Broken Hearts

An Embarrassment of Riches

John Hunter had a problem. A busy surgeon and anatomist, he had little time to care for his own health. By the time he turned 50, he suffered terribly from angina pectoris. His physician, William Heberden, described this syndrome so well that the symptoms are recognizable to any doctor or patient today: a strangling pain in the chest, often spreading down the left arm, brought on by exercise and relieved by rest. As the disease progressed, eating, speaking, or even "any disturbance of the mind" brought on an episode. Heberden warned that the prognosis was poor in such cases: "For, if no accidents intervene, but the disease go on to its height, the patients all suddenly fall down, and perish almost immediately."[1] Hunter had an especially difficult time. Irascible and quarrelsome, he was quick to anger, which then triggered his symptoms. As he frequently complained, "My life is in the hands of any rascal who chose to annoy and tease me."[2]

Living in London in the 1790s, Hunter had few options. He suspected that the problem arose from his heart, for when he performed autopsies on people who had suffered angina, he sometimes found their coronary arteries ossified. But Heberden had little to offer Hunter, only quiet, warmth, and "spirituous

liquors" during attacks, and opium at bedtime to prevent them. The mainstays of eighteenth-century medical practice—bleeding, purging, and vomiting—appeared "to be improper." Hunter's fear of rascals proved to be warranted. Aroused to a "tumult of his passion" at a confrontational board meeting at St. George's Hospital in 1793, he collapsed and died.[3]

Patients today have many more options. Consider the case of former president Bill Clinton. As described by the *New York Times,* Clinton was "a cigar-smoking workaholic whose family history of heart disease and passion for junk food had set him up for a heart attack in the near future." When he developed chest pain and shortness of breath in September 2004, he went to Northern Westchester Hospital. Initial tests were reassuring. However, when he went to Westchester County Medical Center the next morning for a follow-up test—a coronary angiography—the cardiologists discovered "potentially life-threatening accumulations of plaque in the arteries." With four arteries more than 90 percent obstructed, Clinton was at "the edge of the cliff in coronary terms."[4] Three days later, surgeons at New York–Presbyterian Hospital put Clinton on a heart-lung machine for 73 minutes and performed a quadruple bypass of his coronary arteries.

Clinton's treatment did not end with surgery. He began an aggressive medication regimen that included aspirin, a statin, a beta blocker, and an angiotensin-converting enzyme inhibitor. He disciplined his diet and exercised more. In February 2010, while hard at work on the recovery effort after the Port-au-Prince earthquake, he developed recurrent symptoms. When angiography revealed that one of his bypass grafts had become obstructed, an interventional cardiologist performed balloon angioplasty and placed two stents in his coronary arteries. Clinton scarcely interrupted his busy schedule for the procedure.[5] His doctors hoped that their efforts would prevent, or at least postpone, a heart attack.

Even though Clinton's treatment received unusual attention, it exemplifies what has become a routine part of modern health care. Coronary artery disease, a rare curiosity in Hunter's England, is now the leading cause of death in the United States and worldwide. People who have angina or even heart attacks have a cornucopia of options for medical care—perhaps too many options. Hunter and Heberden had to rely on clinical histories and autopsies, but doctors now offer patients an escalating series of diagnostic tests, from the familiar electrocardiograms to stress tests, CT scans, radioisotope perfusion studies, coronary angiography (still the gold standard), intravascular ultra-

sound with near-infrared spectroscopy, and many others. With a diagnosis in hand, patients have to decide among an amazing diversity of preventive and therapeutic strategies, from yoga and diets to aspirin, statins, angioplasty, bypass surgery, and perhaps even gene therapy some day. Hunter had to focus only on avoiding rascals; patients today struggle to choose the right treatment.

Although patients and doctors benefit from the many achievements of scientific research and therapeutic innovation, they do not always know how to use them well. Decision making frustrates patients, doctors, and other caregivers in many areas of medicine. Does a child's earache require antibiotics? Which women should have screening mammograms? Who should undergo coronary angioplasty? This book offers a historical perspective on medical decision making. I focus on the history of cardiac therapeutics, particularly coronary artery bypass surgery and coronary angioplasty. None of what I say, however, applies only to heart disease, cardiology, or cardiac surgery. These techniques simply provide apt examples of the promise and vexations of modern medicine. Patients deciding about medical treatments often have basic questions. Will the treatment work? Is it safe? What else should be considered? Historical analysis reveals why it can be so hard to answer these questions and use our knowledge to make good decisions.

Broken Hearts tells a story about the essential complexity of medical thinking. At times I need to dive deep into medical science to make sense of the controversies and present them fairly. Before I do this, it makes sense to begin with a primer about the clinical syndromes and their treatments. This will provide a foundation for the more detailed histories that follow.

The heart, at a basic level, is one of the easiest organs to describe. It works as a pump, moving blood in a circuit from the right side of the heart, through the lungs, to the left side of the heart, throughout the body, and then back again. Each side has two chambers, an atrium and a ventricle. The ventricles do most of the pumping. The left ventricle, which must pump blood everywhere from the head to the toes, does most of the work. Like all organs, the heart needs a steady supply of blood to provide it with oxygen and other nutrients. It has a dedicated system of arteries for this purpose, the coronary arteries, so named because they encircle the heart like a crown. Although each is important, one major branch, the left anterior descending, supplies most of the left ventricle.

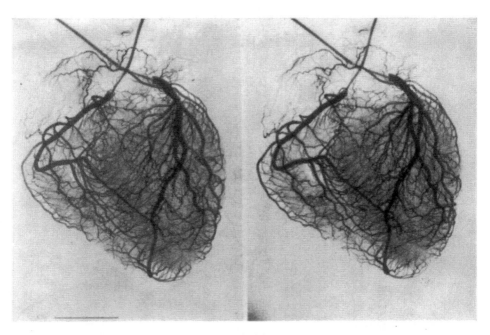

Stereoangiogram of the coronary arteries. If you can cross your eyes, you will see their three-dimensional structure. From William F. M. Fulton, *The Coronary Arteries: Arteriography, Microanatomy, and Pathogenesis of Obliterative Coronary Artery Disease*, 1965, fig. 2.27, p. 46. Courtesy of Charles C Thomas Publisher, Ltd., Springfield, Illinois.

Anything that disrupts the flow of blood through the coronaries can cause problems for the heart. If the heart's demand for oxygen exceeds the supply it receives, its muscle cells run into trouble. Whether the problem is called ischemia (lack of blood) or hypoxia (lack of oxygen), the cells can only tolerate it for a few minutes before they start to die. When an imbalance of supply and demand begins, the cells signal their distress—exactly how remains poorly understood—and the person experiences angina pectoris, the squeezing, suffocating pressure described so clearly by Heberden in 1772. As Cleveland surgeon Donald Effler put it nearly two hundred years later, angina "is the wailing of an anguished heart during that period of stress when it is getting inadequate perfusion."[6] If the ischemia persists and the cells die, the pain becomes more severe, and the affected area of the heart stops working. This is a myocardial infarction, or heart attack. Sometimes the damage strikes a critical part of the heart's electrical system. In the worst-case scenario, the result-

ing arrhythmia can strike a patient dead without warning—an episode of sudden death.

Many things can disrupt coronary blood flow. The most important cause today is atherosclerosis. Pathologists in the nineteenth century first described arteriosclerosis, literally meaning hardening of the arteries. Many diseases could cause this kind of scarring. By the early twentieth century, however, most cases of arteriosclerosis were caused by atherosclerosis, and this became the more commonly used word. The result of a complex mix of environmental exposures (e.g., food, pollution, cigarette smoke), behavior (e.g., diet, smoking, exercise), genes, and other sources of stress, atherosclerosis occurs when cholesterol deposits form in the walls of arteries and trigger a response from the artery's cells. As atherosclerotic plaques evolve over time, they cause inflammation and cell death. Some plaques induce spasms in the artery wall. Others rupture, exposing plaque contents to flowing blood. This can trigger blood clots (technically, thrombi) that obstruct flow through the coronary arteries. Whether called coronary artery disease or ischemic heart disease—one name

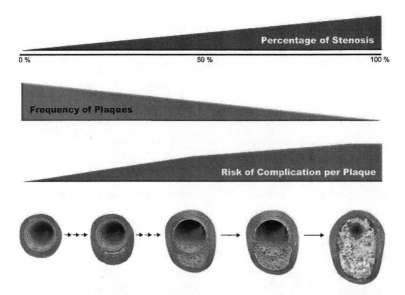

The progressive growth of atherosclerotic plaques. From Morteza Naghavi et al., "From Vulnerable Plaque to Vulnerable Patient: A Call for New Definitions and Risk Assessment Strategies: Part 1," *Circulation* 108 (7 Oct. 2003): 1664–72, fig. 6, p. 1670. With permission from Wolters Kluwer Health.

focused on the cause, the other on the consequence—this process now kills more people worldwide than any other single disease.

Although physicians have probed coronary artery disease in ever-increasing detail, many important mysteries remain. The course of the disease varies enormously from patient to patient, and even from plaque to plaque within each patient. Some people develop symptoms of angina that then remain stable for many years, a syndrome called chronic stable angina. Others have a more dramatic course, with angina worsening over time (more severe, more frequent, more easily triggered). Doctors call this unstable angina. Still other people can have significant coronary artery disease without ever experiencing a twinge of angina. The eventual outcome is equally varied. Some patients endure angina, or treat it successfully, and survive to die from some other cause. An alarming number die from their first heart attack. Others survive one attack, and another, and another, gradually succumbing to heart failure as successive infarctions replace more and more of their myocardial muscle with scar tissue.

Heart disease might seem like such a familiar problem that little more needs to be said about it. After all, coronary artery disease is so common, with such a high mortality rate, that nearly everyone knows someone who died from a heart attack or heart failure. This familiarity, however, conceals a fascinating history not just of the disease and its distribution, but also of its effect on society, its meanings, and the responses it motivates. Any historical study of disease, however, faces a dilemma. Diagnostic terminology changes over time and creates difficult questions of translation and interpretation. If a doctor in 1913 diagnosed a patient with "coronary thrombosis," can a doctor in 2013 be confident that the patient actually suffered a myocardial infarction? The meanings and outcomes of that event would certainly have been quite distinct. Historical epidemiologists, who have tried to reconstruct the contours of heart disease over the twentieth century, know this problem well. Health officials repeatedly redefined the nomenclature used to report causes of death, from "angina pectoris" in the 1910s to "diseases of the coronary arteries" in the 1930s, "arteriosclerotic heart disease" in the 1950s, and now "ischemic heart disease."[7] Approached with appropriate caution, the writings of past doctors yield revealing clues about the emergence of heart disease.

Recognizable cases of angina began to appear in England in the late eighteenth century. John Hunter's illness and death provide one early example. Angina captured little interest, however, until the twentieth century. Atten-

tion turned to heart disease after 1910, when its varied forms—angina, rheumatic heart disease, endocarditis, and others—rose to the top of the list of leading causes of mortality in the United States. Heart disease has held that position to the present day, with the single exception of the flu pandemic from 1918 to 1920. By the 1930s health officials had become particularly concerned about the prominence of coronary artery disease in the United States. Some even named it the American disease. Although skeptics have questioned whether the rise was real or simply an artifact of changing diagnostic practices, most doctors and policy makers took the rise seriously.[8]

Congress made its first move in 1948 when it passed the National Heart Act and established the National Heart Institute (now the National Heart, Lung, and Blood Institute of the National Institutes of Health). By the time of President Dwight D. Eisenhower's heart attack in 1955, few doubted the severity of the problem. But then, at some unnoticed moment in the 1960s, the epidemic reversed course. The decline, first recognized in the 1970s, has been as dramatic as the rise. Heart disease mortality has now fallen more than 60 percent from its zenith. Nonetheless, it remains the leading cause of death in the United States. And even as coronary artery disease has receded in the United States, it has surged elsewhere to become the leading cause of death worldwide.[9]

The rise and fall of heart disease over one hundred years—an epidemic in slow motion—raises a question that has fascinated historians and epidemiologists: how to account for changing patterns of disease over time. The prevalence of a disease can change for many reasons. Sometimes a new cause appears, as happened with radiation poisoning and with HIV over the course of the twentieth century. Sometimes a widespread shift in behavior increases the prevalence of a once rare disease, as happened with cigarette smoking and lung cancer. Medical technology can make occult diseases visible. It can also reduce or even eradicate them. Many factors often work together. Consider high blood pressure. Its prominence in the twentieth century reflects an absolute increase in its prevalence (e.g., because of changes in diet and physical activity), new diagnostic technologies (e.g., blood pressure cuffs), and new medications (e.g., diuretics), all of which made the problem relevant for doctors and patients.[10]

Why did coronary artery disease rise to prominence? Changing social and economic conditions, including everything from the rise in cigarette use to changes in diet, exercise, stress, air pollution, and income inequality, increased

its occurrence. New understandings of its causes made prevention campaigns both possible and essential. New diagnostic technologies and new treatments made it more visible and more relevant for doctors. The tight association between heart disease and the conditions of our modern lives has led scholars to identify coronary artery disease as the emblematic "disease of civilization."[11] Coronary artery disease is not a naturally given fact. Instead, it is exquisitely responsive to how we structure our societies—it is socially produced. This should not be read as a relativist argument that disease is an arbitrary exercise in social labeling. The suffering caused by heart disease, cancer, or AIDS is palpably real. Instead, this argument is twofold. First, the conditions of society powerfully influence the diseases a population experiences. Second, much of what is interesting and relevant about a disease are the social meanings it accumulates. Therefore any understanding of disease must be grounded in careful consideration of its specific social contexts and associations.[12]

The meanings that coronary artery disease accumulated had a powerful impact on how American society responded to it. The symbolic meanings of the heart as the source of vitality, the seat of courage, and the embodiment of emotional experience featured prominently in European medical writing, fiction, and poetry from the seventeenth century through the nineteenth century. With the rise of neuroscience and psychoanalysis in the late nineteenth and twentieth centuries some of this cultural attention shifted from the heart to the brain.[13] But the rising epidemic of heart disease forced the public to remain interested. As Effler wrote in *Scientific American* in 1968, "The heart attack is so common among professional people, executives and men in public office that it has become almost a status symbol. If all the men in these groups who have had coronary attacks were forced to retire (as airline pilots automatically are), the shortage of manpower at the top levels of government, industry, and the professions in the U.S. would cripple the nation. All in all, coronary disease has become so prevalent and ominous in modern society that heroic measures to deal with it seem not only appropriate but essential."[14] The powerful linkage that emerged between coronary disease and those with economic and political power motivated unprecedented efforts to improve its diagnosis and treatment.

Over the course of the twentieth century, doctors developed a series of techniques to detect coronary artery disease and monitor its progress. The first, electrocardiography (EKG), came into wide use in the 1920s.[15] By recording the electrical activity of the heart, an EKG can detect areas that expe-

rience ischemia. This can help doctors confirm that an episode of chest pain arises from the heart. Characteristic changes in the shape of the EKG tracing seen during severe ischemia can also confirm the diagnosis of a heart attack. But EKGs cannot reveal the extent or distribution of disease in the coronary arteries. This only became possible with the development of selective coronary angiography.

In 1929 a German surgeon, Werner Forssmann, first threaded a catheter (a small, flexible, plastic tube) through a peripheral vein and into his own heart. Forssmann took an x-ray of his chest to confirm that the catheter's tip had reached his right atrium. Initially rebuked for reckless self-experimentation, Forssmann developed this idea into a technique that could produce images of living, beating hearts. He was awarded a Nobel Prize in 1956. Visualizing the coronary arteries proved more difficult. Doctors wanted a way to inject the contrast agent into the coronaries without filling the heart's chambers as well, but they feared that doing this would be toxic and would somehow disrupt the heart's delicate metabolism. In the 1940s various teams of doctors experimented with inserting a needle through the base of a person's neck into the aorta and injecting contrast near the opening of the coronary arteries. This produced faint tracings of the coronaries but carried considerable risk to the patients.[16]

The breakthrough came—by accident—in October 1958, when Cleveland Clinic cardiologist Mason Sones injected contrast through a catheter that he had inadvertently threaded into a coronary artery. Although Sones was terrified that the agent would cause an infarction, the patient experienced no ill effects. Even after Sones reassured himself that the technique was safe, it took him another five years to develop a reliable system. He had to collaborate with engineers from the North American Philips Company to design customized x-ray generators, beam splitters, image intensifiers, and synchronized cameras. Their work produced a system that could capture high resolution pictures and movies of the lumen (the space inside) of living coronary arteries.[17] The technique spread slowly in the mid-1960s, but by the early 1970s many large medical centers had access to selective coronary angiography. This innovation transformed both the diagnosis and the treatment of coronary artery disease.

For most of the early twentieth century, coronary artery disease treatment focused on the same two tasks that Heberden had described: provide relief during acute attacks and instruct the patient in preventing them by avoiding

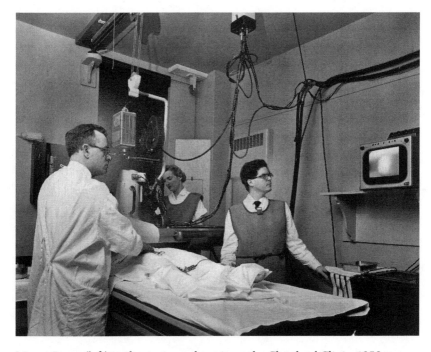

Mason Sones (left) in the angiography suite at the Cleveland Clinic, 1958. Photograph by Hasting-Willinger & Associates. With permission from Cleveland Press Collection, Cleveland State University Library.

physical and emotional stress as well as other triggers. By the 1950s, doctors could recommend a few therapies: nitrates to dilate arteries, opiates to treat pain, and prolonged bed rest—for weeks or even months—to reduce the workload of a damaged heart. As epidemiological studies produced increasing evidence about risk factors, especially diet, exercise, and smoking, doctors began to advise patients about lifestyle changes that might slow the course of atherosclerosis. However, many patients found these changes difficult to make. Doctors and patients desperately wanted a way to treat the disease.

Surgeons had tried since the 1930s to increase blood flow to the heart, but these techniques rarely worked as well as patients or doctors hoped.[18] The advent of coronary angiography in the 1960s helped surgeons determine whether the atherosclerotic plaques were focal or diffuse, and which plaques seemed most worrisome. With access to Sones's technique, surgeons at the Cleveland Clinic redoubled their efforts to provide surgical treatments for coronary artery disease. After a series of partial gains, they succeeded with

Broken Hearts

tigator Award in Health Policy Research, from the Robert Wood Johnson Foundation, Princeton, New Jersey. Additional funding was provided by the Informed Medical Decisions Foundation, the Massachusetts Institute of Technology, and Harvard University. I would especially like to thank Jack Fowler, Albert Mulley (again), Diana Stilwell, and John Wong at the IMDF, and Cynthia Church, Michael Gallo, David Mechanic, and Lynn Rogut at the RWJF.

The editors and staff at the Johns Hopkins University Press, including Sara Cleary, Vanessa Kotz, and Jackie Wehmueller, have earned my lasting gratitude for guiding this work from manuscript to its finished form.

Final thanks go to my own doctors—Jane Fogg, Keith Stuart, and especially Roger Jenkins—who made sure I stuck around to finish the work, and to Liz, Sam, and Natalie, who have endured countless claims of "I'll be there in just a minute" as I squeezed work in amid the joys of family life.

Medical School, I have learned much about medicine and social justice from many colleagues, especially Anne Becker, Gene Bukhman, Leon Eisenberg, Paul Farmer, Jeremy Greene, Salmaan Keshavjee, Jim Yong Kim, Ed Lowenstein, Joia Mukherjee, Scott Podolsky, Joseph Rhatigan, and Sadath Sayeed. Allan Brandt and Charles Rosenberg, at the Department of the History of Science at Harvard University, remain ever generous with their knowledge and wisdom.

An emerging working group on the history of cardiovascular disease, within the American Association for the History of Medicine, has provided useful feedback and leads, especially Henry Blackburn, Joel Howell, Carla Keirns, Aaron Mauck, Todd Olszewski, Gerald Oppenheimer, Sejal Patel, Kavita Sivaramakrishnan, and Sarah Tracy. I have also worked with valued colleagues through the Robert Wood Johnson Foundation, especially Carol Ashton, Charles Bosk, David Meltzer, Keith Wailoo, and Nelda Wray.

Many physicians who lived through the events I describe answered my questions, offered their own memories and perspectives, and reviewed relevant portions of the manuscript, including Anthony Breuer, David L. Brown, Louis Caplan, John Caronna, Anthony Furlan, Sid Gilman, Floyd Loop, H. Royden Jones, James E. Muller, Judith Robinson, David Shahian, Ted Stern, Ton van der Steen, and Stuart Yudofsky. Robert Aronowitz shared his insights as both a historian of medicine and a physician who had observed the transformations I detail in this book. Mary Norman Jones provided both a lay reader's perspective and careful editing. Meredith Bircher and Lori Kelley did invaluable work securing permissions for the figures.

I have presented pieces of this work to many audiences and benefited from their feedback, including the Informed Medical Decisions Foundation; the Robert Wood Johnson Foundation; the annual meetings of the Society for the Social Studies of Science, the History of Science Society, and the American Association for the History of Medicine; the National Committee for Quality Assurance; the Weatherhead Center for International Affairs, Harvard University; the Department of Social Studies of Medicine, McGill University; the Department of Global Health and Social Medicine, Harvard Medical School; the Department of the History of Science, Harvard University; the STS Circle at Harvard's Kennedy School of Government; the Knight Science Journalism Program at MIT; and the Program in Medical Humanities at SUNY Stony Brook.

This research was made possible by the generous support of a RWJF Inves-

Acknowledgments

This book is one piece of a much larger research project, in pursuit of which I have accumulated countless debts. Robert Martensen first set me working on the problem of coronary revascularization when I was a medical student in 1993. Albert Mulley contacted me ten years later and encouraged me to pursue questions I had initially left unanswered. His insight and support have been invaluable ever since. The research, amid teaching commitments at both the Massachusetts Institute of Technology and Harvard Medical School, would not have been possible without the tireless effort of my research assistant, Katherine Shera. Several other students, at MIT, Harvard Medical School, and Wellesley College, have also worked on pieces of this research, including Xaq Frohlich, Sonya Makhni, Catherine Mancuso, Michael Matergia, Benjamin Oldfield, Michael Rossi, Nicolle Strand, and Emily Wanderer.

Many archivists have generously made their collections accessible for study. I would especially like to thank Pam Cornell and Alethea Drexler, at the McGovern Historical Collections and Research Center; Jack Eckert, at the Countway Medical Library; Marjorie Jackson, at the Library and Learning Resource Center at the Texas Heart Institute; Fred Lautzenheiser and Carol Tomer, at the archives of the Cleveland Clinic Foundation; Pamela Miller and Lily Szczygiel, at the Osler Library for the History of Medicine at McGill University; Cathy Pate and Anne Fladger, at the archives of the Brigham and Women's Hospital; and JoAnn Pospipil, at the Baylor College of Medicine Archives.

I have benefited from colleagues with an extremely helpful diversity of interests and expertise in Boston. The Program in Science, Technology, and Society at MIT provided my primary base. There I enjoyed the mentorship of and collaboration with Caspar Hare, Stefan Helmreich, David Kaiser, Vincent Lepinay, David Mindell, Heather Paxson, Natasha Schull, and Rosalind Williams. At the Department of Global Health and Social Medicine at Harvard difficult treatment decisions. Will a treatment work? In what ways? How well? Is it safe? How does it compare to other options? What other factors, whether cost or quality of life, ought to be considered? Uncertainty dogs the effort to answer each of these.

This book offers a historical perspective on the complexity of medical decision making. It uses the history of coronary artery bypass surgery and coronary angioplasty to expose vexing ambiguities that persist at the core of the medical enterprise, in particular the challenge of producing definitive knowledge about the efficacy and safety of medical treatments. This is a story about medical discovery, but one focused on the decision dilemmas created by the emergence of these two new treatments. An appreciation of their complexity, as seen through history, can help patients and doctors make more careful and deliberate decisions. New discoveries in the future may leave patients with easier and better choices. But it is unlikely that decision making will ever be just a question of medical knowledge and practice. Instead, medical decisions always involve priorities, values, and preferences. This history, as a result, can offer valuable lessons to historians, doctors, patients, and policy makers.

Preface

When I had cancer a few years ago, I was lucky in many ways. I faced no difficult decisions. As soon as the radiologist confirmed the existence of the fist-sized tumor I had first felt near my stomach, the course was clear. The tumor had to be removed. A kind surgeon quickly obliged, and I was tumor-free in ten days. When genetic analysis revealed that the tumor was not susceptible to any chemotherapy, I was spared another decision: there was no reason to try an expensive pharmacological treatment of uncertain value. Instead, I watch and wait and hope. If the cancer comes back, my surgeon has promised to try to remove it once again.

Something about all this appealed to the historian in me. Mine was cancer therapy as it existed in the 1890s: find a tumor, cut it out, and hope for the best. That way of therapeutic thinking has an elegant logic and simplicity. Few areas of medicine today retain such clarity of purpose and action. I now face much more complicated decisions as an otherwise healthy 42-year-old than I did as a cancer patient. Fear of heart disease creates most of the trouble. Should I start taking a baby aspirin to prevent a heart attack? If not now, at what age? What about a statin to reduce the risk that one of my atherosclerotic plaques will rupture? I do not actually know that I have any, but as an American male, I almost certainly do. I could go to a local imaging center and purchase a CT angiogram of my coronary arteries, but this might open a can of worms. Am I ready to commit myself to a life of multiple prophylactic medications to preserve my cardiac health? And what if I ever did feel a twinge of angina? Then the dilemmas really become difficult.

The decisions that all people now face, from the vitamins, supplements, and medications we take to preserve our health, to the drama of surgery and intensive care, reflect fundamental changes in medical knowledge and practice over the past half century. The capacity and complexity of medicine have been transformed. Yet the many blessings have brought their own share of curses. Patients and doctors in the early twenty-first century face exceedingly

Figures

Contents

Johns Hopkins Paperback edition, 2014
9 8 7 6 5 4 3 2 1

Johns Hopkins University Press
2715 North Charles Street
Baltimore, Maryland 21218-4363
www.press.jhu.edu

The Library of Congress has cataloged the hardcover edition of this book as follows:
Jones, David S. (David Shumway)
    Broken hearts : the tangled history of cardiac care / David S. Jones.
        p. ; cm.
    Includes bibliographical references and index.
    ISBN 978-1-4214-0801-9 (hdbk. : alk. paper) — ISBN 1-4214-0801-5
(hdbk. : alk. paper) — ISBN 978-1-4214-0802-6 (electronic) — ISBN
1-4214-0802-3 (electronic)
    I. Title.
    [DNLM: 1. Myocardial Infarction—physiopathology. 2. Atherosclerosis—
physiopathology. 3. Cardiovascular Surgical Procedures—adverse effects.
4. Myocardial Infarction—complications. 5. Postoperative Complications—
etiology. WG 300]
    616.1'237—dc23   2012018227

A catalog record for this book is available from the British Library.

ISBN-13: 978-1-4214-1575-8
ISBN-10: 1-4214-1575-5

*Special discounts are available for bulk purchases of this book. For
more information, please contact Special Sales at 410-516-6936 or
specialsales@press.jhu.edu.*

Johns Hopkins University Press uses environmentally friendly
book materials, including recycled text paper that is composed
of at least 30 percent post-consumer waste, whenever possible.

BROKEN
HEARTS

The
TANGLED HISTORY
of CARDIAC CARE

DAVID S. JONES

Johns Hopkins University Press

Baltimore

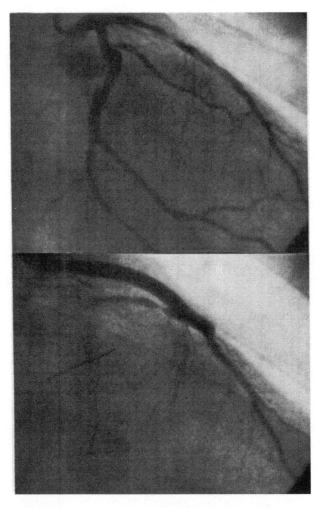

Selective coronary angiography. The coronary arteries
before and after bypass surgery: atherosclerotic plaque is
visible in the upper center of the image (*top*), interrupting
the dark shadow of the contrast-filled left anterior descending coronary artery. The large saphenous vein graft comes
in from the upper left (*bottom*), providing blood (and
contrast) to the artery. René G. Favaloro et al., "Direct
Myocardial Revascularization with Saphenous Vein
Autograft: Clinical Experience in 100 Cases," *Diseases of
the Chest* 56 (1 Oct. 1969): 279–83, fig. 4, p. 282. With
permission from the American College of Chest Physicians.

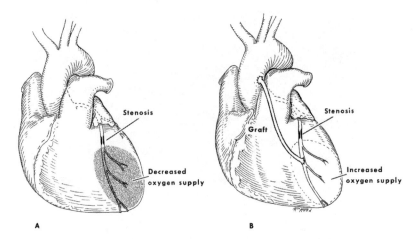

Coronary artery bypass graft surgery. The before (A) and after (B) drawings show how the graft can increase oxygen supply to the heart. Illustration by Robert Reed, 1973. Reprinted with permission, Cleveland Clinic Center for Medical Art & Photography © 1973–2012. All Rights Reserved.

coronary artery bypass grafting. In its basic form, bypass surgery uses a piece of vein taken from the leg to form bypass grafts between the aorta and the coronary arteries downstream of major obstructions. René Favaloro published his first case series in 1968, and the technique spread quickly. By the early 1970s hundreds of surgeons throughout the country performed bypass surgery. The rise of the procedure motivated cardiologists to develop a competing— and less invasive—technique. First described by German angiologist Andreas Grüntzig in 1977, angioplasty (percutaneous transluminal coronary angioplasty) uses a balloon-tipped catheter to stretch open coronary obstructions.[19]

Bypass surgery and angioplasty could increase blood flow to the heart and produce dramatic and immediate relief of angina. Surgeons and cardiologists believed that these procedures could prevent heart attacks and prolong survival. Use of bypass surgery grew quickly during the 1970s. It became one of the most common surgical procedures in the United States in the 1980s. Angioplasty got off to a slow start in the 1980s and remained in the shadow of bypass surgery. However, with the advent of stents—wire mesh tubes that could be inserted by catheters to keep the dilated vessels open—angioplasty took off in the 1990s. Bypass surgery peaked in the United States at 607,000 operations in 1997 and then, under pressure from angioplasty, declined by a quarter over the next decade. Angioplasty continues to grow, with over 1 mil-

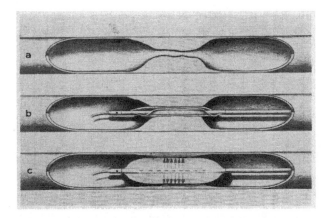

Percutaneous transluminal coronary angioplasty. Large
plaque (*a*); catheter, coming from the right, threaded
through the plaque (*b*); and inflated balloon compressing
the plaque (*c*) to restore full flow through the artery.
From Andreas R. Grüntzig, Åke Senning, and Walter E.
Siegenthaler, "Nonoperative Dilatation of Coronary-Artery
Stenosis," *New England Journal of Medicine* 301 (12 July
1979): 61–68, fig.1, p. 62. With permission from the
Massachusetts Medical Society.

lion procedures each year in the early 2000s.[20] Together these procedures
form a $100 billion industry in the United States alone.

Although the advent of bypass surgery and angioplasty coincided with the
dramatic reduction in heart disease mortality in the United States, analysts
think the techniques played only a minor role in the decline. Pharmaceutical
treatments had also improved dramatically since the 1960s, with new drugs
to control high blood pressure (and thereby reduce the demand on the heart),
to slow and possibly reverse atherosclerosis, and to prevent platelets from
sticking together into obstructive thromboses. And even if public health campaigns have been only partially successful, changes in smoking rates, diet, and
physical activity have likely had the greatest effect on coronary artery disease
incidence.[21]

The attempts by doctors, epidemiologists, and historians to determine the
impact of the many treatments of heart disease demonstrate how tricky it can
be to understand the efficacy of medical interventions. Some people think
that medical therapeutics are straightforward: they either help or they do not.
Simple histories follow this thinking and contrast the wonders of twentieth-

century medicine with the superstition and barbarism of bloodletting, purgatives, and emetics common into the nineteenth century. The celebratory message is clear: now we are scientific; now we have powerful therapies. Although few people today would turn back the clock and exchange antibiotics and anesthesia for mercury and leeches, we should not close the door on subjecting therapeutics to close scrutiny.

Everywhere historians look, they find subtle and informative stories. The efficacy of bloodletting, for instance, cannot be easily dismissed. Understood with a sensitivity to when and why it was used, bloodletting may actually have worked in exactly the way patients and doctors desired. Imagine a person suffering from malaria, flushed and delirious from the raging fevers. Bloodletting could render the patient calm, pale, and cool to the touch. Inspired by this insight, historians have documented delicate interactions between therapies, expectations, and hope. They have looked beyond the internal logic of medical science to consider the contexts in which treatments take place. Medical interventions do not target only the causes or symptoms of disease. Instead they increasingly seek to reduce the risk of future disease, promising control amid fears of mortality. Treatments have broad social functions as well, from comforting the sick to enforcing a particular social order—they have social efficacy. Historical sensibility, however, does not excuse all past therapeutic practice. It is still possible that, by the standards of their day, medical treatments went astray.[22] What can be learned from the history of cardiac therapeutics? Historians have only just begun to scratch the surface.[23]

Making decisions about diagnosis and treatment is the fundamental task of medicine. Despite all of the resources devoted to health care, something has clearly gone awry. Consider bypass surgery and angioplasty. Throughout their history, the procedures have been beset with controversy.[24] Did the techniques work as well as their proponents claimed? Which outcomes could be achieved: just relief of pain, or prevention of heart attacks and death? What were the techniques' complications? Did the benefits justify the risks? What about the costs? What were the most convincing ways of answering these questions? Efforts to resolve the debates have made the techniques of coronary revascularization among the best studied in all of medicine. And yet controversy persists. Skeptics argue that as many as 85 percent of angioplasties are undertaken with unrealistic expectations of benefit. Others argue that only 15 percent are done for inappropriate or uncertain indications.[25] These efforts to determine appropriateness are themselves uncertain because they rely on imperfect knowledge of the effectiveness and safety of the procedures. To the extent that ambiguity ensnares medical decisions, the value of these decisions remains difficult to judge.

The problem is not limited to coronary revascularization. One study found that many patients do not receive all the care that would be recommended based on existing medical knowledge. Only two-thirds of coronary artery disease patients received appropriate care, and fewer than half of all heart attack patients studied received beta blockers, even though these drugs have been shown to reduce subsequent mortality by nearly 25 percent.[26] Other experts focus on the opposite problem and argue that too many patients receive care they do not need or even treatments that do more harm than good. One study surveyed patients who had undergone angioplasty with stenting for stable angina and found that fewer than 20 percent had been told about potential disadvantages of the procedure, and only 10 percent had discussed other treatment options with their doctors. Patients' expectations, physicians' hopes, and fears of malpractice all conspire to push doctors to use every technology that might contribute to a patient's care, especially diagnostic techniques. Diagnostic workups, however, trigger subsequent compulsions. Cardiologists complain that angiography initiates a clinical cascade that forces them to intervene whenever they find a plaque.[27]

Similarly perplexing examples can be found throughout medicine. Patients and doctors alike are often dismayed by end-of-life care. Doctors resent the growing importance of standardized treatment protocols. Patients worry about their physicians' financial conflicts of interest. Comparative effectiveness research, a seemingly laudable endeavor to test which treatments work best, has become a bogeyman of health care politics. Insurers and managed care regulators increasingly constrain the choices of patients and doctors. Those without insurance face an even more dire constraint, the inability to access the health care system for anything short of an emergency.

Gone are the days when paternalistic doctors made decisions on behalf of their patients. Instead, different ideas now exist about the best way to make medical decisions and provide a rational basis for medical practice. Doctors struggle to define what counts as good enough evidence of whether treatments work. Their efforts reveal much about the values of medicine and the obstacles physicians face. Proponents of evidence-based medicine believe that careful scrutiny of clinical research can produce objective and reliable treatment guidelines. But this requires doctors to navigate between general principles and the actual experiences of individual patients. And even when doctors respect the medical literature, they also attend to other sources of authority. Two Australian doctors captured this well in their tongue-in-cheek catalog of how doctors relied not just on evidence-based medicine, but also on eminence-, vehemence-, eloquence-, providence-, diffidence-, nervousness-, and confidence-based medicine. Proponents of shared decision making, meanwhile, believe that patients, once they have been adequately informed about options and evidence, must be able to exert their preference. Decision theorists, medical ethicists, sociologists, lawyers, and economists have all had their say about how medical decisions should get made. Eloquent physicians offer their insider perspectives to demystify these processes for patients.[28]

Despite an abundance of advice, decision making has become a burden. Doctors began to feel this in the years after World War II, as the rapid proliferation of scarcely distinguishable medications challenged them to choose the right drugs for their patients. One marketing expert in 1955 likened a doctor to a housewife in a grocery store facing the "misery of choice." The situation has only gotten worse. By the late 1990s doctors and patients had access to an ever wider assortment of treatments for coronary artery disease, leaving them with what Dartmouth health policy researchers identified as a "dilemma of choice."[29] How did choice become a miserable dilemma? After all, we live in a consumer age that enables and celebrates choice. But framing medical decision making as choice might be misleading. "Choice" carries the connotation of a consumer transaction, of someone picking among desirable commodities. In health care, however, patients do not choose between desirable commodities. Instead they face treatment options that no one would ever have chosen in other circumstances. They must decide between the lesser of two evils.[30]

One thread runs through all medical decisions: uncertainty. What is the cause of a heart attack? Who is at risk? Is angiography reliable? Will a treatment work? Uncertainty undermines answers to all these questions. This might seem inevitable. Complex problems often have difficult answers. But no one should assume that uncertainty is inevitable. At times, uncertainty is sincere, in that some questions have simply defied decades of concerted work. But in other cases, it is the result of a more passive process, such as the consequence of deciding not to study a particular problem. In some situations uncertainty has been crafted strategically. The tobacco industry and skeptics of global warming both cultivated scientific uncertainty in order to forestall science-based regulation.[31] Whenever uncertainty is invoked, we should ferret out its causes. Is it genuine, or is it being accentuated to support a particular interest? If uncertainty does exist, does it reflect lack of scrutiny of the problem or confusion that persists despite scrutiny?

In medical school I was taught that physicians had to learn to tolerate uncertainty. Since clinical medicine is so often uncertain, doctors would do little for their patients if they acted only in moments of absolute certainty. Is it possible to act amid ambiguity without being reckless? It is, but doing so requires doctors to develop a self-conscious attitude toward the imperfections of medical knowledge. Many scientists seek to find the answer to a particular problem and hope that their methods will yield certainty. Social scientists often realize that the complexity of the phenomena they study precludes single answers. They grapple constantly with questions of interpretation in face of ambiguous information. By bringing social science perspectives to the problems of disease and therapeutics, we can acknowledge uncertainty, analyze it, and sometimes even move toward new policies despite it. That is a goal of this book.

Accounts of decision making must take seriously the dilemmas and uncertainties and create portraits in which both patients and doctors can recognize themselves. Decisions are never made in a vacuum. Instead, patients and doctors bring their own interests and values to the table. They navigate constraints introduced by the values and priorities of families, health care systems, and their social and cultural contexts. This confluence can generate substantial conflicts.[32] The challenges could be illustrated in any area of medical practice. I focus on coronary artery disease and the techniques of coronary revascularization for many reasons. First, coronary artery disease killed more Americans by far than any other disease in the twentieth century and is now the leading cause of death worldwide. Second, cardiology and cardiac surgery have played a dominant role in the health care economy. The technologies of coronary revascularization—angiography, angioplasty, and bypass surgery— provide up to 30 or even 40 percent of the revenue of some hospitals and medical centers.[33] Third, cardiac therapeutics have been scrutinized more carefully than any other area of medical practice by many of the best minds in science and medicine. As a result, it is possible to document the medical theories, practices, and conflicts in remarkable detail.

The title of this book, *Broken Hearts*, has multiple meanings. The book is about heart disease, the patients who experience it, and the treatment decisions they must make. The book also explores the disappointments that follow when treatments do not work as well as patients and doctors expected or when unanticipated complications occur. The subtitle, about the tangled history of cardiac care, highlights the aspect I am most interested in: how history can reveal the complicated and often overlooked subtleties that make medical decision making so problematic. I explore this complex history through two stories, each of which explores a different phase of the decision-making process. The first part examines what doctors and patients think about the causes of disease and the potential effectiveness of treatments. The second part moves from the benefits of treatments to their risks and studies how doctors detect unwanted complications and assess a treatment's safety. A concluding chapter raises the issue of geographic and racial variations in medical practice to illustrate the many factors beyond knowledge of risk and benefit that influence medical decisions. Together these analyses offer a framework for understanding the key components of any decision about health care.

Part I, "Theory and Therapy," begins with a surprising puzzle. As heart attacks took their toll on American society in the twentieth century, doctors and medical researchers worked hard to understand their causes. But as recently as the 1970s, physicians disagreed about the most basic facts. How did this situation come to pass? As doctors studied heart attacks from the 1910s into the 1960s, their usual methods of animal research and postmortem analysis suggested many possibilities but did not foster a strong consensus. Only as surgeons began to intervene acutely against heart attacks in the 1970s did researchers produce strong evidence that coronary thrombosis was the cause of heart attacks, and that one particular mechanism—the plaque rupture hypothesis—was the cause of coronary thrombosis. This theory, however, had little initial appeal for doctors because it did not provide them with new therapeutic interventions. As new treatment options emerged from the 1970s into the 1990s, competition between therapies influenced medical theory. New medications, especially statins and platelet inhibitors, fostered interest in plaque rupture, while the continuing popularity of bypass surgery and angioplasty provided support for older models of progressive obstruction. Confusion about the relative importance of plaque rupture and progressive obstruction continues to generate unrealistic expectations of what treatments can accomplish.

The tangled history of the plaque rupture hypothesis demonstrates an important aspect of discovery and knowledge production in medicine. The familiar "bench to bedside" story of doctors working in laboratories to generate new knowledge that they then translate into new therapies is incomplete. Knowledge also flows in the opposite direction, from bedside to bench. What doctors do—how they examine, describe, and manipulate bodies—powerfully influences what doctors think. The complex interplay between practice and knowledge is especially pronounced with surgery and other mechanical interventions, in which doctors interact directly with diseased tissues and bodies.[34] This situation, in principle, is not a problem. A healthy give-and-take between theory and practice is necessary if medicine is to be a self-correcting enterprise. It does, however, leave medical knowledge vulnerable to a wide range of influences. Doctors become enthusiastic about new therapies for many reasons: because treatments seem to work, because they offer hope where none had existed before, because they reward physicians financially, or even because they have been marketed successfully. This enthusiasm spills over and recrafts the fabric of medical knowledge. The history of the plaque rupture hypothesis helps explain why it can be so hard to generate reliable knowledge in medicine.

As new treatments become available, initial enthusiasm almost always arises from the anticipated benefits the treatment will provide. Patients, doctors, and researchers often focus less on the potential complications. Part II of this book, "Complications," follows knowledge of the cerebral complications of cardiac surgery to show how complications become marginalized in medical research and practice. The story begins in the 1890s and traces the history of cardiac surgery from its origins through its rise to preeminence in the 1960s and 1970s. From the outset, surgeons knew that their manipulations of the heart would put the brain at risk. They developed heart-lung machines to protect brain function while they stopped the heart to repair damaged valves and congenital defects, but they quickly realized that their devices were an imperfect substitute for the natural function of the heart and lungs. As patients suffered terribly from strokes, delirium, and other cerebral complications, physicians from many specialties deployed a remarkable diversity of techniques to describe and understand the problem. And yet in the first decade of coronary artery bypass surgery, scarcely anyone paid any attention at all to its cerebral complications.

The initial inattention to the cerebral complications of bypass surgery, in the 1970s, and the subsequent minimization of them in the 1990s and 2000s, offer an opportunity to study the obstacles to certain kinds of medical knowledge. Methodological challenges, competing priorities, and an enduring faith that the benefits of cardiac intervention always justify its risks, have left doctors uncertain about the dangers of cerebral complications, and patients surprised when complications occur. This is a fundamental problem. When doctors devote more energy to proving that treatments work than they do to ascertaining complications, they produce an asymmetrical knowledge base, one with better knowledge of efficacy than of safety. The asymmetry introduces a bias in favor of medical intervention. If doctors and patients know more about benefits than about risks, then the calculus of risk and benefit will always favor proceeding with the treatment. Questions of responsibility are crucial here. Who or what is responsible for treatment complications? Who shares responsibility for eliminating them? The history of cerebral complications explains why the adverse effects of medical treatments can be so difficult to study and so easy to explain away.

Although I focus on questions of efficacy and safety, these are never the full story of a medical decision. If doctors had perfect knowledge of safety and efficacy, would they make the right decisions? Perhaps. But doctors have long known that many other variables influence their decisions. The conclusion to this book, "Puzzles and Prospects," introduces this dilemma. As bypass surgery and angioplasty spread in the 1980s, researchers realized that the use of these interventions varied from place to place and depended, in part, on whether the patients were white or black. This was part of the newly recognized problem of geographic variation in medical practice. Reports of "unwarranted" variation suggested that something irrational influenced medical practice despite doctors' commitments to evidence-based medicine. Researchers set out to understand and eliminate unwarranted variation. They documented an astonishing range of possible influences only to find that the causes of practice variation themselves may vary from place to place. Amid concerns about the rising costs of health care over the last two decades, doctors redoubled their efforts to define standards for appropriate practice. As hard as this has been in the United States and Europe, it has proven even more troubling elsewhere. Profound inequalities in cardiovascular practice exist between rich and poor countries. With heart disease the leading cause of death in most countries worldwide, these inequalities pose fundamental challenges for health and social justice in the twenty-first century.

Several themes run throughout this book. The promise and limits of visualization technologies appear often. The "Theory and Therapy" chapters focus on the most familiar forms of visualization, on how pathologists and radiologists produced images of coronary arteries and their plaques. The "Complications" chapters show how neurologists, psychiatrists, and psychologists worked to visualize something more abstract—brain function and injury—in an attempt to produce concrete representations of the cerebral complications of cardiac surgery. These modes of visualization are all part of the scientific drive to see and thereby understand.

The case studies also explore the myriad meanings of risk. Doctors and patients struggled to decipher the unpredictable risk of atherosclerosis and heart attacks. Surgeons and neuroscientists documented the risks posed by medical treatment. Analyses of geographic variation grappled with a less intuitive idea, the risk of unnecessary treatment. Other questions recur as well. What is the nature of a fact in medicine? When and why do specific concerns become relevant for physicians? Which technologies are appropriate in particular societies, and why? How does uncertainty undermine physicians' efforts to answer these questions and provide the best care for their patients? Answers to these questions are essential to medical theory and practice.

In this book, I explore a history that remains very much alive. Studies of recent history can be especially informative, but they present specific challenges. Any history of cardiology and cardiac surgery faces an embarrassment of historical riches. The vast resources invested in coronary revascularization generated an immense record. Doctors have written tens of thousands of articles, editorials, reviews, and textbook chapters. Their efforts have attracted ongoing attention—from adulation to condemnation—from journalists. The clinical records of tens of millions of bypass surgery and angioplasty patients reside in hospital file rooms. Hospitals, medical schools, and professional societies have their own records as well. No one could read it all.

In my effort to get the story right, I have read thousands of journal articles, by physicians famous and obscure, in journals that remain influential or have gone defunct. I have also tried to get beneath the published accounts. Physicians have been surprisingly forthright in their interviews with reporters, making media coverage of cardiology and cardiac surgery a treasure trove of revealing opinions. Nostalgic physicians have collected and published oral histories from their colleagues. These are rich in narrative detail, even though they rely on recollections of events long ago. The archives of hospitals, medical libraries, and local medical history collections preserve the files of some past (and even some current) physicians. Many of these are invaluable, even if often frustratingly incomplete, while others remain inaccessible. Finally, many of the key players in cardiac medicine over the past fifty years are still active. I interviewed a dozen or so, but most agreed only to speak off record. This is revealing of tensions that have long existed in the field. These interviews color my thinking even if I cannot rely on them for documentation.

The enormous scope of cardiac therapeutics involves a daunting cast of characters. Patients, richly described in physicians' narratives but less often in their own words, offer a vivid sense of what it's like to experience heart disease and its treatments. Most of my story comes from the work and research of cardiologists and cardiac surgeons, with a supporting cast of internists, neurologists, psychiatrists, psychologists, pathologists, research assistants, and policy experts. I have focused on a few sites. The Cleveland Clinic is the most important. Doctors there developed selective coronary angiography and published the first successful case series of bypass surgery. They nearly completed the trifecta, by recruiting Andreas Grüntzig to bring angioplasty to the clinic in 1979, but visa challenges intervened.[35] Clinic neurologists conducted one of the first rigorous studies of the complications of bypass surgery. The site also offers historians an extensive archive and published histories of its hospital, cardiologists, and neurologists.[36]

Many other places play important roles, especially in Boston and Houston. Ambitious doctors at the leading academic medical centers often made the most aggressive efforts to develop new techniques and engaged actively in debates about them. But cardiovascular medicine is a vast enterprise. Relevant contributions have come from all corners of the United States, as well as from overseas. Hundreds of doctors contributed to the stories I tell. It is not always important to keep careful track of who is who, but it is worth noting what perspectives and interests a particular actor might have had in specific contexts. I have tried to strike a balance between providing detailed portraits of the lived worlds of the doctors and patients without overwhelming readers with countless details about people and places. Some stories that I tell only in passing deserve full investigations. Many sources of information remain to be mined, especially medical school curricula, the archives of professional societies, and patient records. I have already made good progress on one follow-up project, a detailed analysis of the prehistory of coronary artery bypass surgery.

No one book can cover all the available material, and this book is necessarily selective. It highlights specific problems and suggests ways of thinking about them, but it does not offer a systematic review of the current state of the medical literature—cardiologists, cardiac surgeons, and others write these. Nor is it an advice manual for patients and their families about making the right decisions. Anyone faced with decisions about angiography, bypass surgery, or any medical treatment should look elsewhere for the latest updates. Instead, my approach is historical, inflected by my training as a physician. It is often easier to subject the past to careful scrutiny than to do so with the present: the time has passed, and the stakes now seem lower. Distance from the past fosters both criticism and sympathy, a balance that can be hard to achieve in the present, when hearts and minds are at stake. Hindsight often makes it possible to discern the subtle influences on how medical theory and practice developed. But strong continuities exist between past and present. Historical perspective can help open the present to thoughtful analysis and critique. Readers of this book should emerge better equipped to make good decisions.

Whether history generates wisdom hinges on what we do with the knowledge. Thoughtful historical analysis can provide clues about what we should carefully consider today as we work to improve the quality of medical care and health policy. The stories herein highlight the barriers that stand between patients, doctors, and perfect medical knowledge and practice. Everyone needs to be aware of the imperfections in what physicians know about efficacy and safety, and of the limits of what they can do. I cannot claim to have all of the right answers, but I do hope that this book will help readers find them.

I am sometimes critical of medical knowledge and practice, but this book should not be read as an attack on doctors. My years as a medical student and a psychiatrist taught me two lessons that are relevant here. First, medicine needs thoughtful critique. It remains an imperfect art and science. Only through careful reflection by physicians and others can continuing progress be made. Second, doctors and patients face extraordinarily difficult problems. The tragedy of human disease, as seen in the suffering caused by heart disease, is vast. The best medical treatments often provide only partial solace. Patients and doctors struggle to do the right thing in difficult situations. I hope that the stories and critiques I present will make readers more sympathetic to these challenges, even as the narrative strives to temper medical hubris with historical perspective. It is important to know not just what we believe, but also why we believe it.

PART I / Theory and Therapy

CHAPTER ONE

The Mysteries of Heart Attacks

Every day in the United States, thousands of men and women submit to angioplasty and bypass surgery. Each technique is a marvel of cardiac science and ingenuity. Nonetheless, the treatments require patients to make a dramatic leap of faith. Look closely at the details of the bypass procedure: after an anesthetist renders the patient unconscious, the surgeon removes arteries and veins from the patient's arms, legs, or chest and then saws open the rib cage to stitch the bypass grafts to the coronary arteries. Angioplasty involves only slightly less bravado. The cardiologist snakes a three-foot-long balloon-tipped catheter into a coronary artery. The balloon fractures the plaque, stretches the artery wall, and deploys a wire mesh stent to prop it open. The newest stents gradually release drugs, including one developed from Easter Island bacteria, to prevent the formation of disruptive scars.

Over one million Americans consent to such procedures each year. Why? Patients, and their doctors, expect bypass surgery or angioplasty to relieve their symptoms and, in some cases, extend their lives. Faith in the procedures is backed up by decades of clinical experience and research showing the many ways in which these treatments can help patients. But careful studies of patients and doctors have also shown that expectations often exceed what experience and research reasonably support. One survey, for instance, found that patients with stable coronary disease expect a ten-year gain in life expectancy from angioplasty, even though no such benefit has ever been demonstrated. Internists have more modest expectations, but even these exceed what evidence-based medicine has found.[1] What has generated excess faith in coronary revascularization?

The simple, intuitive logic of bypass surgery and angioplasty plays a crucial role. When patients and doctors believe that atherosclerotic plaques grow progressively until they cause angina and then heart attacks, the procedures make perfect sense. *New York Times* magazine recently had a back cover advertisement for angioplasty at Mount Sinai Hospital that capitalized on this belief. Atop an image of an evidently fit man hiking across a limitless prairie, the text described how one patient, a plumber, returned to work just two days "after having his own pipes cleaned out." The ad concluded, "Ironic that a plumber came to us to help him remove a clot."[2] The plumbing analogy makes a powerful claim: since angioplasty or surgery can relieve the obstruction or bypass it altogether, they must work. The pairing of theory and therapy is so persuasive that excellent outcomes seem inevitable.

But this faith is only as good as our knowledge of what causes heart attacks. One of the best-kept secrets in medicine is that as late as the 1970s, even though heart attacks killed one-third of all Americans, doctors had not reached consensus about their cause. Harvard pathologist Stanley Robbins vented his frustration in 1974: "No pathologist needs to be told that the more we study myocardial infarction, the less certain our understanding of it." Ongoing research had "only penetrated to deeper levels of ignorance."[3] The persistent debate would seem surprising given that the cause of heart attacks ought to be a matter of simple anatomic fact. Many scientific facts now seem so well established that we take them for granted, for instance, that the earth orbits the sun and blood circulates throughout the body. The history of heart attacks, however, demonstrates something that many historians and philosophers of science have long understood: there is nothing simple about matters of fact. Despite their seeming self-evident obviousness, the facts about the earth and blood only gained status as facts after a long, complex process of observation, debate, revision, and consensus. Scientists for centuries have struggled to determine whether their methods are sound, whether their instruments are reliable, and even who has the authority and credibility needed to make such claims about the natural world. Scientists' efforts to resolve such disputes inevitably reflect social, economic, and political interests that reach far beyond hospital and laboratory walls.[4]

Why did the cause of heart attacks remain mysterious for so long? One of the most important answers was that the two stalwarts of medical research—animal experiments and autopsy studies—offered imperfect representations of the mechanisms of heart attacks in living humans. Armed only with the inconclusive evidence these methods generated, doctors struggled for decades to understand the role of competing theories.

Though the heart attack has a short history in medicine, the phenomenon is not a new one. Humans have presumably suffered from heart attacks for centuries, if not millennia. Archeologists, for example, have probed Egyptian mummies and found evidence of atherosclerosis. Whether ancient Egyptians suffered heart attacks is harder to tell. From Roman times through centuries of European history, wealthy elites lived lives of feasting and indolence that would worry a modern cardiologist. Many of them could afford the luxury of medical care. Despite this, heart attacks made only rare appearances in the medical literature before the twentieth century. William Osler devoted 104 pages to diseases of the circulatory system in the 1892 first edition of his influential *Principles and Practice of Medicine*. Of these, just two pages cover anemic necrosis, the "white infarct" found in some patients who suffered sudden death. Osler also included a section on angina, with his suspicion that coronary arteriosclerosis caused most serious cases. But he followed an old tradition and classified the syndrome as one of the "neuroses of the heart."[5]

In the 1890s the phrase "heart attack" began to pop up in the medical and popular literature, as well as in obituaries, suggesting a certain familiarity with the concept. In 1895, for instance, the *New York Times* ran a story about a gold miner who feigned a heart attack after hitting a rich ore deposit. He wanted to be evacuated from the mine so that he could race off to his broker and place a stock order based on his insider's knowledge before news of the strike spread publically. Such colloquial uses, however, reveal little about what the author meant by the concept. Part of the challenge, as in the history of any disease, is that disease names and diagnostic practices change over time. What we commonly call heart attacks, physicians would now term more formally a myocardial infarction. When the phenomenon gained attention in the early twentieth century, doctors generally labeled it a coronary obstruction or coronary thrombosis. I refer to the phenomenon with our phrase, heart attack, but take care not to confuse past and present understandings of the problem.[6]

Even though the labels have changed over time, any physician practicing now would recognize the signs and symptoms described by doctors in the early twentieth century. Chicago physician James Herrick published an influential account in 1912. One of his patients, a 42-year-old physician, previously in good health, "was seized . . . with a sudden, excruciating pain in the lower sternal region, which pain radiated to the arms and to the epigastrium. He was profoundly shocked, very weak and nauseated, the skin cold and clammy, the pulse rapid and thready. His colleagues who saw him thought he would die in a very short time." Another patient, vacationing at Atlantic City, developed "terribly severe pain" at the boardwalk. When he returned to Chicago, his doctor found his heart "dilated, weak, and rapid." He died two weeks later.[7]

Although the clinical symptoms were clear, the cause was not. When Herrick diagnosed his patients with "coronary thrombosis," his diagnosis made a causal claim. He believed that his patients had died because of a blood clot obstructing a coronary artery. Autopsy of the 42-year-old doctor, for instance, revealed a large thrombus in two of the major coronaries arteries, with smaller plaques throughout the coronary tree, and several large scars in the cardiac muscle. But Herrick knew that there was no perfect relationship between cause and outcome: "While such obstruction by a thrombus was very often suddenly fatal, it was not necessarily so." Some patients died of heart attacks but had no coronary disease at autopsy. Other autopsies found extensive coronary disease in patients who had never felt a twinge of angina. When Herrick presented his initial discussion of coronary obstructions in 1912, he summarized seventy years of medical research about coronary arteries and had no trouble explaining the variable outcomes. Extensive research on animals, supplemented by occasional clinical reports, suggested that interconnections sometimes existed between the coronary arteries. If one artery became obstructed, then blood could flow through the collaterals and supply the heart tissue downstream of the initial obstruction.[8] The outcome of any particular occlusion depended on details of the patient's coronary anatomy.

Physicians dislike such uncertainty. For decades after Herrick's work, they continued to revisit the question. They studied patients who had died of heart attacks, hoping that if they looked closely enough at the autopsy specimens, they would detect subtle but consistent findings that linked a pathological process to the clinical outcome. The vagaries of human pathophysiology dashed such hopes time and time again. One team of Boston doctors learned this the hard way. In the 1930s they developed a meticulous technique to study the heart. They removed the heart from the deceased's chest, injected a colored lead gel into the coronary arteries, cut the heart open so that it would lie flat on an x-ray plate, took an x-ray of the injected tissues, and then did a careful dissection of the arteries themselves. All for nothing. Their results suggested that the "site of the occlusion or occlusions in the coronary arteries bears no necessarily constant and immediately obvious relationship to the location of an infarct which may be found in the heart." After completing a painstaking analysis of 335 autopsies, they concluded, "There is no characteristic syndrome necessarily associated with coronary arterial occlusion."[9] Like Herrick before them, they could find patients with heart attacks, and they could find obstructions in coronary arteries, but they could not find a clear relationship between heart attacks and obstructions.

Frustrated physicians wanted to know why heart attacks occurred when and where they did. Atherosclerosis was a chronic process that evolved over a patient's lifetime. What happened at a particular moment at a specific plaque to transform atherosclerosis into the catastrophe of a lethal heart attack?[10] Doctors have several basic options when they set out to explore the underlying causes of the diseases that their patients suffer.

The oldest, and often still effective, approach relies on clinical history. Doctors can talk to patients to learn about the timing and experience of the symptoms. Sometimes the clinical history alone is sufficient to clinch a diagnosis, but rarely does knowledge of the history and diagnosis reveal the cause of what went wrong. Coronary artery disease, for example, did not give up its mysteries easily. The classic symptom, the squeezing chest pressure called angina pectoris, does not occur in all patients. Some have symptoms that resemble heartburn, others develop fatigue or shortness of breath, and still others—especially women—have no obvious symptoms at all. Even when doctors correctly relate a symptom to the heart, it can be difficult to determine where in the heart the problem arises or what the underlying cause might be. Angina warns patients and doctors that something bad is happening, but it does not give away its cause.

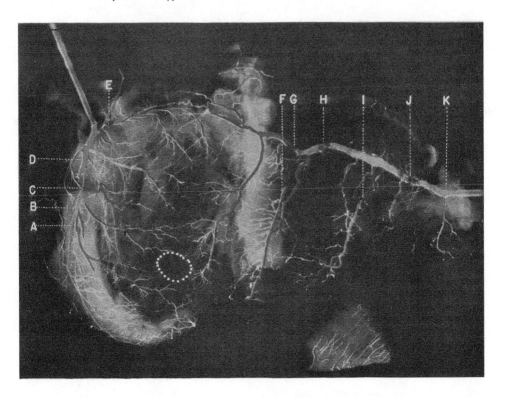

Electrocardiograms, which came into wide use in the 1920s, gave cardiologists new ways to study the heart. The characteristic tracings, now a staple of television medical dramas, reveal some of the heart's underlying electrical activity. Doctors have acquired great skill at reading the tracings to detect a range of cardiac disturbances. Careful analysis can detect subtle signs of coronary disease. Elevation of the ST-segment, for instance, signifies acute myocardial ischemia, a sign that a heart attack might be underway. Pronounced elevations, so-called tombstone changes, have an especially dire prognosis. Electrocardiograms can also pinpoint the disease location in a region of the heart and sometimes in a particular coronary artery, for example, an anterolateral infarct often associated with obstruction of the left anterior descending coronary artery. But as with the clinical history, an electrocardiogram reveals little about the specific cause of a heart attack.

To solve the mystery of heart attacks, doctors realized that they might need to study a living human heart at the moment of an attack, but such a direct approach was not possible until the advent of selective coronary angiography

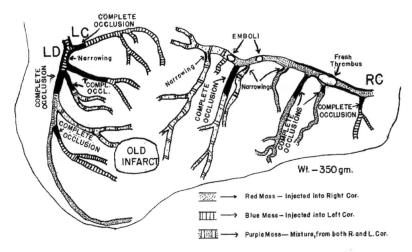

Postmortem angiography (*opposite*) and interpretation (*above*) of coronary artery occlusions, thrombus, and emboli. Using such techniques, researchers sought to map the distribution of coronary atherosclerosis. From Herrman L. Blumgart, Monroe J. Schlesinger, and David Davis, "Studies on the Relation of the Clinical Manifestations of Angina Pectoris, Coronary Thrombosis, and Myocardial Infarction to the Pathologic Findings, with Particular Reference to the Significance of the Collateral Circulation," *American Heart Journal* 19 (Jan. 1940): 1–91, fig. B, p. 5, fig. C, p. 6. With permission from Elsevier.

in the 1960s and the subsequent emergence of coronary artery bypass surgery. In the meantime, doctors had to make the most of the traditional approaches, animal research and postmortem studies. Each, however, was an imperfect approximation of a living human. Neither proved capable of providing definitive answers.

Physicians had long turned to animal research to study both physiology and pathophysiology. Ancient Greek physicians, largely forbidden from dissecting humans, dissected dead and living animals to probe the body's secrets. In the late nineteenth century, French physiologist Claude Bernard made animal research the centerpiece of his vision of modern medicine. Countless medical scientists followed his lead throughout the twentieth century. Cardiologists in particular made extensive use of laboratory animals. One 1966 review described a menagerie of animal models and animal organ preparations: cows, rabbits, dogs, cats, frogs, and humans, as well as normal hearts, isolated hearts, enzyme-digested embryonic hearts, transplanted hearts, resting and beating guinea pig atria, cat and dog ventricles, and acutely failing dog hearts.[11]

Such animal research, however, provided little insight into heart attacks. Animals in the wild rarely suffer atherosclerosis. Scientists who wanted to study this disease had to work assiduously to induce atherosclerotic lesions in laboratory animals. In 1908 Russian physician A. I. Ignatowski first reported that rabbits fed a nonvegetarian diet would develop the disease. Nikolai Anichkov picked up this lead. Working in Russia and then Germany between 1909 and 1913, he accomplished the feat by feeding rabbits sunflower oil mixed with cholesterol. Microscopic examination revealed that the basic components, structure, and distribution of laboratory atherosclerosis resembled those seen in humans. Other researchers, however, could not replicate this finding in rats or dogs. Moreover, the clinical relevance of his initial work was unclear because Anichkov had raised the rabbits' cholesterol levels well beyond those seen in humans. Doubts about whether animal atherosclerosis provided a useful proxy for human atherosclerosis undermined the usefulness of this research for decades.[12]

Animal models of heart attacks proved even more problematic. It was one thing to induce atherosclerosis, but quite another to provoke a full-fledged heart attack. Spontaneous heart attacks in nonhuman mammals were exceedingly rare (and often poorly documented). In 1959 researchers at Northwestern University finally succeeded at inducing a heart attack in a laboratory animal. They were surprisingly frank about the rigors of this twelve-year project. They had imported fifteen rhesus monkeys from India and fed them a diet of monkey chow mixed with cholesterol, butter suspended in water, and bread soaked in cream. They lost several to fulminant tuberculosis, which swept through the animal lab. The monkeys, "completely undomesticatable," ate only some of their food and threw the leftovers around their cages. Some escaped from the cages, and researchers spent hours chasing them as they swung among the hanging light fixtures. Years of frustrating work eventually paid off. One female monkey struggled mightily with the researchers as they removed her from her cage to photograph her. The strain was too much: half an hour later, she collapsed and died. When the researchers performed the autopsy, they saw three thrombotic occlusions and a massive myocardial infarction. They concluded that they had achieved the first heart attack in an experimental animal.[13] The extensive work required to induce a single heart attack in an animal experiment prevented researchers from making widespread use of this approach.

While some researchers worked to develop animal models, others relied on the tried-and-true method of clinical-pathological correlation. Since the nineteenth century, doctors had built an edifice of clinical knowledge by comparing clinical studies of patients to the pathological findings from autopsies. Boston pathologist Timothy Leary (no relation to the counterculture guru) described this in 1939: "Careful observation of series of cases under hospital conditions and expert postmortem study of the changes produced in the bodies of victims of fatal forms of disease furnish the basis of knowledge which may suggest procedures for experimental research, and may lead to knowledge of the causation, treatment, or, most important, the prevention of diseases."[14] But Leary, who grappled with many medical mysteries in a remarkable forty-two-year career as Boston's medical examiner, knew that this was only a best-case scenario. Initial studies of heart attacks, such as those by James Herrick, yielded inconsistent results. Many competing theories soon appeared.

Doctors knew well that attacks often occurred during emotional excitement. Leary, for instance, described how one father, "much stirred emotionally" in expectation of hearing his son's violin recital, died suddenly. Another young man, a bookie, dropped dead after being harangued by his boss for arriving at work intoxicated. Physical exertion or anything else that increased demand on the heart could also cause problems. Leary emphasized the particular dangers of indigestion, physical stress, and exposure to cold. Two other Boston doctors produced a more detailed list of triggers, including "alcoholic debauch," travel, gorging, gardening, swimming, running to catch a train, golf, watching exciting sporting events, and "excessive sexual activity."[15] Doctors hoped that by cataloging the triggers of heart attacks, they might be able to discern the underlying causes.

This initial study of triggers, however, provided only another partial answer to the question of cause. What did the triggers do that stopped the heart? Leary often wondered if a triggering event could induce spasm in an artery that was already narrowed by an atherosclerotic plaque. An autopsy of the bookie revealed no evidence of atherosclerosis. Was it possible that fatal spasm could occur in healthy arteries? Other doctors, however, dismissed the importance of triggers. After all, many patients died of heart attacks in their sleep. These doctors argued that coronary artery disease progressed according to its own inexorable logic, with the fatal culmination occurring at its own pace.[16]

Pathologists proposed many mechanisms to explain how slow, but relentless, progression could occur. Their studies suggested that atherosclerotic lesions accumulate progressively with age, from the earliest traces seen in young children and teens to the extensive disease seen in the elderly. Microscopic analyses showed that such growth came from the continuing accumulation of cholesterol deposits and the continuing proliferation of muscle cells and fibroblasts. It required little imagination to see how such growth could narrow a blood vessel until no blood flowed.[17] After all, this often happened with household pipes as debris accumulated slowly over time. This simple theory appealed to both common knowledge and common sense.

A related theory invoked a slightly more dramatic mechanism. As atherosclerotic plaques grew, they disrupted the normally smooth lining of blood vessels. These rough patches could activate the clotting mechanisms in blood and become the point of attachment for a small blood clot that would coat the inner surface of the vessel. As atherosclerosis continued, it triggered repeated bouts of clotting and thrombus deposition. Over time layers of thrombi accumulated, like layers of silt in a river channel, until flow slowed and then ceased.[18]

The theories of progressive obstruction and thrombus deposition emphasized gradual processes occurring slowly over many years. Other theories had a more catastrophic perspective. Some researchers believed that as plaques formed in the walls of blood vessels, they induced the growth of tiny capillaries into the vessel wall to supply blood to the growing plaque. These fragile intramural capillaries were a disaster waiting to happen. Whether spontaneously or in response to a blood pressure surge, they could rupture and bleed into the wall of the coronary artery. If they bled forcefully enough, the artery wall would swell and pinch off flow through the artery itself. J. C. Paterson, a Canadian pathologist who found evidence of such intramural hemorrhages in most of his autopsies, defended this theory vigorously in the late 1930s.[19]

Timothy Leary came to favor a different, though still dramatic, theory. In his work as medical examiner, he had the chance to examine many victims of sudden death. Often, especially in older patients, he found atheromatous cysts in the coronary arteries: toxic collections of fat, cholesterol crystals, and dead cells. When these ruptured, they spilled their debris into the blood vessel. This toxic brew triggered the clotting cascade, and blood quickly congealed into an occlusive thrombus. Leary also undertook animal studies. Like Anichkov a generation earlier, he fed rabbits diets high in cholesterol to induce atherosclerosis in their short rabbit lives. He even succeeded, where Anichkov

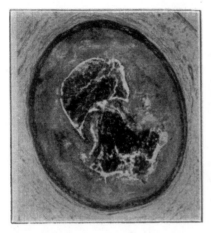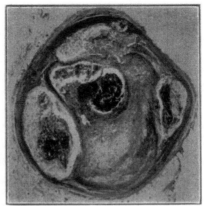

Histological specimens of plaque rupture. An atheromatous abscess (*left*), in the bottom half of the coronary artery wall, that has ruptured and spilled its contents into the artery lumen (upper left), taken from autopsy of a 66-year-old man. Four abscesses (*right*), one of which has ruptured into the lumen, taken from autopsy of a 62-year-old man. Timothy Leary, "Pathology of Coronary Sclerosis," *American Heart Journal* 10 (Feb. 1935): 328–37, figs. 7 and 8, p. 334. With permission from Elsevier.

did not, in producing advanced lesions of atherosclerosis, especially atherosclerotic abscesses. One of his rabbits may have even died from this induced coronary artery disease. Although Leary saw atherosclerosis as a chronic disease of disrupted cholesterol metabolism, death could occur suddenly, "usually due to the rupture of an atheromatous 'abscess.'" The drama of these ruptures might have appealed to Leary's morbid sensibilities. As he continued to work as Boston's medical examiner, he became fascinated by disasters, such as the 1943 Cocoanut Grove nightclub fire and the explosion at the East Ohio Gas Company in Cleveland in 1944.[20]

Leary's theory won him praise from colleagues near and far. One South Boston physician wrote to Leary in 1934 to say, "Nothing should give more worth-while fame and personal satisfaction than the intellectual feat of clarifying the awful mystery of coronary disease." F. G. Banting, renowned for discovering insulin, thanked Leary for participating in a 1934 conference: "The symposium on arteriosclerosis was one of the finest discussions that I have ever attended. I was tremendously interested in your observations, and particularly in the experimental work which was so splendidly presented."

Leary also critiqued the other hypotheses, especially the claims about intra-mural hemorrhages, writing in 1938 that conclusions about the significance of vascularization were not justified.[21]

Debate nonetheless raged from the 1930s through the 1960s. As one pathologist later recalled, "While the nation waged periodic wars, leading cardiologists and pathologists did likewise." The hemorrhage hypothesis became increasingly popular, especially among pathologists. The more carefully they looked for pockets of blood or hemorrhage in coronary artery walls, the more often they found them. Arthur Master and his colleagues at New York's Mount Sinai Hospital even recommended, in 1944, replacing the term "coronary thrombosis" with "coronary occlusion" because the intramural hemorrhages could cause heart attacks without a thrombus.[22] Their insistence on this change demonstrates an important point. The names given to a disease by physicians can have profound implications for how they think about the disease. Although Herrick thought in 1912 that "coronary thrombosis" described the problem well, Master believed that the phrase made unwarranted assumptions. As long as the role of thrombosis remained unclear, these physicians thought a more neutral term might be more appropriate. A rose by any other name might smell the same, but in medicine physicians need to worry about the connotations and consequences of the labels they assign.

Resolution eluded investigators through the 1960s. One problem was the complexity of human disease. When pathologists examined autopsy specimens, they often found evidence not just of hemorrhage, but also of rupture and thrombus. Which was the cause, and which were the red herrings? Researchers from the warring camps came together under the auspices of the American Heart Association in 1962 to attempt a consensus. They could agree only on the vaguest of conclusions: "Coronary atherosclerosis is a progressive process; at any stage, factors, as yet unknown, may precipitate either sudden death or the patterns of myocardial ischemia or necrosis." This was the state of affairs that so maddened pathologist Stanley Robbins, who wrote in 1974 that the only conclusion that could be drawn about the relationship between thrombosis and heart attack was that "the issue continues to be vexed."[23] Robbins and others could only hope that a unifying theory would someday emerge.

CHAPTER TWO

The Case for Plaque Rupture

In March 1844 sculptor Bertel Thorvaldsen attended a performance at the Royal Theatre in Copenhagen. We do not know what he thought of the show, but we do know what happened to him that night: he died suddenly at the theater. An autopsy performed at his home revealed the likely cause. As reported in the *Journal of the Danish Medical Association*, Thorvaldsen's coronary arteries contained "several athromatous lesions, of which a rather significant one was ulcerated and the atheromatous mass extruded into the arterial lumen." This report earned little attention because Thorvaldsen's ruptured plaque was but one of the flood of findings of unknown significance that emerged from the morgues and dissecting rooms of European hospitals in the mid-nineteenth century. Nearly one hundred years later, as we have seen, Boston pathologist Timothy Leary found ruptured plaques in his own autopsy series, although plaque rupture was one of several competing hypotheses about the cause of heart attacks. In the following decades, Leary's work received less attention than the rival theory of intramural hemorrhages. Only in the 1990s, nearly one hundred and fifty years after its first clear description, did plaque rupture capture the medical imagination.[1]

How and why does a theory gain influence, or not, at a particular time and place? Physicists write poetically about how the elegance or beauty of a theory gives it credibility, while biologists cite the sheer weight of evidence that supports many of their theories. Historians have looked elsewhere for answers, describing how social and political interests influence the uptake and even the development of scientific ideas. The plaque rupture hypothesis demonstrates how medicine's engagement with the meaning-laden world of disease and its commitment to therapeutic intervention influence the ways in which doctors make the case for specific ideas. The case for plaque rupture depended on vision and labor, specifically the different techniques pathologists used to visualize the micro-anatomy of human bodies and the amount of painstaking work required to do so. Two tensions, both well known to medical science, colored the debates. First, researchers pushed animal research to the limits of its plausibility as a model for human disease. Second, they navigated the intersection of laboratory theory and clinical practice, specifically the debates between cardiologists and pathologists about the best way to treat a heart attack.

The search for adequate animal models of atherosclerosis continued into the 1960s. Studies of animals living in the wild and in captivity revealed that animals sometimes, if rarely, developed atherosclerosis. One 1964 conference assembled experts on atherosclerosis in different animals. Over four hundred species were discussed, including humans, marsupials, anteaters, monkeys, dingoes, ocelots, tapirs, yaks, penguins, crows, toucans, hippos, sperm whales, a 20-year-old monitor lizard, red pandas, spectacled bears, immature bluefin tuna, and broad-breasted bronze turkeys. The research, however, often revealed as much about the researchers' assumptions as it did about the animals. Orcas, according to one small study, developed advanced atherosclerotic lesions, while dolphins lacked such "stigmata of severe disease." The researchers attributed this difference in part to the cetaceans' personalities. Pacific striped dolphins, who cavorted playfully in aquaria, were "very docile." Killer whales, in contrast, were "aggressive, malicious, man-eating beasts, wily and willful, with complex, highly developed brains." Similarly, when the Northwestern researchers described their successful efforts to induce a heart attack in a rhesus monkey (as described in the previous chapter), the monkey's personality figured prominently in their accounts. Yes, they had put her on a diet of 42 percent butterfat to induce atherosclerosis, and yes, the fatal attack had occurred as she had grappled with the researchers. But the researchers were quick to add that the monkey had always been "hyperactive, nervous, and highly emotional."[2]

One researcher, Vancouver pathologist Paris Constantinides, set out to overcome the limits of observational studies and create a reliable animal model of advanced atherosclerosis. The work required a delicate balance. Humans develop atherosclerosis over decades, but laboratory animals do not live that long (and few scientists would have had such patience). Researchers could accelerate the process by feeding the animals extremely high-fat diets, but if they pushed the diet too hard, the animals died quickly from cholesterol poisoning. Constantinides found a solution. By putting rabbits on a diet that alternated between high-fat and low-fat months, he could keep them alive long enough—ideally more than two years—for them to develop advanced atherosclerotic plaques, with atheromatous abscesses, necrosis, cholesterol crystals, and calcifications. The rabbits' plaques resembled human lesions "down to their last anatomic detail."[3]

But Constantinides wanted heart attacks. After experimenting with various drugs, he learned that he could reliably induce a coronary thrombosis in rabbits with advanced atherosclerosis by injecting them with a two-drug cocktail, one to increase blood pressure and one to make blood clot more easily. A combination of epinephrine (for the pressure) and Russell's viper venom (for the hypercoagulability) did the trick. This reliable animal model of heart attacks allowed him to do a systematic pathological analysis to find out what had happened. Epinephrine caused the rabbit's blood pressure to surge, the "violent stretching" of the vessel wall cracked open an atherosclerotic plaque, and reaction of the plaque with the snake venom triggered a thrombus every time.[4] But even with this success, Constantinides faced the question that had forever plagued animal researchers. Did the animal model adequately represent what happened in humans? He had shown that ruptured plaques caused coronary thrombosis in rabbits. Did this happen in humans as well? Since his rabbit model and its viper venom hardly reflected human conditions, he could not say.

Intrigued and frustrated that the cause of heart attacks remained "wrapped up in mystery" and "wide open to speculation," Constantinides sought evidence of plaque rupture in humans. Having relocated to Barnes Hospital in St. Louis by 1963, he studied in excruciating detail coronary artery specimens from twenty patients who had died with coronary thromboses. Instead of slitting open the coronary arteries and eyeballing them for evidence of thrombosis, he used a rigorous and standardized protocol. He or his assistant cut the coronary arteries into serial cross-sections, each only 0.007 milli-meters thick. He mounted thousands of such sections onto microscope slides, stained them, and examined them under his microscope. In case after case, he found a fissured plaque at the base of every thrombosis. When he reported his initial results at a Chicago meeting of experimental biologists in April 1964, a reporter from the *Journal of the American Medical Association* wrote an enthusiastic account: "While fissures in the atherosclerotic lining of throm-bosed coronary arteries have been observed occasionally in the past and have therefore been considered an infrequent factor, his own findings suggest that such breaks are a constant causal factor." Constantinides published a full ac-count of his human findings in 1966, confident enough in his methods and their consistent findings to make a strong causal claim: "Since we did not find any thrombi without fissure or any fissures without thrombi, we are forced to conclude that there was a causal association between these two findings."[5] Even though he deployed the rhetoric of a cautious scientist forced by his data to accept a difficult conclusion, he had tenaciously built his case from animal research and clinical observation. He believed that he had found the cause of heart attacks.

San Francisco cardiologist Meyer Friedman converged on the same answer from a different direction. Friedman is best known for his ideas about the type A personality. Starting in the 1950s he had popularized the idea that aggres-sive, hard-driving people (and apparently orcas) brought coronary artery dis-ease on themselves. This idea captured the public and medical imagination for decades, enshrined in the many images of workaholic executives dropping dead from a disease they had brought on themselves.[6] He spent much of his career exhorting his patients to adopt healthier approaches to life. Friedman had more conventional interests as well. In the early 1960s he grew perplexed by a disconnect between heart attack theory and therapy.

As heart attacks killed more and more Americans, doctors desperately sought new treatments. Many turned to heparin. First isolated from dog liver in 1916 and then purified in 1929, heparin had anticoagulant properties and could prevent blood from clotting (hence its common description as a "blood thinner"). Starting in the 1950s, cardiologists enthusiastically adopted hepa-rin for treatment of heart attacks. The idea was simple. If coronary thromboses (blood clots) had anything to do with heart attacks, which had been suspected since Herrick's day, then a drug such as heparin, that prevented blood from clotting, would surely help. But even as cardiologists welcomed heparin, pathologists increasingly accepted the intramural hemorrhage hypothesis— that heart attacks began when capillaries within coronary arteries ruptured and bled into the artery wall. When Friedman surveyed pathologists in 1962, he found that 70 to 90 percent of them believed that hemorrhages caused coronary occlusions. If they were right, then heparin made no sense: the problem with heart attacks was not the thrombosis but the hemorrhage that had caused the vessel wall to swell and pinch off the coronary artery. Heparin might even be dangerous. Its anticoagulant action would exacerbate the bleeding and occlusion caused by intramural hemorrhages.[7] Friedman realized that cardiologists were pursuing a path most pathologists considered foolhardy.

Frustrated by the schism between theory and practice, "between the opinions and conclusions of the majority of pathologists and those of the physiologist, clinical investigator, and clinician," Friedman set out to determine who was right. He had to find the cause of heart attacks. Unlike Constantinides, he did not believe that animal research would yield an adequate answer. Existing animal models, as he explained only half in jest, had too many limitations: "We have not yet been able to replicate completely the human situation in the laboratory. We have yet to study cigarette-smoking, poorly exercising, obese, high blood-pressured, diabetic, 'mesomorphic' genetically predisposed animals, who also ingest cholesterol and animal fats, and compare them with animals devoid of these traits and habits." Dismissing the value of animal studies and unable to study heart attacks in living people, he jumped directly to autopsy studies. He obtained fifty-seven coronary artery segments from patients who had seemingly died from coronary artery disease and, like Constantinides, cut them into exceedingly thin sections, 0.050 to 0.100 millimeters each. In forty-one cases he found an occlusive thrombus, and in thirty-nine of these, he found evidence of plaque rupture. Although he often found blood within the artery wall (the sort of observation that had motivated the intramural hemorrhage hypothesis), he realized that this blood had entered the wall not from a ruptured capillary but from the coronary lumen through the ruptured plaque. Exposure of the blood to the atheromatous abscess and its contents had triggered the thrombus. Friedman concluded that plaque rupture, a "luminal catastrophe," was thus the cause of both the thrombi and the intramural bleeding seen by previous investigators.[8]

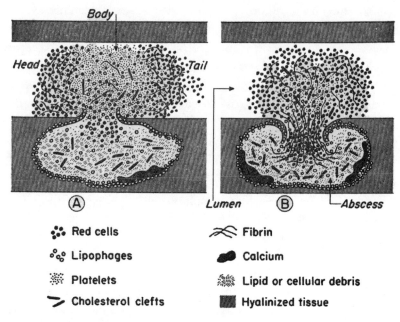

Diagrams of plaque rupture. What happens when a plaque ruptures and discharges its contents (e.g., lipophages, cholesterol clefts) into the lumen, forming a blood clot of red cells, platelets, and fibrin (*left*). What happens when a plaque collapses and fills with blood (*right*). Friedman believed the left-hand mechanism caused heart attacks. From Meyer Friedman and G. J. Van den Bovenkamp, "The Pathogenesis of a Coronary Thrombus," *American Journal of Pathology* 48 (Jan. 1966): 19–44, fig. 1, p. 25. With permission from Elsevier.

Two aspects of the work of Constantinides and Friedman deserve special comment: the role of labor and the importance of visualization. Their meticulous analyses of coronary arteries required back-breaking work. Constantinides, who scrutinized over forty thousand microscope slides, often described the unusual effort he had made. As he told the *Journal of the American Medical Association*, "Only a few cases had been studied in this manner in the past." Of course, as is so often true, the work of the successful researcher was made possible by a long-suffering assistant who bore the brunt of the labor. For Constantinides, this was Sharon Miller, whom he thanked "for her patient technical assistance with the preparation of the serial sections of the study." Friedman similarly explained that few people had undertaken such diligent studies because of "the sheer tediousness entailed in viewing thousands of sections and making the drawings that are necessary for successful reconstruction of the total thrombosed segment under scrutiny. Such work is undramatic and, only too often, downright boring." Only a few other groups invested the enormous effort required. When they did, they too found evidence of plaque rupture.[9]

The effort invested by Constantinides, Friedman, and their indefatigable laboratory technicians had a clear payoff in that it allowed them to claim the methodological high ground. When skeptics did not find ruptures in their own series, Friedman and Constantinides could say that their critics had just not searched hard enough. One particular exchange is revealing. William Roberts, who led the pathology division of the National Heart and Lung Institute (formerly the National Heart Institute) in the 1970s, cast doubt on plaque rupture research. He agreed that researchers would need to cut serial sections in order to detect thrombi and, beneath them, ruptures. However, he was skeptical of the value of this approach: "Even by this means interpretation is often difficult." The detailed structure of the artery and its plaques, for instance, could be damaged by the fixation and sectioning needed to prepare the tissue for microscopic analysis. As a result, "Interpretation of whether a plaque is cracked or, if present, an artifact is real is fraught with too much difficulty, in my opinion, to give this possible mechanism of thrombosis undue weight." He cited his own work as evidence against plaque rupture. In his study of 107 patients, he found thrombi in only 42 of them. Moreover, "Cracks in plaques were infrequently observed beneath thrombi in the patients studied by us, and probably most thrombi form over plaques with intact surfaces."[10]

The details here are important. Roberts and his collaborator did not look as closely as Constantinides and Friedman had done. Roberts had divided the coronary artery into 5 millimeter segments and then cut and examined two or three cross-sections from each, an average sampling interval of 1.7 to 2.5 millimeters. Constantinides and Friedman had sectioned every 0.007 to 0.10 millimeters, as much as two hundred times more densely. Roberts was confident that his technique was adequate: "The chance of missing a thrombus by this technique is unlikely." If his method had not revealed thrombi and ruptures, then they did not exist. Friedman had harsh words in response. He described how Roberts had disparaged the careful work that he, Constantinides, and others had done, the kind of work "of course, he himself had failed to do."[11] Friedman gave Roberts's critique no weight. For Friedman, only researchers who had matched his rigor—and effort—deserved to join the debate.

Scientific ideas usually gain credibility through replication. Someone makes a discovery, others repeat the experiment or observation, and the more often this happens, the more confident they can be that the discovery is correct. Friedman used the rhetoric of replication to bolster his case. As he pointed out, plaque rupture had been described independently by three sets of researchers and confirmed by two others. But Friedman also exploited a rhetoric of nonreplication: since the skeptics, such as Roberts, had used less fastidious methods and had not really attempted to replicate his own work, they could be ignored. With so few other groups willing to expend the effort, there would be less competition about plaque rupture in the field of scientific ideas.

Friedman's response to Roberts's critique raises one other important issue: visualization. Ever since Galileo turned his telescope to the stars, scientists have sought to use new technologies to render the invisible visible. And ever since Galileo demonstrated his telescope to the courtiers and clergy of Florence, skeptics have wondered whether visualization technologies could be trusted. Scientists, and the historians, sociologists, and philosophers who study them, have realized that telescopes, microscopes, and the myriad other ways of seeing do not capture an unmediated image of nature. Instead, scientists and their devices produce carefully crafted representations that reveal some aspects of the object being studied but obscure others, with the images often designed to strengthen a particular argument.[12] Physicians, for instance, have long worked to make the causes of disease visible. With the rise of physical examination in the nineteenth century, doctors developed a new "clinical gaze" that led them to see the disease that lay underneath a patient's signs and symptoms. Autopsies opened the body and its organs for direct visual inspection. The word itself, "autopsy," is derived from an older word related to eyewitness testimony, to knowledge derived from actually being there.[13] Microscopes allowed pathologists to zoom in and see the fine structures of bodies and their diseases at a resolution invisible to the naked eye. Light microscopes could magnify by a power of ten, one hundred, or even one thousand. Electron microscopes could go even further.

A specific assumption motivated this push from autopsy to light microscopy and on to electron microscopy: more magnifying power and higher resolution would bring the observer closer to truth. Friedman invoked this explicitly against Roberts's critique, noting that his "reasoning, I believe, is no more valid than that of a light microscopist who, unable to detect the cristae of mitochondria with his own instrument, depreciates the validity of the discovery of these structures by the electron microscopist."[14] The trade-off was clear. The more thinly pathologists sliced tissues, the more they could learn, but at the cost of potentially limitless labor. Friedman pushed the rigor to one level, 0.10 millimeter sections examined with light microscopy. Constantinides went to 0.007 millimeter sections. Someone could have gone further: 0.001 millimeter sections with electron microscopy. But no one did. This left Friedman confident that he, along with Constantinides and the three other groups, had produced a definitive enough answer. Their claims about plaque rupture had unmatched methodological credibility and rhetorical power. Their confidence, however, proved premature.

CHAPTER THREE

The Case against Plaque Rupture

Paris Constantinides and Meyer Friedman had great confidence that their methods had produced definitive evidence in support of the plaque rupture hypothesis. The three other groups who employed comparably rigorous techniques converged on plaque rupture as the most likely cause of heart attacks. One might expect that the results of such methodical scientific research would speak for themselves. In practice, however, this does not always happen. Almost any phenomenon in science or medicine is open to multiple interpretations.[1] Nature is complicated, scientists are creative, and many factors influence how scientists resolve debates.

The search for the cause of heart attacks demonstrates this well. When Timothy Leary and others first made serious attempts to decipher the cause in the 1930s, they proposed a series of theories, from progressive obstruction to intramural hemorrhage, spasm, plaque rupture, and others. But they could not produce evidence for any one theory that convinced everyone. A similar story played out in the 1960s. Even as Constantinides and Friedman developed one line of evidence—autopsy analyses—to a high state of refinement, other scientists introduced other ways of studying and visualizing the pathogenesis of heart attacks. These competing methods supported different theories and prevented closure of the heart attack debate in the 1960s and 1970s. Even as bypass surgery became more and more popular in the 1970s, physicians remained uncertain about the cause of the problem that bypass hoped to prevent.

The challenges to the plaque rupture hypothesis came from many directions. The primary challenge arose from old doubts about the relevance of coronary thrombosis. Even if Friedman and Constantinides had produced compelling evidence that a plaque rupture was at the root of every coronary thrombosis, they had not convinced their colleagues that a coronary thrombosis caused every heart attack. The rigor of their method proved to be an obstacle here. Because it took so long to study the hundreds or thousands of cross-sections from a single patient, the researchers could examine only a small number of patients. Did their small samples fairly represent the entire population of people who suffered heart attacks? Did thromboses cause the attacks in all of them? Those questions lingered for decades.

Many of the competing theories from the 1930s remained ongoing concerns in the 1960s. Although Leary came to favor the plaque rupture hypothesis, he had also described cases in which spasm appeared to play the dominant role. Remember the violinist's father who dropped dead from excitement before his son's recital, or the bookie who died after a reprimand at work? When Leary dissected their bodies, he found no remarkable disease in their coronary arteries and concluded that the emotional arousal had triggered coronary spasm. He acknowledged that his evidence was circumstantial: "The clinical and pathological material here presented can be, of course, only suggestive as to the possible role of coronary spasm leading to attacks of angina or in producing sudden death."[2] In the absence of positive evidence that spasm had occurred, the idea gained little traction.

Coronary spasm hypotheses made a comeback in the 1950s. In 1959 Los Angeles cardiologist Myron Prinzmetal described what he called "variant angina." Unlike classic angina, it was not triggered by exertion and relieved by rest. It occurred and resolved spontaneously. Having seen spasms in the arteries of rabbits' ears, he suggested that spontaneous changes in vascular tone could cause angina and even infarction. This fit well with popular notions—ironically made popular by Meyer Friedman's type A personality hypothesis—that linked stress to coronary disease. Supportive evidence came in the 1960s as doctors described good clinical results using nitrates and other drugs that relaxed the muscle cells in artery walls and relieved vascular spasms.[3]

In one remarkable case doctors realized that the spasm theory could explain an outbreak of angina and heart attacks among workers at a Wisconsin munitions factory. In 1966 a factory near Milwaukee reopened to make nitroglyercin-cellulose fuel, used as a solid rocket propellant. Although employees wore gloves and aprons to minimize their contact with the chemicals, some exposure occurred. Workers suffered from headaches, weakness, dizziness, and nausea, all symptoms of nitroglycerin poisoning. Many had to quit their work at the factory. Others acclimated to the exposure: their bodies increased their vascular tone to balance the vasodilating effects of the chemical. This adaptation, however, had a cost. On weekends some workers developed symptoms of nitroglycerin withdrawal, with rapid heart rates, high blood pressure, and chest pain. One worker even suffered a heart attack at six o'clock on a Monday morning, 62 hours after her last "dose" of nitroglycerin during her Friday shift. She awoke to chest pain radiating to her neck and arms, along with shortness of breath, pallor, and sweats—the classic symptoms of an acute infarction. Even though a doctor arrived within minutes, she lapsed into a coma and died quickly. Other workers suffered nonlethal heart attacks. In none was there any evidence of significant coronary atherosclerosis. The doctors who investigated the outbreak instead concluded that nitroglycerin withdrawal on weekends triggered rebound vascular spasm that could be severe enough to kill.[4]

Many other researchers agreed that coronary thromboses had no monopoly on heart attacks. Even though Constantinides and Friedman had seen a thrombus in nearly every one of their cases, other researchers reported different experiences. From the 1930s into the 1970s, team after team studied victims of heart attack and sudden death and reported what percentage of them had coronary thromboses. Results ranged an astonishing amount, from zero to 100 percent. Everyone knew that the answers depended in part on the details of the researchers' methods and their assiduity. Two Edinburgh pathologists, for instance, noted that "the incidence of coronary thrombotic occlusion is a direct function of the care of the technique employed in examination of the heart." What technique did they trust? They used a butcher's meat-slicing machine to cut 2-millimeter sections of frozen hearts and coronary arteries. Occasional, more detailed, checks convinced them that their method "appears a legitimate and safe variant of the immensely laborious task of preparing serial sections of several cms. of large numbers of coronary arteries." They examined one hundred hearts and found occlusive thrombi in only thirteen. Two Oxford cardiologists, however, condemned this method. Although they also used a bacon slicer, they examined the coronary arteries carefully before sectioning the heart and found occlusions in 70 percent of their patients.[5]

The debate was not just about the best way to slice a heart. Researchers introduced many new methods in their effort to resolve the debates. Constantinides and Friedman had taken one method—examination of serial cross-sections of coronary arteries—to a rigorous extreme. But there were other ways to visualize coronary arteries and their plaques. An Italian researcher, Giorgio Baroldi, injected a plastic resin into postmortem coronary arteries. Once the plastic had fixed, he could remove the tissue and reveal a plastic cast of the entire arterial tree. When he used this technique to examine 522 hearts from Milan and 499 from the Armed Forces Institute of Pathology, he found occlusions in fewer than 20 percent of people who suffered sudden death. Thinking that he had produced definitive evidence, Baroldi, like Arthur Master before him, called on researchers to replace the misleading phrase "coronary thrombosis" with the more etiologically neutral "idiopathic coagulation necrosis," a description of the pathology of a myocardial infarction that made no claims about possible causes.[6]

Another pair of researchers focused on the time course of coronary pathology. They charted the prevalence of thrombosis in relation to how long the patient survived after the onset of the heart attack. If thromboses caused heart attacks, then researchers should find them whenever they dissected patients who died in the early hours of an attack. The prevalence of thromboses in survivors would decrease over time as the body gradually broke down the clot. They found the opposite. In a first study, published in 1960, they found thromboses in only 16 percent of patients who had died within an hour of the onset of symptoms. This increased to 37 percent in patients who had died after one to twenty-four hours, and 54 percent after one day. A 1970 follow-up confirmed this finding: 15.7 percent at one hour, 36.4 percent at one to eight hours, and 57 percent after eight hours. Since the prevalence of thrombosis increased with longer survival after onset, the researchers concluded that the thrombus was a consequence, and not the cause, of the heart attack. Some aspect of the attack must "disturb the hemodynamics of an already embarrassed coronary circulation to favor the formation of a thrombus." They used this finding to critique the value of autopsies: "Autopsies unfortunately provide only one still picture of a process which during life is rapidly evolving, changing and dynamic. The cause often cannot be distinguished from effect."[7]

A third group, from Stockholm, introduced yet another method into the mix. Prior work with peripheral blood clots had shown that thrombi incorporate fibronectin and other blood proteins as they grow. If researchers injected a patient with fibronectin tagged with a radioisotope, such as iodine-125, then the radiation would concentrate in the growing thrombus. The Stockholm researchers adapted this insight to time the appearance of coronary thromboses. When heart attack patients were admitted to their coronary care unit, researchers injected them with iodine (to saturate the thyroid gland and protect it from radiation), waited four hours, and then injected the radioisotope-labeled fibronectin. Seven of these patients died during the hospitalization. The researchers searched for blood clots during the autopsy, placed them on scintillation counters, and studied the distribution of the radiation within the thrombus. In six cases the radiation was found throughout the clot, suggesting that the thrombus had formed after the infarct had begun, possibly hours after. Remember, the patients received the tracer four hours after admission to the hospital. If any part of the thrombus had formed earlier, then this portion would have been tracer-free. In fact, the seventh patient, who did not receive the tracer until forty-eight hours after the onset of symptoms, had such a tracer-free core.[8] Thrombus once again seemed to be a consequence and not a cause of infarction.

Cardiac researchers exhibited great ingenuity in their efforts to crack the case of heart attacks in the 1960s. In addition to pushing the limit of traditional pathological methods, they developed novel approaches with plastic casts, careful timing, and radio-labeled blood clots. These techniques produced different representations of pathophysiology. Although it was possible that all could have zeroed in on a common answer, this did not happen. Instead, different methods supported different theories, leaving many researchers frustrated into the 1970s.

Despite the diversity of methods, they all shared the limitation that had flummoxed pathologists for decades. They involved after-the-fact analyses of people who had died from heart attacks. As the Swedish team complained in 1973, "We know of no way of telling whether a living patient has a coronary artery thrombus or not."[9] This, however, had already started to change.

Ever since Wilhelm Röntgen first produced an x-ray image (now formally called a roentgenogram) of his wife's hand in December 1895, physicians had worked to find more and better ways of using x-rays to visualize the structures of living bodies. Unlike microscopes, which allowed scientists to make the miniscule visible, x-rays allowed doctors to make opaque structures transparent so that they could see the inner structures of the body without having to open it. The simplest x-ray images worked well for some structures of the body. Bones show up clearly, as anyone who has ever broken one knows. Lungs can also be studied easily because the edema, inflammation, and masses of many lung diseases contrast with the radiolucent air space of healthy lungs. But even with these exemplars of radiology, it took physicians time to learn how to make use of x-rays. Accustomed to diagnosing fractures and lung disease by the traditional techniques of a physical exam, doctors had to restructure their practices in the 1910s and 1920s to make x-rays useful. Over time, this happened. Physicians and patients have come to take x-rays for granted, despite many ambiguities that persist in their production and interpretation.[10]

Visualizing the heart proved to be a more difficult challenge. Although the shape of the heart could be seen in contrast to the translucent lungs, the muscular chambers and flowing blood did not provide any contrast to allow visualization of the heart's inner structure. With the work of Mason Sones in the 1960s, physicians gained access to selective coronary angiography, but even when angiography worked well, it still provided only a partial representation of coronary atherosclerosis. Surgeon Donald Effler realized this as he worked with Sones to develop coronary artery surgery at the Cleveland Clinic in the 1960s. His work allowed him to study the coronary arteries in several ways. He could examine Sones's preoperative angiogram, and then he could directly touch and see the coronary arteries when he opened the chest during surgery. The modalities did not always align. Sometimes angiography gave a misleadingly severe impression of coronary artery disease. In one patient, for instance, angiography suggested "severe involvement of both coronary arteries with multiple segmental occlusions." When Effler opened the chest and touched the heart's surface, the coronary arteries felt beaded, with atherosclerotic deposits interrupting the smooth artery walls. But then he looked closely with his own eyes: "Visualization of these vessels does not show remarkable pathology." As Effler and his colleagues gained more experience with angiography and surgery in the 1960s, they found that angiograms more often underestimated the full extent of the atherosclerosis. This sometimes caused trouble in the operating room. One of their techniques, the patch graft repair, worked well only on localized atherosclerosis. If angiography misled, they had to reassess their operative plan: "Frequently, the localized segmental obstruction visualized at operation is more extensive than suggested by the preoperative coronary arteriograms."[11]

Other doctors, especially pathologists, set out to test the reliability of coronary angiography by comparing angiographic and postmortem findings. They found similar problems. A 1962 Mayo Clinic study, from the earliest years of coronary angiography, found that angiography and pathology agreed 61 percent of the time but that angiography underestimated the extent of disease in 22 percent of cases and overestimated in 17 percent. By 1973 the technique was far better established, but radiological-pathological correlation had not improved significantly. A study from the University of Minnesota found agreement in 64 percent of cases, underestimates in 33 percent, and overestimates in 3 percent. Two doctors from St. Vincent's Charity Hospital in Cleveland captured the concerns about angiography well: "To say that the above-mentioned technique relegates the electrocardiograph to a place in the Smithsonian Institute, as was said at a recent panel discussion, should not be allowed to pass without comment."[12] Angiography, after all, could reveal only the silhouette of the contrast-filled artery. It revealed nothing about the status of atherosclerosis in the smaller branches of the coronary tree, or about the quality of blood flow through the vessels.

Despite awareness of the technique's limitations, cardiologists and surgeons were beguiled by the images it produced, in part because angiography enjoyed the same self-evident realism and objectivity that photography and conventional radiography had long enjoyed. Doctors could look at the images and think they were seeing the arteries and their obstructions. Sones heightened the realism by using his image amplifiers and cameras to make movies—cineangiograms—of living coronary arteries. These movies animated the tissues, allowing viewers to watch contrast flow around the plaques and through the arteries of a beating heart. Sones played these movies at national meetings. One even won first prize from the American College of Chest Physicians.[13]

The comparisons of angiography to familiar, trusted, self-evident technologies were explicit. Sones described angiography as a "photographic safari through the heart." Effler, writing in *Scientific American,* described how angiography provided "visual diagnosis" and transformed the prospects for cardiac therapeutics: "Sones's technique of selective coronary arteriography produced a leap forward in our ability to read coronary disease that can be fairly likened to the impact of the invention of the printing press on the written word." This technology "made it possible to select patients intelligently and to tailor the operative procedure to the needs of the individual patient and his specific disease." When Sones received the prestigious Albert Lasker Clinical Medical Research Award in 1983, Houston surgeon Michael DeBakey praised the technique: "The resultant moving pictures clearly show the precise location of any significant obstruction and lesion . . . [Angiography] has helped to save thousands of lives throughout the world." One book, written with the help of Cleveland Clinic physicians to educate patients about cardiac disease and therapeutics, described how angiography provided "a literal 'road map' of the heart's blood supply."[14] Whether they compared angiography to a book, a map, a photograph, or a movie, the implication was clear: seeing should lead to believing.

Eager to believe what they saw, researchers quickly put angiography to use in the debates about heart attacks. The technique initially added to the doubts about coronary thrombosis. In a 1967 study, researchers examined fifteen patients with symptoms of angina and electrocardiographic evidence of damaged cells but found no angiographic evidence of coronary disease. A 1975 study found normal appearing arteries in patients who had had heart attacks. In this study, however, the catheterizations were done six weeks to one year after the attack; the researchers could not identify what had caused the attack at the time. Other angiographers got lucky. One Denver cardiologist studied a 46-year-old woman who suffered hourly angina attacks. Although initial angiography revealed normal vessels, she suffered a series of episodes during the procedure. The cardiologist identified a coronary spasm each time. Subsequent studies found spasm in five of six patients tested within six hours of a heart attack. In each person the spasm occurred at the site of an existing plaque, which supported a specific theory about what causes heart attacks. Spasm, superimposed on an existing atherosclerotic lesion, caused total occlusion of a coronary artery. As flow through that artery slowed to a stop, the heart attack began. A thrombus then formed—the consequence, not the cause, of the attack—and sealed the heart's fate.[15]

Such compelling visual evidence in support of spasm and against coronary thrombosis posed a challenge to Constantinides, Friedman, and other researchers who thought they had found the cause of heart attacks. As uncertainty persisted into the 1970s, researchers grew increasingly embarrassed.

Ontario pharmacologist Plinio Prioeschi described how understanding of heart attacks had passed from a phase of oversimplification, in which everyone had been happy with the coronary thrombosis hypothesis, to one of "anarchy." New York cardiologist Charles Friedberg decried the "disturbing discord" in the irreconcilable autopsy results. Noted pathologist Stanley Robbins complained that the vexing results had only deepened pathologists' ignorance. When a British pathologist mocked skeptics of coronary thrombosis for failing to recognize a simple anatomic fact, Baroldi, an advocate of the spasm hypothesis, had a self-righteous response: "For centuries it was also ostensibly simple that the earth was the center of the universe, and those who postulated otherwise were similarly dismissed as radicals and heretics."[16] In this crucial period, as coronary artery bypass surgery launched as a treatment that promised to prevent heart attacks, physicians grudgingly admitted that they still did not understand the mysteries of what caused them.

CHAPTER FOUR

Learning by Doing

Into the 1970s autopsy studies, animal research, and coronary angiography continued to generate results that lent support to competing theories about the cause of heart attacks. Even the basic question—was a coronary thrombosis the cause or consequence of the attack?—remained contentious. Uncertainty about this point undermined the claims of researchers who saw plaque rupture as the ultimate cause.

By the end of the 1970s, however, the discord had largely disappeared, primarily thanks to new methods of classifying and intervening. New ways of classifying heart attacks and other acute coronary syndromes simplified the task of deciphering the cause: physicians hoped to find consistent causes of specific subtypes of coronary phenomena. More important, physicians developed new ways—and new courage—to treat heart attacks. Some things in medicine can be learned only through intervention. Does a treatment work? What does the response reveal about the cause of the disease? The only way to know is to try; a therapeutic gamble becomes a means of knowledge production. What doctors saw during heart attack angiography and surgery changed their thinking about the causes of heart attacks.

Many scientists see taxonomy and other classifying endeavors as simple-minded descriptive work, no more profound than stamp collecting. This is wrong. How scientists classify and name the phenomena of nature has real consequences for how they think about and understand the natural world. Classification can define objects as worthy of scientific attention, influence what aspects of them are studied, and bestow a range of meanings and associations on them. In medicine, for instance, diagnostic categories make judgments about what is or is not a disease. Is alcoholism a disease? What about homosexuality, hysteria, or depression? The status of each has changed over time. Such distinctions have immense impact on everything from patients' existential experience of disease to bureaucratic determinations of which diseases merit treatment or insurance reimbursement.[1]

Diagnostic categories can also influence how doctors study and explain disease. Heart attacks demonstrate this well. In his two reports James Herrick had described the wide range of outcomes he believed could follow a coronary thrombosis. The unluckiest patients experienced "instantaneous death . . . in which there is no death struggle, the heart beat and breathing stopping at once." Others died within a few minutes or hours, "found dead or clearly in the death agony by the physician who is hastily summoned." Not all victims died. In some "death is delayed for several hours, days or months, or recovery occurs." The last group had just "mild symptoms, for example, a slight precordial pain ordinarily not recognized." Herrick assumed that such variable outcomes depended on the details of the thrombus: "The clinical manifestations of coronary obstruction will evidently vary greatly, depending on the size, location and number of vessels occluded."[2] As doctors continued to study heart attacks over the decades, however, they realized that important distinctions existed beneath these different clinical syndromes.

When pathologists studied the actual damage to a heart's muscle after a heart attack, they found two basic patterns. In the most severe attacks, the damage spanned the full thickness of the ventricle's wall, a transmural infarction. In milder cases the damage involved only the innermost layer of the heart, a subendocardial infarction.[3] Researchers suspected that these two patterns reflected slightly different mechanisms of injury. In a transmural infarction, blood flow to a section of the heart ceases completely, killing all cells downstream of the obstruction. In a subendocardial infarction, some blood gets to the tissues, so only the innermost layers, which experience the highest pressures during a heart's contraction, die as a result.

Cardiologists, interested in diagnosing living hearts, learned to use electrocardiograms to classify the severity of the damage during or after a heart attack. When they interpreted EKGs, they distinguished several phases of the tracing. These corresponded to different parts of the cardiac cycle: the atria contracted, filling the ventricles with blood (the P-wave); the ventricles contracted, pumping blood through the body (the QRS-complex); and the ventricles repolarized in preparation for the next beat (the ST-segment). Instead of simply diagnosing a heart attack or a myocardial infarction, cardiologists in the 1970s specified whether the patient had ST-segment elevations (an early sign of myocardial ischemia and hypoxia) or pathological Q-waves (a sign that part of the ventricle wall had infarcted and become a scar). They also learned to correlate these classifications with those used by pathologists. Patients with Q-waves and ST-segment elevation almost always had transmural infarctions.

Armed with these distinctions, cardiologists and pathologists revisited data on the prevalence of thrombosis in heart attacks. In the late 1960s and early 1970s, pathologists had defended the full range of possible claims, with thrombosis found in anywhere from zero to 100 percent of heart attacks. Many groups set out to see if this variance depended at all on the type of infarction. Michael Davies and his colleagues in London produced the most influential studies. They, like so many other physicians, had become fed up with the lingering controversy about a seeming matter of fact: "It is hardly credible that there should be a continuing debate about what is ostensibly so simple a morphological problem, the relation of coronary thrombosis to myocardial infarction." As part of a series of 6,400 routine autopsies between 1964 and 1974, they found 500 cases of fatal heart attack. Of these, 469 had at least one transmural infarct, and 31 had only diffuse subendocardial damage. Looking carefully at the coronary arteries, they found occlusive thrombi in 95 percent of the transmural infarcts but in only 13 percent of the subendocardial infarcts. They also found plaque rupture underneath nearly every thrombus. The different patterns of damage reflected different pathological processes. A follow-up study of 100 sudden deaths found occlusive thrombi in 74 and evidence of plaque rupture in another 21. The conclusion, published in the influential *New England Journal of Medicine*, seemed clear: "Virtually all patients who die suddenly of ischemic heart disease have actively progressing arterial lesions that are predominantly thrombotic."[4] By limiting their analysis to the most severe cases—transmural infarctions and sudden deaths—these physicians identified consistent findings about the presence of coronary thromboses that had eluded earlier researchers, who had relied on a less differentiated concept of heart attack.

The most compelling evidence of coronary thrombosis, however, came not from autopsy studies of newly classified syndromes but from new kinds of therapeutic intervention. Ever since Herrick had drawn attention to heart attacks, doctors had sought ways to treat them. Traditional management—weeks of bed rest and supportive care—left much to be desired. Acute management of heart attacks began to change in the 1960s, with the development of coronary care units. Using electrocardiograms to identify the arrhythmias that frequently killed people after heart attacks, and using new drugs and defibrillators to treat patients, doctors could save some lives. [5] But these methods treated only the complications of heart attacks and not their underlying causes. In the 1970s doctors attempted to target the underlying causes with new treatments: bypass surgery, angioplasty, and new medications, such as clot-busting drugs. These interventions transformed not only patients' prognoses but also physicians' understandings of the attacks themselves.

New surgical procedures, for instance, proved to be surprisingly informative. Even as debate persisted about whether thromboses, spasms, or intramural hemorrhages caused heart attacks, many doctors believed that increasing the blood flow to the heart during or after an attack could only help. But could they actually do it? Surgeons had experimented with treatments for coronary artery disease since the 1910s and tried an astonishing diversity of techniques to improve the blood supply, creating scars, grafting tissues, and implanting arteries into the heart. One technique used a patch to widen a coronary artery at the site of an atherosclerotic plaque. Initially these procedures were limited to patients with stable disease. Into the 1960s physicians did not dare to operate on a patient in the throes of a heart attack. But as surgeons gained confidence and speed with these techniques, they set out to rescue patients during acute attacks. The first surgical treatment of a heart attack came in January 1966, when surgeons at the Cleveland Clinic used a patch graft to perform emergency revascularization. The patient survived. After René Favaloro developed coronary artery bypass surgery there in 1967, he first tried it during an acute attack in April 1968. In the initial series of sixty patients, only five died. Other surgeons quickly followed suit.[6] The satisfaction must have been immense. Imagine the scene. The surgeon opens the patient's chest to reveal a heart, beating feebly, visibly pale from impoverished blood flow. As soon as the grafts are placed and the clamps removed, the surgeon watches as blood returns to the heart, the tissues flush, and the contractions increase their vigor.

Surgeons soon took this one step further. Instead of waiting until the attack had begun, they began to operate to prevent an imminent infarction. This was another consequence of physicians' changing classifications of acute coronary syndromes. Cardiologists had long known that between the extremes of chronic stable angina and heart attack, a third syndrome existed, one characterized by increasing frequency and severity of angina attacks. Over the decades they had given this syndrome many names: intermediate syndrome, crescendo angina, unstable angina, or impending acute infarction. In the early 1970s, the term "preinfarction angina," with its implication of an imminent calamity, dominated the medical literature. The phrase imposed a linguistic compulsion on physicians and patients to respond. Although some doctors found the phrase misleading, "since it presupposes knowledge of a future event not certain to occur," some prominent surgeons believed that the "clinical state foreshadowing imminent infarction can be defined with a high degree of certainty." This risk state demanded intervention. As Favaloro explained at a Los Angeles meeting of thorascic surgeons in 1972, "it is best to operate before myocardial infarction occurs." In fact, of the Cleveland Clinic's first sixty emergency revascularizations, forty were in patients with "impending myocardial infarction."[7]

Something similar happened with the "widow-maker lesion," a large plaque in the left main coronary artery revealed by angiography. Effler called this the "lesion of sudden death." The mortality rate among people with such a plaque was so high that surgeons believed surgery could only help. They considered the widow-maker to be an appropriate indication for bypass surgery, regardless of whether the patient had symptoms of coronary artery disease.[8]

Surgeons' eagerness to attempt bypass surgery for acute and impending infarctions put pressure on cardiologists. Even as coronary angiography became more widespread in the late 1960s, many cardiologists remained reluctant to perform the procedure in patients with heart attacks or even unstable angina. Angiography was not without risk. They feared the catheter might critically impede flow through an already compromised coronary artery. The manipulations and contrast agents might stress the ailing heart and trigger lethal arrhythmias. Cardiologists did not consider this risk worthwhile if the angiography was simply a diagnostic exercise in the absence of specific therapy. The situation changed, however, when surgeons began to attempt emergency revascularization. For surgeons such as Effler and Favaloro, angiography was the "sine qua non" of successful revascularization, identifying the most serious lesions and the best opportunities for surgical repair. The hope that they might save lives by operating on patients during acute attacks changed the dynamic of risk and benefit for angiography. It was no longer simply a diagnostic test. Instead, it was the first step—a precondition—to a potentially life-saving therapy. This motivated cardiologists to take the risk. The Cleveland Clinic team started slowly, reporting their initial results on twenty-eight patients with heart attacks or impending infarction in 1971. The procedure proved safe enough, entailing only "minimal risk." Most of the patients survived the subsequent surgery. By 1972 their series had increased to sixty. Other centers quickly followed Cleveland's lead and gained confidence with both angiography and surgery during acute coronary events.[9]

Physicians finally had something they had long wanted: access to the inside of a living heart at the moment—or at least within a few hours—of a heart attack. Cardiologists could perform emergency angiography and gain an initial picture of the coronary arteries. Surgeons could open the chest and see and feel the arteries directly. What did they find? Thrombi, and lots of them. In one influential study, cardiologists in Spokane, led by Marcus De-Wood, set out to resolve the ambiguities that had plagued past studies of heart attacks. Between 1971 and 1978, they performed angiography on 322 patients suffering acute transmural infarctions. When they were able to get their catheters into the patients within four hours of the onset of the attack, they found total occlusion in 87 percent of them. Nearly 70 percent had clear angiographic evidence of coronary thrombi. Demonstrating careful scientific humility, the cardiologists admitted that they could not prove that thrombi had caused the attacks, only that there was likely a "dynamic interaction" between spasm, platelets, and plaque that led to occlusion.[10]

DeWood's group did not stop with angiographic visualization of occlusive thrombi. Instead, when conditions allowed, they took their patients to bypass surgery. Of the 59 patients who had "definite angiographic features of thrombus," surgeons successfully retrieved the thrombus in 52.[11] This experience must have been doubly gratifying. Surgeons did not just have the satisfaction of placing a bypass graft and providing a new source of blood—and potentially life—to an infarcting heart. They also restored flow through the native vessel

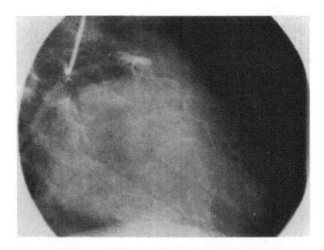

Angiographic demonstration of thrombus. Presumably
this is legible to a trained observer. From Marcus A.
DeWood et al., "Prevalence of Total Coronary Occlusion
during the Early Hours of Transmural Myocardial
Infarction," *New England Journal of Medicine* 303 (16 Oct.
1980): 897–902, fig. 1, p. 898. With permission from the
Massachusetts Medical Society.

by removing the clot. They apprehended the prime suspect, the coronary
thrombus, at the scene of the crime.

Other treatments for acute infarction proved equally revealing. Even as surgeons worked to remove or bypass occlusive thrombi, cardiologists tried to
dissolve them, a process call thrombolysis. A Baltimore researcher had identified the first clot-busting drug, streptokinase, in the 1930s. Clinical trials
using streptokinase to treat various diseases began after World War II. One
1959 study, for instance, gave intravenous streptokinase to twenty-two heart
attack patients. Although the researchers worried that they might exacerbate
intramural hemorrhages—this was the heyday of that theory—the treatment
proved safe. Eight "gravely ill" patients, and others less sick, "improved considerably." However, the drug did not always work as well as physicians hoped.
Some patients developed severe allergic reactions. Unreliable efficacy and
concerns about safety conspired to keep streptokinase from widespread use
through the 1960s.[12]

Cardiologists worked to revive thrombolytic therapy in the 1970s. Angiography proved to be a major asset in this work. Researchers could now inject streptokinase directly into the coronary arteries, under angiographic guidance, and watch the results in real time. One study successfully opened eighteen of twenty-nine occluded arteries. The researchers saw this as support for the pathogenic role of the thrombus: the results were "in agreement with those of most pathologists who found fresh thrombosis in the majority of patients with acute myocardial infarction." However, debates about patient selection, mode of administration, and dose generated continuing uncertainty. Resolution did not come until 1986, when researchers in a large Italian trial reported that intravenous streptokinase reduced mortality after a heart attack by 18 percent, with the greatest benefit seen in patients who received the drug most quickly. The results were encouraging, even if imperfect: thrombi played enough of a role in heart attacks that dissolving thrombi provided substantial benefit. Therapeutic response supported the underlying theory. As one streptokinase researcher concluded, the success of the drug "provides further evidence for the concept that thrombosis is the mechanism of coronary artery occlusion in acute transmural myocardial infarction."[13] This was a way of reasoning backward from therapy to cause. Since a clot-lysing drug helped, a clot must have been at least partly to blame.

The imperfect efficacy of streptokinase suggested that something other than a dissolvable thrombus played a role. Angiography made it easy for cardiologists to see what this something was. As they watched streptokinase break down clots, they saw that a residual obstruction often remained, even after the thrombus dissolved. Many thrombi had formed at the site of atherosclerotic plaques, and these were not affected by streptokinase.[14] The drug could only dissolve clots. It could do nothing about the atherosclerotic plaques themselves. Another new treatment, balloon angioplasty, quickly filled the breach.

German cardiologist Andreas Grüntzig first described the technique of balloon angioplasty in 1977. The technique had a deceptively simple logic: by inflating a cylindrical balloon at the site of an atherosclerotic obstruction, angioplasty could compress the plaque and stretch the vessel's wall, producing a larger space through which blood could flow. Although questions lingered about the details of these mechanisms—would the stretching last? exactly what happened when you squeezed a plaque?—cardiologists in Europe, the United States, and elsewhere raced to develop the technique. Within two years they had adapted it for use against heart attacks. One study randomly assigned patients to angioplasty or thrombolysis. Treatment was successful in both groups: flow was restored through the occluded artery in 83 and 85 percent of the patients. But streptokinase left patients with an 83 percent residual obstruction; although the clot had been dissolved, the original plaque persisted. Angioplasty managed to both displace the thrombus and compress the plaque, leaving only a 43 percent residual obstruction. The doctors hoped that angioplasty, by restoring a greater volume of flow, would provide "more effective preservation of ventricular function."[15]

The use of angiography in conjunction with bypass surgery, thrombolysis, and angioplasty produced visible and tangible evidence of the importance of coronary thrombosis in the severest heart attacks. But what about spasm and other possible causes of infarctions? In the 1960s cardiologists had reported good outcomes with vasodilators. These physicians would detect spasm in patients with angina pectoris, treat it with dilating drugs, and relieve angina attacks. They had concluded from this therapeutic success that spasm likely played a role in the pathophysiology of heart attacks. However, as the fortunes of therapies directed against thrombi rose in the early to mid-1970s, the fortunes of therapies directed against spasm fell. The basic therapeutic test was an obvious one: give a patient a dilating drug and see if it altered the course of a heart attack. Many doctors tried this. Two cardiologists from Massachusetts General Hospital gave nitroglycerin in varied forms—a pill placed under the tongue, an intravenous injection, or an intracoronary injection during angiography—to thirty patients with acute transmural infarctions. Only four patients showed any response at all to the vasodilator, and one of those responses was slight. A second arm of the study, meanwhile, found that eight out of ten patients had a significant response to streptokinase. The conclusion, as far as the researchers were concerned, was inescapable. Spasm, if it occurred, was not decisive: "The dominant occlusive process is coronary thrombosis."[16]

Many cardiologists and surgeons came to celebrate how their new therapeutic capabilities had transformed their understanding of coronary artery disease. As Chicago pathologist Seymour Glagov described in 1990, cardiologists could intervene "directly on the plaque": "we may now disrupt, sear, aspirate, shear, and displace plaques and lyse occluding thrombi." Whereas pathologists once had the final word about the ultimate causes of death and disease, cardiologists and surgeons used their new therapeutic techniques as assays to test different theories about the causes of heart attack. The contribution of treatment to discovery was explicit. As Atlanta physician Laurence Harker wrote, "Most of what we have learned about the role of thrombosis in relation to atherosclerosis has been a result of intervention."[17] New modes of practice produced new ways of knowing.

The self-conscious use of therapeutic intervention as a means of producing knowledge about the causes of disease raises interesting questions about the relationship between therapy and experiment. Doctors do not simply generate new knowledge in laboratories and then apply it in clinical practice. They have always learned from their patients. When doctors use new treatments against diseases that they only partially understand, they can learn something useful. By choosing patients carefully, or by studying enough of them, physicians can isolate and probe specific aspects of the disease process. Even though the goal of a medical intervention might be treatment and not experiment, the outcome of the intervention might be new knowledge, as if it had been an experiment.[18]

This dynamic does not mean that doctors are involved in continuous human experimentation. The situation is subtler in important ways. Patients, who come to doctors in pursuit of treatment, are often the driving force behind an intervention. With some therapies, the mechanisms are so well understood that little is left to chance. If you give appropriate antibiotics to a patient with a bacterial infection, the infection should resolve in a consistent and predictable way. In these respects medical practice is not a conventional experiment. But doctors often act in the setting of uncertainty, about disease theories, therapeutic mechanisms, or the specific circumstances of an individual patient. As a result, the outcome of a particular therapy in a particular patient cannot always be predicted. A wise physician pays close attention to the outcomes in hope that future treatments will be better understood. Does this make many treatment episodes experiments? Yes and no. Outcomes are uncertain, and knowledge might be produced. But as long as there is a justifiable expectation that the patient will benefit, then the encounter can fairly be considered therapeutic.

The use of therapeutic intervention to generate knowledge about disease inserts an often unrecognized complexity into the center of the medical enterprise. Much of the pride and success of modern medicine stems from a fundamental tenet: decipher the cause of disease in the laboratory and then develop appropriate treatments. This vision took hold in Europe and then the

United States in the late nineteenth century, pioneered by famed scientists such as Louis Pasteur, Robert Koch, and Claude Bernard. As Elisha Gregory described in his presidential lecture before the American Medical Association in 1887, "Knowledge must come first, then wisdom brings its practical application."[19] Many innovations in medicine followed this path from bench to bedside.

But in the search for the cause of heart attacks, information moved in the other direction. Pathologists had attempted to decipher the cause of heart attacks, but their methods had lent support to competing theories. Building on those leads, surgeons and cardiologists developed various fundamentally different treatments, tried them out in patients, and then worked backward to convince themselves of what the underlying cause must be. This was not new or unique to cardiac therapeutics in the 1970s. In ancient Greece, doctors recognized different paths to medical knowledge. Rationalists developed theories about disease and then sought appropriate treatments. Empiricists disavowed theory. Instead, they simply tried treatments, observed the effects, and used this experience to guide future decisions. But these distinctions were more hypothetical than actual. Over the centuries, many physicians and other practitioners freely integrated rationalist and empiricist approaches, for instance, using outcomes of treatment to revise their theories.[20] A certain amount of therapeutic feedback is needed to produce robust medical knowledge. If doctors developed disease theories in the lab and then refused to update them in light of their clinical experience, they would be isolating themselves from a valuable source of information. Reciprocal interactions between theory and therapy lead to the improvement of each.

The give-and-take is an inevitable consequence of medicine's pragmatic orientation. Although laboratory research has been an essential part of medicine, medicine is ultimately about practice, about doing things on behalf of patients. And the things that physicians do influence how they think. One philosopher captured these dynamics well in an ethnographic study of peripheral vascular disease in a Dutch hospital. She began with a deceptively simple question: What is atherosclerosis? At first pass this seemed like another simple matter of anatomic fact. She found something far subtler. When she asked internists, surgeons, radiologists, pathologists, and patients about atherosclerosis, she received different answers. Each group engaged with the disease through different practices. Patients felt pain when they walked long distances. Radiologists visualized arteries with ultrasound and angiography.

Pathologists kept limb specimens in refrigerators and studied slides under microscopes. Each different way of doing and experiencing the disease produced a different disease. There was not just one atherosclerosis, but many atheroscleroses, not one body, but multiple bodies, each different in some way, depending on how atherosclerosis was experienced, studied, or enacted. But these divisions are not insurmountable. Even as different practices generate different conceptions of the body, specialists from different disciplines often collaborate on patient care. They create a "trading zone," where adjustments of language and practice allow collaboration despite differences in perspective.[21]

When collaboration works well, physicians integrate what they learn from the clinic and laboratory to produce a composite theory of disease. But even though it is understandable how and why therapeutic practice shapes medical knowledge, it can be problematic. If physicians want to use treatment response to revise their pathophysiological theories, then they must be confident that they truly understand how the disease responded to the treatment. How good is such medical knowledge? Pathologists have long recognized the limits of their methods. They've known, for instance, that autopsy specimens are imperfect representations of what the living body had been like. How reliable a representation could doctors produce by studying how bodies responded to treatments? Important details often remain unclear. For example, angiography produces an image of the silhouette of blood flowing through the coronary artery. When doctors saw streptokinase reduce a 100-percent occlusion to an 83-percent residual obstruction, could they conclude that the benefit came wholly from clot dissolution? Might spasm have played a role? Did the residual obstruction reflect atherosclerotic plaque or some thrombus that had not dissolved? Furthermore, when streptokinase worked in some trials but not others, was the problem the drug, the theory, or the particular details of the clinical trials?

Other therapeutic inferences raise similar questions. When vasodilators did not work, doctors had to judge whether the spasm theory was wrong or the drugs had simply failed to relieve a spasm that was actually there. This ambiguity rose to the fore recently with a debate about the possible role of infection in atherosclerosis. Many studies in the 1980s and 1990s had suggested that bacteria could colonize atherosclerotic plaques and that chronic low-grade infection exacerbated atherosclerosis. Skeptics argued that even when bacteria were present, they played little role in the course of the disease.

How could the role of bacteria be tested? A therapeutic trial could answer this question. Doctors could give antibiotics to patients to kill off the bacteria. If the bacteria contributed to atherosclerosis, then this treatment should moderate the course of the disease.

Doctors in North America, India, Europe, and Argentina put the theory to the test. They recruited 7,747 patients who had survived a heart attack and gave half of them an antibiotic for twelve weeks and the other half a placebo. The researchers then followed the patients over time to see whether antibiotic therapy would reduce the risk of a subsequent heart attack or death. The study found no difference between the groups.[22] Was the theory wrong? Or was the course of antibiotics too short? The researchers could not say. To be confident when inferring pathophysiology from treatment response, physicians must be certain that the treatment did all that was expected of it, something that cannot always be ascertained. Sociologists of science have described this as the "experimenter's regress." The entanglements of theory and experiment can complicate the interpretation of any action.[23]

None of this is unique to cardiology. Psychiatry also demonstrates the influence of practice on theory. Psychiatrists, more so than other physicians, remain unsure about the underlying causes of the diseases they treat. The brains and bodies of patients with depression, schizophrenia, and other mental illnesses bear no obvious signs of the disease. Neither autopsy studies nor imaging techniques have revealed decisive markers. So how can psychiatrists learn what causes the diseases? Response to medications has been crucial. Many of the drugs developed in the 1950s and 1960s targeted the neurotransmitter dopamine, so psychiatrists developed dopamine-based theories of depression and psychosis. When selective serotonin reuptake inhibitors, most famously fluoxetine (Prozac), became popular in the 1980s and 1990s, depression became a disease of serotonin deficiency. If new experimental drugs that alter glutamate activity prove worthwhile, then glutamate theories will surely follow. This is a credible approach to an extremely difficult problem. It does, however, expose psychiatric science to pharmaceutical marketing and enthusiasm in a way that has raised many concerns.[24]

The experiences of psychiatrists point to distinct questions physicians need to consider whenever they use therapeutic outcomes to influence their pathophysiological thinking. The first is whether the therapeutic trials produce evidence that is as definitive as doctors and patients desire. The second is whether other influences, including financial conflicts of interest or other sources of unwarranted enthusiasm, color how doctors interpret therapeutic trials. Given the enormous revenue generated for physicians and hospitals by angiography, angioplasty, and bypass surgery, this is a reasonable question to ask. But that's getting a bit ahead of the story.

CHAPTER FIVE

The Plaque Rupture Consensus

A proliferation of new autopsy techniques, diagnostic technologies, and therapeutic interventions from the 1960s through the 1980s provided many new ways to study the causes of heart attacks. Methodological diversity initially perpetuated the discord that had dominated the field since the 1930s. In the 1980s, however, the diverse techniques began to converge toward a unifying theory, one that accepted the causal role of coronary thrombosis and, underlying that, plaque rupture. In this chapter, I describe the consensus that took shape, showing how researchers fit the pieces of the intellectual puzzle together into a coherent story about the causes of heart attacks. My use of the term "consensus" here should be read as descriptive, not prescriptive. The plaque rupture hypothesis is, at present, widely agreed on by medical experts. Does this mean that it is an accurate account of nature? That is a much more difficult question.

In the plaque rupture model, as in the traditional model of progressive obstruction, the first steps are insidious. The atherosclerotic process begins with the silent growth of fatty streaks, shallow deposits of fat-laden cells that can

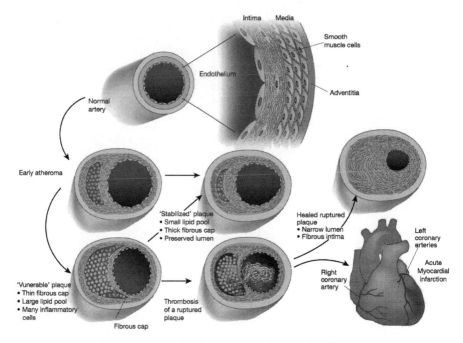

Development of stable and fragile plaques. According to the plaque rupture consensus, some plaques (*middle*) grow progressively large and fibrous; these cause angina. Others become vulnerable plaques (*bottom*), which rupture, thrombose, and cause heart attacks. From Peter Libby, "Inflammation in Atherosclerosis," *Nature* 420 (19–26 Dec. 2002): 868–74, fig. 4, p. 872. Reprinted by permission from Macmillan Publishers, Ltd. (Nature Publishing Group).

be found in the arteries of American teenagers. As fatty streaks grow and then consolidate into atherosclerotic plaques, they follow one of two diverging paths. Some become fibrous plaques, composed mostly of muscle cells and connective tissue that grow slowly but stably over time, as envisioned in the model of progressive obstruction. When they grow large enough, they limit blood flow, especially during exercise, as cardiac workload increases. These are the lesions of chronic stable angina. Other plaques, however, follow a different path. They become unstable, the sort that killed Bertel Thorvaldsen in 1844 and that Timothy Leary described in detail in the 1930s. Composed of fat-laden macrophages (the scavengers and recyclers of our cellular worlds), these plaques develop into toxic collections of dying cells, cholesterols crystals, and calcified debris. Regardless of size, they can erode, fissure, or rupture and release their debris into the bloodstream, which triggers a blood clot. The clinical outcome depends on the location, extent, and duration of the rupture, thrombosis, and ensuing occlusion.[1]

Plaque rupture provided a unifying mechanism, "the missing link" of acute coronary syndromes.[2] Some plaques underwent repeated cycles of rupture, thrombosis, clot lysis, and rupture once again. Patients with these plaques might present with the sputtering or crescendo symptoms of unstable angina. If a thrombus formed, persisted briefly, and then lysed spontaneously—as thrombi often did—then the patient might suffer a subendocardial infarction. Persistent thromboses produced transmural infarctions, with the size and outcome of the heart attack depending on the location of the plaque. Even a small plaque rupture in just the wrong place could trigger the fatal arrhythmias of sudden death. The full phenomenology of coronary artery disease, from stable angina to sudden death, became explicable under this simple hypothesis.

The plaque rupture hypothesis tied together the findings of autopsies, animal studies, and clinical interventions. Cardiologists and other researchers celebrate it as a success story of bench to bedside and back again. Not only did the hypothesis explain a range of existing observations, it also accounted for observations that emerged in the 1980s and 1990s as cardiologists developed new imaging and diagnostic technologies. For instance, one team developed a thermal probe that could study the temperature profile of atherosclerotic plaques. The researchers suspected that the increased metabolic activity of the macrophages that infiltrated plaques would yield heat. Sure enough, by studying plaques that had been removed from patients, researchers showed that increased temperatures correlated with the density of the inflammatory cells.[3]

Additional support for the plaque rupture hypothesis came from another new technology, angioscopy. Further miniaturization of electronic and optical devices in the 1980s allowed engineers to mount tiny fiberoptic cameras onto the tips of coronary artery catheters. Unlike angiography, which produced x-ray silhouettes of blood vessels, angioscopy allowed cardiologists to see the inside of living arteries. It provided a more immediate experience of the disease. Doctors rhapsodized about the wonder of this new technology: "In the past year we have taken a metaphoric voyage through the coronary arteries of living man, using fiberoptic angioscopes." This voyage of discovery confirmed many features of the plaque rupture synthesis. Plaques, ulcers, surface thrombi, progressive thrombi, and occlusive infarctions could all be seen.

The correspondence of symptoms and endothelial surfaces, a new kind of clinical-pathological (but living) correlation, was no less than a "systematic description—a paradigm—of the pathophysiology of acute and chronic coronary artery disease in man."[4]

The authors' word choice here is revealing. Ever since Thomas Kuhn's use of "paradigm" in his 1962 *The Structure of Scientific Revolutions,* the concept has captured the imagination of everyone from scientists to business executives to the cartoonists who lampoon them. Kuhn's idea was that scientists have organizing frameworks (paradigms) not just for our explanations of natural phenomena, but also for our perceptions of those phenomena. Scholars continue to debate exactly what Kuhn meant, whether a classic Kuhnian "scientific revolution" has ever taken place, and what drives the dynamics of scientific change (i.e., a paradigm shift).[5] Amid this controversy, one aspect of Kuhn's work maintains its intuitive appeal: the idea that a new paradigm can often explain things that had seemed anomalous under the prior conventional wisdom. The plaque rupture hypothesis served this function for cardiologists.

The new consensus about plaque rupture, for instance, provided an explanation for many long-standing puzzles. Ever since V. P. Obraztsov and N. D. Strazhesko first described the clinical syndrome of heart attacks in 1910, doctors have wondered whether something triggers the events. The Russian doctors believed that the attacks were triggered. One of their patients was struck by an attack during a competitive card game, another after an unpleasant conversation, and a third after climbing stairs. This idea became a focus of fierce debate in the 1930s, in tandem with the controversies over the role of coronary thrombosis. A team of New York cardiologists led by Arthur Master provided an influential analysis. He admitted that angina had many well known triggers, including heavy exertion, large meals, excitement, or cold air. But what about heart attacks? Reviewing data on the onset of 930 heart attacks, he found that one-third occurred at rest, one-quarter in sleep, and one-fifth during only mild activity. Only a few had traditional triggers: moderate or heavy exertion (10%), walking (16%), or eating (10%). Master revisited this question twenty years later. His 1960 review, based on even more examples, upheld his initial conclusion: there was no clear relation between exertion and the onset of heart attacks. Master's conclusion held sway for a generation, and researchers lost interest in triggering.[6]

In the early 1980s, however, a team of researchers in Boston returned to the question of triggers, quite by accident. The team had begun a study to investigate whether two particular drugs could limit the size of the infarct during a heart attack. They assumed that the ability of the drugs to do this would depend on how soon they were given after the attack had begun. As a result, the researchers kept close track of exactly when the attacks started in hundreds of patients. When one of the researchers, James Muller, reviewed the data, he was surprised to see a dramatic daily pattern in the timing of heart attacks. Their prevalence was three times higher in the morning than in the evening. What could account for this circadian rhythm? As Muller and his team explained in the *New England Journal of Medicine* in 1985, even though it was "now recognized" that heart attacks were caused by plaque rupture and the resulting coronary thrombosis, "the circumstances that trigger these pathologic events remain unknown."[7] They set out to identify other circadian rhythms that could credibly be involved. They learned that there were daily rhythms in blood pressure and in the level of stress hormones—both cortisol and catecholamines—in the blood. All were higher in the morning than at night. There were also changes in blood viscosity and tendency to clot, also higher in the morning.

Muller's analysis echoed, in a way, what Paris Constantinides had done to induce plaque ruptures. Constantinides had used epinephrine and snake venom to trigger rupture and thrombosis in his rabbit model. Muller speculated that natural rhythms in vascular tone and clotting tendency accounted for the morning preponderance of heart attacks. The plaque rupture hypothesis helped explain this diurnal rhythm.

The theory offered an explanation for some treatment effects as well. Doctors had learned by the 1980s that beta blockers and aspirin could each reduce the risk of heart attacks. Perhaps, Muller suggested, this happened because the beta blockers blocked the morning surge in catecholamines and the aspirin offset the effects of morning hypercoagulability. This suggested a new therapeutic goal: not to prevent the triggers, but to prevent their consequences, "to sever the linkage between a potential triggering activity and development of a catastrophic coronary thrombosis."[8]

What about heart attacks that occurred without obvious triggers? Plaque rupture shed light on those as well. Early statements of the plaque rupture consensus emphasized that rupture could strike out of the blue. Erling Falk, an early proponent of the hypothesis, explained that even though rupture could be catastrophic to patients, it was "probably a random event," little more than "an incidental event in the evolution and growth of the atherosclerotic plaque." Muller echoed this, but with a subtle and significant twist: "The onset of infarction, in some cases, may be the result of the unfortunate simultaneous occurrence of several events, each by itself of little consequence, but catastrophic when occurring together." In his version of plaque rupture, many in this series of unfortunate events could potentially be explained. Studies in the 1980s and 1990s had identified several proximate causes of plaque rupture: inflammation, plaque necrosis, weakening of the plaque's cap (the breakdown of the collagen fibers that hold the cap together), and the mechanical strain of an artery affixed to the surface of a beating heart. Attention increasingly focused on macrophages, whose enzymes broke down connective tissues. With increasing precision, researchers identified specific cytokines, growth factors, and other molecular mediators of inflammation. Heart attacks were brought into the age of modern molecular biology.[9]

Equipped with gene expression arrays, genetically modified mice, and sophisticated new techniques for imaging cells and subcellular components, researchers could visualize coronary artery disease in new ways. Knowledge of the molecular details of the inner workings of plaques allowed physicians to understand, in principle, which plaques were most at risk of rupture. Based on this model, Muller described the idea of a "vulnerable atherosclerotic plaque."[10] The holy grail, for Muller, was a way to detect or, even better, to detect and rehabilitate vulnerable plaques before they ruptured.

As the consequences of this new vision became clear, doctors broadened the concept beyond vulnerable plaques. Clinical outcomes depended not just on the plaques but also on vulnerable myocardium and vulnerable blood. What happened if a plaque ruptured? It depended. A thin thrombus might seal the rupture without disrupting flow. But if the person's blood were in a hypercoagulable state, then a large, occlusive thrombus could form. If this happened, the heart might experience a substantial infarction, but it might not, depending on whether interconnections between the coronary arteries provided sufficient collateral flow. In 2003 cardiologists articulated this new consensus in *Circulation,* the premier journal of cardiovascular medicine: "The term 'vulnerable patient' may be more appropriate and is proposed now for the identification of subjects with high likelihood of developing cardiac events in the near future."[11]

Patients in this expanded at-risk pool needed careful management. The consensus report recommended a wide range of tests to identify vulnerable patients, including electrocardiograms, CT scans, c-reactive protein assays, and catheterization. Other authors added MRI, intravascular ultrasound, near-infrared spectroscopy, and coronary thermography to the mix. The hypothesis thus justified the use of a wide range of techniques to visualize risk, not just in the coronary arteries but in the patient as a whole. Some day doctors might even be able to harness "future genomic and proteomic techniques." All would be integrated by "state-of-the-art bioinformatics tools such as neural networks" into composite risk scores that would guide clinical decisions.[12] Just as genomics made many promises in the 1990s and 2000s about the bounty it would provide to science and to society in general, so too did molecular cardiology become a "promissory science."[13]

By the early 2000s, the plaque rupture hypothesis was ascendant. It drew support from many sources. Empirically minded physicians could explain that they accepted the hypothesis because of the weight of evidence supporting it. Others appreciated the elegance with which plaque rupture tied together so many threads. Professional and institutional status also played a role. The early adopters of the plaque rupture hypothesis in the 1990s were leading figures in academic medicine. Once their experience— in labs, clinics, or drug development—had convinced them of plaque rupture, they set out to convince others. In charge of cardiovascular programs at the Mayo Clinic, Massachusetts General Hospital, Mount Sinai Hospital, or the Cleveland Clinic, they published enthusiastic reviews in prominent journals, especially the *New England Journal of Medicine* and *Circulation*. The strategic deployment of authorial authority reached its zenith in the 2003 *Circulation* consensus statement. Signed by fifty-eight prominent cardiac researchers, this statement validated the concepts, diagnosis, and treatments of vulnerable plaques.[14]

Is this new theory the right theory? Historians often take a long perspective. Over the twentieth century, doctors offered many hypotheses about the causes of heart attacks. Plaque rupture was one of the first but only recently became the most prominent. In view of this history of shifting theories, it would not be especially radical to predict that new theories will inevitably appear and that the consensus will shift once again. Many cardiologists and pathologists might disagree. They could cite as evidence how their understanding of heart attacks underwent a remarkable transformation in under a generation. Frustration about even the simplest facts of infarction in the

1970s gave way to enormous confidence in the 2000s. Many avenues of research contributed to the transformation, with puzzling anomalies falling into place under the new paradigm. The ability of the plaque rupture hypothesis to synthesize so many types of evidence fuels much of the faith in it. Is this faith justified? Modern physicians have an uncanny ability to be aware of how quickly medical science can transform and yet remain confident that they, unlike all their predecessors, finally have the story right. History might argue for more caution.

CHAPTER SIX

Rupture Therapeutics

The previous chapter told a conventional intellectual history of the plaque rupture hypothesis. It traced how the idea became popular as doctors fit more and more observations into the plaque rupture framework. But there are other facets of this history. From the 1930s into the 1960s, as doctors debated the role of plaque rupture and intramural hemorrhage, the morbid pathology of myocardial infarction was largely an academic question. Doctors at that time had no way of intervening directly to alter the course of coronary artery disease. But this changed between the 1960s and the 1980s. As doctors developed new drugs and new surgical and nonsurgical interventional techniques, the cause of heart attacks changed from a matter of fact to a matter of concern. Plaque rupture became relevant in the setting of new therapeutic interventions that promised to prevent or mitigate the effects of a heart attack.[1]

A second history of the plaque rupture hypothesis can therefore be told, one that looks not at the evidence that contributed to the plaque rupture consensus but at the relationship of plaque rupture to changing therapeutic practice in the 1980s and 1990s. In 1966, 1976, or even 1986, the hypothesis offered physicians and their patients little therapeutic value. It had not yet produced new treatments and did not rise to prominence until the late 1980s and 1990s, in conjunction with new drug therapies, specifically statins and platelet inhibitors. This chapter shows how the advent of these new drugs fostered interest in plaque rupture.

As cardiologists and pathologists debated the role of coronary thrombosis and plaque rupture in the 1960s and 1970s, some researchers explored a specific question: What was it about ruptured plaques that caused blood to clot? Scientists quickly zeroed in on platelets. Our blood contains elaborate systems that limit bleeding and trigger the healing process whenever skin or other tissues are injured. Certain blood proteins, once activated by exposure to damaged tissues, in turn activate other proteins, a process known as the clotting cascade, which culminates in a thrombus, or blood clot. The thrombus provides a scaffold on which the damaged tissue can regrow. Blood platelets play a crucial role in this process. They respond to tissue injury by amplifying the clotting cascade and stabilizing the resultant thrombus. The system works well when it seals off bleeding cuts and scratches. When the clotting cascade gets triggered by the damaged cells of an atherosclerotic plaque, however, the result can be catastrophic—the coronary thrombosis dreaded by physicians since the time of Herrick's initial articles.

In the 1960s researchers discovered that aspirin, long used to treat pain and fevers, also inhibited the activation of platelets. This made aspirin a potential antithrombotic agent, something that could possibly be used to prevent or treat coronary thromboses and, if coronary thromboses played a causal role, heart attacks. Clinical trials in the 1970s produced disappointing results. Some found that aspirin helped; others did not. The National Institutes of Health launched a large experiment in 1975 to resolve this issue. The clinical trial had all the bells and whistles of cutting edge clinical research: it was large, randomized, double blind, and placebo controlled. Doctors at thirty clinical centers throughout the United States enrolled 4,524 patients who had survived a heart attack. Half were given daily aspirin (500 mg twice per day), and half were given a placebo. The researchers followed the patients for several years and found, in the end, that total mortality was higher in the patients who received aspirin (10.8% vs. 9.7%). They looked at various subgroups (defined by sex, weight, family history, smoking history, and so forth), but in no group did aspirin produce significantly lower mortality. Although aspirin slightly reduced the incidence of nonfatal infarction (6.3% vs. 8.1%), it was associated with significantly higher rates of gastritis and gastric ulcers (23.7% vs. 14.9%). Disappointed by these results, the researchers did not recommend using aspirin to prevent a second heart attack.[2]

By the late 1980s, however, subsequent trials had produced evidence that aspirin actually did provide benefit, first shown in patients with unstable angina, then in treatment of heart attacks, and finally in prevention of heart attacks. By the early 1990s data from clinical trials consistently supported the use of aspirin across the full range of coronary artery disease treatment and prevention.[3] The efficacy of aspirin made little sense if heart attacks were caused by progressive growth of atherosclerotic plaques. Slowed flow, caused by a large obstruction, would not be enough to activate platelets and generate a thrombus. The efficacy of aspirin made perfect sense, however, in light of plaque rupture. Platelets adhered to ruptured plaques, activated, and then triggered spasm, blood clotting, and thrombosis. The outcome of a specific rupture—whether asymptomatic healing, unstable angina, infarction, or sudden death—depended on the vigor of platelet activity, something that aspirin could inhibit.

Aspirin's many side effects dampened initial enthusiasm. Many patients who took it for long periods—which would be required if used to prevent heart attacks—developed stomach irritation. In retrospect this comes as little surprise: the dose used in the NIH study, 1000 mg per day, is far higher than the dose now recommended for cardiac protection, just 81 mg per day. And even the low dose causes gastritis in some people. The side effects of aspirin motivated a search for drugs that had the antiplatelet effects of aspirin without its side effects. One line of work led to the development of clopidogrel, a drug that blocked a crucial platelet enzyme. An influential clinical study suggested that clopidogrel was safer and possibly more effective than aspirin. Marketed as Plavix since 1997, the drug has remained one of the best-selling drugs in the United States, with over $6 billion in sales in 2010 and nearly $150 million spent on marketing.[4] Advertisements explicitly invoked the drug's mechanism: "Help stop a clot before a clot stops you."

Aspirin made sense as a treatment, moreover, for another reason. As researchers in the 1980s and 1990s tried to understand why plaques ruptured, they became interested in inflammation. Since the work of Rudolf Virchow in the mid-nineteenth century, pathologists had wondered whether inflammation played a significant role in atherosclerosis. For most of the twentieth century, however, inflammation was overshadowed by interest in cholesterol, until the plaque rupture hypothesis resurrected the old interest in inflammation, and especially in the role of macrophages and other inflammatory cells.[5] The well-known anti-inflammatory effects of aspirin contributed to its ability to mitigate this aspect of atherosclerosis, prevent rupture, and improve clinical outcomes.

One other class of medications provided support—quite unexpectedly—for the plaque rupture hypothesis: lipid-lowering medications. For as long as a link between cholesterol and atherosclerosis has been hypothesized, clinicians have targeted cholesterol to halt or reverse the course of the disease. In the 1960s researchers tried to do this by encouraging patients to change their diets, but rigorous efforts to test the efficacy of this approach ran into trouble. A perfect study would have needed to enroll huge numbers of patients and somehow get them on blinded diets for many years, which proved impossible to coordinate and implement. When large trials were attempted, most famously the Diet Heart Study, they proved complex, controversial, and inconclusive. Subsequent studies targeted multiple risk factors for atherosclerosis, for instance, by encouraging patients to quit smoking, eat a low-fat diet, and control blood pressure. These studies showed how hard it could be to alter the course of the disease. One study obtained angiography in twenty-five patients before and after one year of aggressive medical management to reduce cholesterol levels. Of eighty-eight plaques identified in these patients, sixty-nine remained stable, fourteen grew larger, and only five shrank.[6]

Increasingly aggressive treatment, however, eventually yielded more impressive results. The most persuasive of these came with a new class of medicines, the statins. Discovered in the 1970s by a Japanese researcher, who screened over 6,000 drugs derived from fungal broths in search of one that would disrupt cholesterol metabolism, statins became available for clinical use in the 1980s. Research produced evidence that statins could reduce cholesterol levels, slow progression of some plaques, induce regression of others, reduce the need for revascularization procedures, and, most important, reduce the number of heart attacks and deaths. These results led doctors to prescribe statins for increasing numbers of Americans, and this, in turn, has raised interesting questions. The condition statins treat, high cholesterol, usually has no symptoms. Although statins reduce the risk of a heart attack, many of the people who take them would probably have never had one. Like other treatments for asymptomatic conditions, statins change what it means to be healthy, contributing to our present situation in which nearly all older Americans, regardless of how they feel, take multiple medications.[7]

Statins also created puzzles for medical researchers. Even in studies that showed dramatic clinical benefit, the plaque regression was so subtle that researchers could only detect it through computer-assisted analysis of angiographic images. In one influential study published in the *New England Journal of Medicine* in 1990, researchers had randomized 146 patients to three different treatments. The statin group experienced far fewer clinical events (death, heart attack, or revascularization) than the conventional therapy group (3 of 46 patients vs. 10 of 52 patients). Those taking statins were also more likely to experience plaque regression (seen in 32% vs. 11% of patients). This regression, however, was subtle, a mere 0.3 percent change in obstruction, which corresponded to a 0.002 millimeter change in the minimum lesion diameter.[8] This posed a mystery: If the plaques were only slightly smaller, what had produced the substantial reduction in heart attacks?

Such findings made little sense under the model of progressive obstruction. With that model, the efficacy of a medication would depend on its ability to reduce the size of the obstructive plaque. Plaque rupture, however, offered an answer. Statins helped not by shrinking plaques but by reducing the risk of plaque rupture. Physicians remain uncertain about how statins accomplish this. They might decrease the lipid content of plaques or transform cholesterol into crystals, producing a stiffer lipid core, which is more resistant to rupture. Or they might have some unanticipated anti-inflammatory effect, for instance, modifying the artery lining so that circulating white blood cells do not enter plaques and transform into macrophages. Uncertainty over statins' mechanism, however, has done little to hinder sales. From Mevacor to Pravachol, Zocor, Lipitor, and Crestor, statins have yielded one blockbuster after another. Lipitor, one of the best-selling drugs in the United States, topped even Plavix, with $6.9 billion in sales and nearly $300 million spent on advertising in 2010.[9] Simvastatin, the generic version of Zocor, rode on these coattails and generated over $6.2 billion in sales without any marketing at all.

The efficacy and increased prescribing (and advertising) of aspirin, clopidogrel, and statins facilitated the emergence of the plaque rupture hypothesis. Plaque rupture, unlike progressive obstruction, explained how these drugs might reduce the frequency and lethality of acute coronary events. The drugs, in turn, solved the problem of the therapeutic imperative.[10] Medicine is an applied science: patients come to doctors asking them to do something. While a diagnosis and prognosis can provide reassurance, doctors must offer treatment. In the nineteenth century, new clinical and laboratory sciences had made doctors increasingly skeptical of their traditional therapeutics, which were based on bloodletting, emetics, and purgatives. But even when they doubted these treatments, they had to use them until they had something different to offer—new knowledge is only useful if it allows physicians to do something new. Plaque rupture demonstrates this well. Before doctors recognized the value of statins and platelet inhibitors, they could not put the plaque rupture hypothesis to work for patient care. They could describe the risk of rupture but offer no specific treatment. Once the drugs became available, however, the dynamic changed. Doctors could describe to patients how the rupture of fragile plaques caused heart attacks and, for the first time, offer treatments that intervened against the causes and consequences of plaque rupture.

The case I've made may be circumstantial, but the timing of the relevant events is uncanny. Plaque rupture only slowly became the conventional wisdom in medicine. The idea had circulated since the 1930s and was carefully documented by teams of researchers in the 1960s. Textbooks in the 1970s had little to say about the causes of heart attacks, and that was noncommittal. Textbooks and journal reviews didn't begin to devote substantial space to plaque rupture until the 1990s. When they did, they emphasized the newness of the theory. Peter Libby, a prominent advocate of the plaque rupture hypothesis, described it as "recent" and "emerging" in 1998. The *New York Times* described it as a "new and emerging understanding" in 2004.[11] After decades of debate and neglect, plaque rupture finally became popular in the 1990s, at a time when it became linked to popular and profitable medicines. The efforts of pharmaceutical companies could only have helped. They spent billions of dollars marketing statins and platelet inhibitors. These drugs became the most profitable blockbusters known to medicine. The money spent marketing these drugs and funding research on them marketed the disease model as well. No such marketing initiative existed for plaque rupture before the late 1980s. However, once the hypothesis came to be linked to promising and heavily marketed drugs, plaque rupture became a promising and widely accepted theory.

The therapeutic appeal of the plaque rupture hypothesis has even extended beyond actual therapies: doctors expect a new therapeutic bonanza. During the 1990s, molecular biologists elucidated the mechanisms of plaque rupture in ever greater detail. This inspired physicians to move beyond macroscopic, mechanical interventions like bypass surgery and angioplasty, and even beyond statins and platelet inhibitors, to seek new ways of intervening at the molecular scale. Each newly recognized pathway became the basis for a potential new treatment. Researchers sought drugs that would inhibit interstitial collagenase, stromelysin, gelatinase B, and the other bewilderingly named macrophage-derived matrix-degrading metallo-protease enzymes. They tested whether angiotensin inhibitors, and possibly even fish oils, could protect the cells that line our arteries. They imagined how transforming growth factor-ß might enhance matrix production and strengthen the plaque, while platelet-derived growth factor might prevent the death of smooth muscle cells; they did worry, however, that these growth promoters might foster excessive and harmful plaque growth. And they sought antioxidants, cytokine modulators, and gene therapy to increase the expression of cholesterol receptors in the liver. One cardiologist described five basic strategies for stabilizing plaques, from endothelial passivation to cholesterol reduction and platelet inhibition. As he described in 2002, "Cell biology has created a new paradigm for prevention of acute coronary syndromes resulting from plaque rupture."[12] Plaque rupture, productive of drugs, sales, and research funding, has become the basis for a new molecular cardiology.

A biography of the plaque rupture hypothesis thus takes shape.[13] Plaques may have ruptured for as long as there have been long-lived, over-nourished, and sedentary humans, but the idea of plaque rupture only emerged in the 1930s, as a burgeoning epidemic of coronary artery disease led pathologists to focus their microscopes and attempt to discern the cause of the disease. The plaque rupture hypothesis remained marginal through the 1960s, in part because it did not allow doctors to do anything new for their patients. In fact, plaque rupture theory may have suffered because it was such a frightening concept in the absence of specific therapeutics, as I describe in chapter 8.

Only with the advent of statins and platelet inhibitors in the 1990s was the promise of the plaque rupture synthesis fully recognized. In this respect, the theory is like many other ideas in the history of medicine. Germ theory, which contends that many diseases are caused by bacteria, took shape in the late nineteenth century, but competing theories remained influential until new therapies, especially antibiotics, appeared. High blood pressure was recognized when physicians engineered sphygmomanometers and other devices to detect it, but it did not become interesting until insurance actuaries recognized its ability to predict mortality rates, and it did not become relevant to doctors until they had treatments to manage it.[14] Medicine is in essence about prognosis and treatment. Medical ideas gain traction according to their prognostic and therapeutic utility.

If it is true that enthusiasm for statins and platelet inhibitors contributed to the consolidation of the plaque rupture consensus, then some wariness might be in order. The use of medical treatments to resolve debates in medical theory ties medical knowledge to medical practice. The pragmatic focus of medicine makes this inevitable, appropriate, and even essential. However, whenever medical therapy informs medical theory, the theory becomes subject to the vagaries of therapeutic enthusiasm. Enthusiasm has long been a bugbear for physicians, something to be avoided and replaced with objective reason. Doctors in the 1970s repeatedly voiced their anxieties about their susceptibility to unwarranted enthusiasm. The concern was in part about financial conflicts, with the idea that physicians' assessments were colored by the potential profitability of a medical practice. But it was also more general, about the seemingly irrational ways in which doctors become excited about a treatment in the absence of credible evidence. Enthusiasm once existed for everything from bloodletting and purgatives to lobotomy, radical mastectomy, and hormone replacement therapy. Those treatments have now largely been abandoned.[15] Will current enthusiasm for statins and platelet inhibitors also fade? No one can tell yet. The existence of enthusiasm itself should not condemn a treatment. And much evidence does support the use of these drugs. But patients, doctors, and researchers must keep a close eye on how therapeutic enthusiasm penetrates medical knowledge. Doctors, confronted with sick patients, must act. This therapeutic imperative, for better or worse, can change what doctors think. We must be aware of when and how this happens.

CHAPTER SEVEN

Therapeutic Ruptures

The consolidation of the plaque rupture consensus demonstrates how medical practice can influence medical theory. Only when doctors began to intervene acutely during heart attacks in the 1970s did they become convinced that coronary thrombosis caused the attacks. Only with the increasing use of statins and platelet inhibitors in the 1990s did the plaque rupture hypothesis become popular. Something else can happen as well: established therapies can persist even after the theories that initially justified them have been abandoned. The plaque rupture hypothesis led many doctors to question the value of bypass surgery and angioplasty, in particular the ability of these techniques to prevent heart attacks. The resilience of both procedures against these critiques offers a revealing display of the potential autonomy of medical practice from medical theory.

When René Favaloro described bypass surgery in 1968, and when Andreas Grüntzig described angioplasty in 1978, their hopes for the procedures depended on the then-popular theory of progressive obstruction. The basic idea was that plaque accumulated over time, like deposits in old pipes. As these grew, they caused angina and eventually infarction. Everyone has had this experience with household plumbing: first a sink drains slowly, and then one day it is so obstructed that it overflows. Many doctors and patients mapped the familiar and intuitive model of pipes onto the problem of coronary artery disease. For instance, when doctors at the Cleveland Clinic described bypass surgery to one patient in 1973, they detailed the intricate procedure but ended with a reassuring note: "It's a plumbing job." This imagery persisted in medical and popular culture for decades. Even though one cardiologist issued a biting critique of cardiology's plumbing imagery in *Slate* in 2007, the idea was still invoked explicitly in the 2011 *New York Times* advertisement for angioplasty services at Mount Sinai Hospital.[1]

The logic of therapeutic intervention here is clear. Arteries provide conduits for blood flow. If a plaque obstructs flow in a patient with chronic angina, then the plaque must be reduced or bypassed. Managing the plaque in this way would restore flow, reduce the frequency and severity of angina, and prevent damage to the heart's muscle. Such plumbing metaphors, with all their intuitive appeal, were not simply a way of describing and marketing the procedures to patients. They also influenced the thinking of doctors in powerful ways.

In 1969 Effler expressed his belief that the logic of bypass surgery and the angiographic evidence of restored flow justified his faith in the operation: "The cardiologist who would loudly deny the existence and validity of such factual evidence by refusal to examine it is, in my opinion, allowing emotion to prevail over scientific evaluation." When a team of researchers completed the first randomized clinical trial of bypass surgery in 1977 and found that most surgical patients did no better than the patients treated only with medications, many surgeons dismissed the results. Their critiques ranged widely, from how patients were selected to the skill and experience of the study's surgeons. But much of the critique reflected the conflict between statistical and mechanical ways of knowing. Even if actuarial analyses of survival rates showed no benefit, surgeons believed—they *knew*—that the treatment had worked because it had restored flow. In a model that defined disease as obstructed flow, bypass surgery and angioplasty had to work.[2]

The logic also made good sense for acute heart attacks. If a new total occlusion could be removed within two or three hours, it should be possible to rescue dying tissue and limit the extent of infarction and scarring. As one team of Boston surgeons described in 1972, "the surgeon has demonstrated capability of restoring blood flow to threatened areas with reversal of symptoms and stabilization of the electrocardiogram."[3]

But these were not the only reasons many bypass surgeries and angioplasties were done. In many instances surgeons hoped that by operating on a stable patient, they would prevent a future heart attack, an act of prophylactic revascularization. As Favaloro asserted, "it is best to operate before myocardial infarction occurs." A *Cleveland Magazine* article that celebrated the work of Favaloro and Effler in 1973 captured this well in its subtitle: "Don't Wait Around to Get Zapped by a Heart Attack: Head It Off at the Coronary Artery." This sort of enthusiasm might have been unusually pronounced in Cleveland. After all, Favaloro, Effler, and others at the Cleveland Clinic first brought bypass surgery to national attention in 1968. But the hope spread quickly, to Minnesota, New York, Canada, and even Brazil.[4]

Early attention focused on large plaques, but surgeons and cardiologists soon set their sights on smaller and smaller lesions. In 1981, for instance, Cleveland Clinic surgeons made their case for bypassing small plaques. They cited "the unpredictability of progressive coronary atherosclerosis" and "the relentless natural history of progression of atherosclerosis," as well as their desire to "provide the maximum duration of palliation" and "eliminate the need for future reoperation."[5] All these factors, in their eyes, justified intervention against plaques that had not yet produced heart attacks. In some patients, they even targeted lesions that had not yet produced angina.

Angioplasty pioneers employed almost exactly the same logic. In his initial account of the procedure, Andreas Grüntzig praised how it reduced obstructions: "The atheroma can be compressed leaving a smooth luminal surface." It was "comparatively simple and has the advantage of providing instantaneous revascularization without the need for open-heart surgery." As he and his colleagues gained confidence, they broadened their assessment of when angioplasty should be done. Their early experience supported "its preventive application for mild stenosis at an earlier stage of coronary artery disease." As they explained, "timely early intervention by PTCA [angioplasty] before double- and triple-vessel disease develops and irreversible myocardial damage occurs might appear logical and justified, even though symptoms are mild."[6] Almost anyone could be a candidate for intervention.

Motivated by these hopes, bypass surgery and angioplasty both expanded rapidly in the 1980s and 1990s, even as plaque rupture theory rose to prominence. Bypass surgery peaked at over 607,000 operations in 1997. Its subsequent decline—to 405,000 operations in 2007—reflected the rise of angioplasty, not loss of faith in the principle of bypass surgery. In 2007 interventional cardiologists performed nearly 1.2 million angioplasty procedures, with balloons alone or with stents in addition.[7] Prophylactic revascularization, which motivates many of these procedures, is alive and well in America. But should patients and doctors expect it to work?

For prophylactic revascularization to deliver on its promise, two things must be true. First, plaques need to develop predictably over time so that the ones most likely to cause future heart attacks can be recognized before they rupture. Second, angiography, near-infrared spectroscopy, or some other diagnostic technique must be able to identify the high-risk plaques. Only if these conditions are met can a preemptive strike work and prevent a heart attack. Critics of prophylactic bypass and angioplasty have argued that the plaque rupture hypothesis undermines both.

The first signs of trouble appeared when cardiologists realized that coronary artery disease was unpredictably progressive. The Cleveland Clinic team, for instance, collected data on patients who had undergone angiography twice, with several months or years between the two tests. They found that some lesions progressed slowly, while others remained stable. Still others changed rapidly and dramatically. Even worse, significant new lesions appeared in arteries that had looked normal before. Dutch cardiologists, meanwhile, monitored patients on wait lists for bypass surgery, hoping to detect those with rapidly progressive disease who needed to be prioritized in the queue. They found that "information derived from sequential coronary arteriograms is of little value in predicting future progression."[8] The variable trajectories of atherosclerosis, as revealed by conventional angiography, undermined cardiologists' ability to predict which plaques would cause future problems.

A Mayo Clinic group was frank. Sizing up the literature in 1981, they described how progression was "almost entirely unpredictable." They concluded that their findings "cannot support the decision for surgical bypass of coronary arteries with less degrees of stenosis on the assumption that they will inevitably progress." As discussed in chapter 3, some doctors had wondered whether angiography was a reliable technology and had examined the extent of the angiography-pathology correlation. The skepticism that emerged in the 1980s stemmed from a different anxiety. Even when doctors set aside their doubts about procedure, they faced a problem: angiography provided a portrait of a capricious disease, one that defied their expectations of reliable progression. A fickle disease made for a poor candidate for prophylactic intervention. This was a frustratingly familiar problem for physicians. Despite the increasing marvels of medical technology, many diseases, especially heart disease and cancer, defy physicians' efforts to predict how they will develop.[9] Prognostic uncertainty always complicates treatment decisions.

Pathologists and cardiologists also realized that the plaques that caused heart attacks were often not the largest, most dramatic plaques on an angiogram. Michael Davies and Anthony Thomas, in their influential autopsy series, found that one-third of the lesions responsible for sudden death were small, filling less than half the diameter of the vessel. Such lesions had traditionally attracted scant attention from angiographers: "Thrombosis can develop in the presence of a lesion that may not be regarded as important when it is detected through angiography during the patient's life." This was not just an artifact of clinical-pathological correlation. When cardiologists looked at angiograms of patients who subsequently suffered heart attacks, they discovered that nearly half (48%) of the culprit lesions had been small at the time of the prior study. The conclusion seemed inescapable: "It is often difficult to predict the location of a subsequent infarct from analysis of the first coronary angiogram." This did not mean that large plaques were benign. Cardiologists assumed that larger plaques were higher risk on a plaque-by-plaque basis. But since small plaques were so much more common, infarctions most often came from them.[10]

As doctors recognized the consequence of this finding in the mid- to late-1980s, they looked for a way to identify and treat the small plaques that caused many heart attacks. Here cardiologists ran into the limits of angiography. Since the 1960s, comparisons of angiography and autopsy studies found that the former often underestimated the extent of atherosclerosis. As cardiologists continued to develop new ways to visualize coronary arteries, the more sophisticated techniques confirmed the limits of conventional angiography. One group put ultrasound probes on the surface of the heart during open-heart surgery to examine the coronary arteries. They found evidence of diffuse atherosclerosis where angiography had revealed only discrete lesions. Intravascular ultrasound, conducted with ultrasound probes on the tips of coronary catheters, revealed another problem. Atherosclerotic plaques often grew outward first, away from the blood flowing through the artery, and did not encroach on flow until quite advanced. Since angiography could detect only changes to the lumen, a patient could have many large and small lesions, invisible to angiography, each with a substantial and unknowable risk of rupture. As Cleveland Clinic cardiologists Eric Topol and Steven Nissen wrote in 1995, "the silhouette or 'luminogram' is a relatively poor representation of coronary anatomy and a limited standard on which to base therapeutic decisions."[11]

Even when plaques were identified, angiography could not determine their physiological significance. In one Iowa study, researchers used Doppler probes to measure coronary blood flow during bypass surgery. Looking at obstructions that ranged from 10 to 95 percent, they found no correlation between the obstruction that angiography had revealed and the maximal flow they measured: "Our findings severely undermine the confidence with which the physiologic effects of most coronary obstructions can be determined by conventional angiographic approaches."[12] This discovery cast further doubt on the relevance of angiography results and the ensuing therapeutic rationale.

Taken together these findings posed a serious challenge for cardiologists. Their hopes for prophylactic revascularization relied on their confidence that they could identify and predict the plaques that would some day spawn heart attacks. But as they studied coronary artery disease, they found that atherosclerosis progressed unpredictably and that many heart attacks arose from small plaques that could not even be seen with angiography. Cardiologists could forecast neither the site nor the time of future events. As two experts concluded in 2003, "it is impossible to predict whether structurally vulnerable plaques may become unstable weeks, months, or years after their detection."[13] Without a better crystal ball, patients and doctors had to make their decisions amid uncertainty and hope.

As awareness of the capricious nature of atherosclerosis and the limited ability of angiography to detect small lesions spread in the 1980s, doctors recognized the consequences for prophylactic revascularization. One North Carolina team noted in 1988, "Because it was difficult to predict the site of the subsequent occlusion in our patients from the initial coronary angiogram, coronary bypass surgery or angioplasty appropriately directed only at the angiographically significant lesions initially present in almost all our patients would not have been effective in preventing the majority of myocardial infarctions."[14] In other words, cardiologists and surgeons could manage all of their patients' large, visible plaques only to have heart attacks arise from the many less-evident lesions.

This finding was consistent with that of the randomized trials that had so vexed surgeons. In the 1970s and 1980s, teams of American and European researchers had performed large randomized trials of bypass surgery. Each had shown that although bypass surgery relieved angina in nearly all patients, it provided a mortality benefit only in the sickest minority of patients. A similar story played out as angioplasty boomed in the 1980s and into the 1990s. Although angioplasty could relieve angina, trial after trial showed that patients with stable angina received no survival benefit from the procedure. The plaque rupture hypothesis, combined with growing awareness of the limits of angiography, explained why. Surgeons and interventional cardiologists focused their efforts on the largest, most obstructive plaques, which caused the worst angina but were not the plaques responsible for most future heart attacks.[15]

Critics of rampant revascularization trumpeted the limits of angiography and prophylactic treatment with increasing vigor and wit. When the Iowa group found no correlation between stenosis and flow, they warned that their results should be "profoundly disturbing to all physicians who have relied on the coronary arteriogram to provide accurate information regarding the physiologic consequences of individual coronary stenoses." Peter Libby, a prominent Boston cardiologist and promoter of the plaque rupture hypothesis, proclaimed that the "central dogma" of clinical cardiology had to be reassessed. Spencer King, who had brought Grüntzig and angioplasty from Switzerland to the United States, used his stature as an elder statesman of cardiology in 1998 to warn cardiologists that they needed to move beyond "the interventionalist's fixation on simply mechanical and geometric improvements." Critics did not mince their words: "'Nothing is more sad' says Arthur Koestler, 'than the death of an illusion,'" began two Los Angeles cardiologists. "We believe that the illusion of cardiology is that bypass or dilatation of coronary stenoses reduces the risk of myocardial infarction."[16]

Two Cleveland Clinic cardiologists often led the charge. Concerned about the rapid spread of angioplasty in the 1980s, Eric Topol coined the phrase "oculostenotic reflex" to describe cardiologists' rush to intervene on every significant plaque. By 1995 he and colleague Steve Nissen decried the "irresistible temptation" and its "ritual of reflex angioplasty." When cardiologists ignored the results of a clinical trial that found little value for one angioplasty variant, atherectomy, Topol and Nissen let loose. They claimed that cardiologists were motivated by "'angiographic gratification,' the allure of a better, more gratifying image after atherectomy." This had to stop: "Procedures should not be performed solely to improve the luminal appearance—so-called coronary 'cosmetology.'"[17] Cardiologists needed to move past superficial concerns and focus their attention on outcomes that mattered.

Enthusiasts responded to the challenge by seeking better ways to visualize diseased coronary arteries. The proliferation of techniques gave researchers hope that "it may be possible to develop catheter-based and non-invasive means of imaging and treating these potentially life-threatening lesions." Computers and image amplifiers could improve the sensitivity and accuracy of lesion detection by angiography. Analyses of plaque morphology improved on traditional measures of plaque diameter and area. Careful measurements suggested that irregular, eccentric lesions had a higher risk of rupture than smooth, concentric lesions. Intravascular ultrasound, advocated by Nissen, had "the potential to be the most sensitive technique by which to define the progression or regression of atherosclerosis." Angioscopy, which provided a direct look at the plaques, distinguished the types associated with different coronary syndromes. Near-infrared spectroscopy revealed plaques' lipid composition. Always hopeful about the promise of new technology, teams of prominent cardiologists founded competing startup companies—Volcano Corporation and InfraReDx—to develop and market the new technologies that might detect high-risk plaques and reinvigorate the practice of prophylactic revascularization. Some promising results have appeared. Even when heart attacks did arise from plaques that had previously appeared innocuous, a closer look showed that those lesions often had been thin-capped plaques, the sort most at risk of rupture.[18]

But even if these improved imaging techniques live up to their promises, they will still raise one last question about the therapeutic logic of prophylactic revascularization: Is plaque vulnerability a focal or diffuse process? Although this might seem like a subtle distinction, it is of great importance in medicine. Debates about whether diseases were focal or diffuse preoccupied physicians throughout the twentieth century. Many of the great debates in cancer treatment, for instance, have hinged on this assessment. This played out most publicly with breast cancer.[19] Is it sufficient to remove the most obvious nest of tumor cells with a lumpectomy? Should a surgeon perform a radical mastectomy, removing nearby muscles, lymph nodes, and even bones in an effort to root out local spread of the disease? Or do patients need systemic chemotherapy to kill whatever cells might have already spread to the far corners of the body?

Surgery, by its very nature, relies on the assumption that the disease process is focal enough to justify targeted intervention. This assumption made assessment of the focality of coronary artery disease a crucial question during the development of coronary artery surgery. As cardiac surgeons developed their initial treatments for coronary artery disease in the 1950s and 1960s— artery implants, patch grafts, endarterectomy, and finally bypass—they disagreed about even the simplest anatomic questions of whether lesions were focal or diffuse.[20] Even as attention focused on discrete vulnerable plaques in the 1980s and 1990s, debates about the diffuseness of risk remained.

One Michigan study examined coronary angiograms in 253 patients who had suffered heart attacks. While most had a single complex plaque, 100 had multiple plaques, and these patients were more likely to have worse outcomes. The researchers concluded that "plaque instability is not merely a local vascular accident but probably reflects more generalized pathophysiologic processes with the potential to destabilize atherosclerotic plaques throughout the coronary tree." As researchers learned about the importance of inflammation in atherosclerosis, especially the ways in which macrophages contributed to plaque rupture, they studied macrophage activity in different parts of the coronary tree. They found evidence of widespread activation of macrophages in patients with unstable angina, a finding that "challenges the concept of a single vulnerable plaque in unstable coronary syndromes." Speaking to the American Heart Association in 1995, association president Sidney Smith warned that "new understandings from vascular biology and molecular biology now suggest that therapy must address the entire atherosclerotic process, not just severe obstructions."[21] Increasingly seen as a diffuse disease, atherosclerosis required something more than focal therapies.

A new round of debate sprang up around this question. One study of fifty patients found vulnerable or ruptured plaques in only twenty of them and, in those twenty hearts, only an average of 2.1 "hot spots" per heart. If this finding is generalizable, then vulnerable plaques are rare, localized, and amenable to focal therapies (angioplasty and bypass surgery). Moreover, since statins and other systemic therapies required months of treatment before their benefit appeared, "the most vulnerable plaques may merit stenting or some other form of local therapy to 'buy time.'" Others disagreed and argued that doctors needed to treat "the entire coronary tree," by lowering cholesterol and stabilizing plaques or platelets. A *New York Times* reporter captured the critique well: since angioplasty and stenting targeted only a few of the countless at-risk plaques, interventionalists faced "a losing game of medical whack-a-mole."[22]

The challenge plaque rupture posed to the rationale of revascularization must be understood precisely. No one disputed the ability of bypass surgery or angioplasty to relieve angina in most patients. No one disputed the benefit provided by surgery to patients with the severest forms of coronary artery disease. No one disputed the value of surgery or angioplasty when provided within an hour of onset of a heart attack. The concern instead was that many procedures, especially angioplasty, which had become a burgeoning business, were done for different reasons: to prevent heart attacks and prolong lives in patients with stable, mild to moderate coronary artery disease. It was this indication that the advent of plaque rupture theory undermined.

What percentage of angioplasty procedures might fall under this cloud? Few questions in cardiology remain more contentious. When cardiologists debate the appropriateness of angioplasty, they cite a wide range of estimates of what share is done emergently (and thus presumably appropriately) and what share electively (and thus questionably). In one exchange, different cardiologists asserted that 31, 50, and 85 percent of angioplasties are done electively in stable patients who would not be expected to receive a survival benefit from the procedure. The highest estimate—85 percent elective, non-emergent—came from a 2006 study of angioplasty performed in New York State. In contrast, in a 2011 study, researchers looked at a national registry and concluded that 71.1 percent were done for "acute indications" and were, as a result, largely appropriate.[23] The inability of researchers to reach consensus on this, another seeming matter of fact, demonstrates how hard it can be to establish the evidence base needed to motivate a rational health care policy. Researchers, however, preserve their faith that definitive answers are just a careful clinical trial away. Launched in 2011, the ISCHEMIA trial (International Study of Comparative Health Effectiveness with Medical and Invasive Approaches) hopes to settle, once and for all, whether revascularization can reduce the risk of heart attack or death in patients with angina.[24]

Amid the controversies about the logic of prophylactic revascularization or the share of angioplasties done inappropriately, angioplasty has remained popular among clinicians and patients, many of whom believe that the procedure could extend lives. Despite the barbed wit of revascularization's detractors, prophylactic angioplasty has exhibited considerable therapeutic autonomy and endured substantial challenges to its physiological rationale.[25] Is it possible to explain this resilience? Plaque rupture did not become relevant for clinicians until it was linked to powerful and popular new treatments. This makes sense in light of the therapeutic imperative, which assigns value to medical theories according to their ability to support therapeutic practice. A different explanation is needed for the continuing popularity of prophylactic revascularization in the era of plaque rupture consensus. Financial interests, defensive medicine, and the fear of heart attacks have all played important roles.

Bypass surgery and angioplasty certainly made many people rich. By 1973, just five years after the Cleveland Clinic had launched coronary artery bypass surgery, Donald Effler billed over $1 million each year. Since his salary was under $100,000, money poured into the clinic's coffers. By 1977, one hundred thousand bypass operations were performed annually in the United States. At a cost of $10,000 to $15,000 each, with typically one-quarter of that the surgeon's fee, both doctors and hospitals profited handsomely. Although salary data can be difficult to find, one famous case illustrates what could be accomplished. Denton Cooley, the charismatic founder of the Texas Heart Institute, earned $9,865,000 from his medical practice in 1987. This fueled a lifestyle that caught the eye of *Architectural Digest*. The magazine ran a profile of Cooley and his two Texas homes in its May issue that year.[26]

Cardiologists also reaped the windfall. For every bypass surgery, there was a cardiologist who had provided the diagnostic catheterization. The advent of angioplasty offered nearly unlimited opportunity. With a typical charge of $800 for each procedure, a cardiologist in 1984 could perform angioplasties just one day a week and earn nearly $200,000 from that procedure alone. In the 1990s the rising salaries of interventional cardiologists surpassed the declining salaries of cardiac surgeons. Coronary angiography, angioplasty, and surgery now form a multibillion-dollar industry. By some estimates, the procedures account for one-third of the revenues at medical centers that offer them. Prominent cardiologists profited not just from clinical care but also from consulting. Gregg Stone, a pioneer of catheter technologies, received $10,000 or more from each of twelve companies in 2006 and had equity stakes in another eleven.[27]

The temptations of wealth proved irresistible for some doctors. Cardiac surgeons and cardiologists have been investigated repeatedly for profiteering from unnecessary procedures. Two surgeons in Redding, California, were investigated by the FBI on allegations of unnecessary bypass surgery and Medicare fraud. A Louisiana cardiologist has been sentenced to prison for falsifying medical records to justify unnecessary angioplasty.[28] A Baltimore cardiologist, who inserted thirty stents in a single day in 2008, has been investigated by the Senate Finance Committee for Medicare fraud. Other investigations have been launched against doctors in Austin, Texas, and Salisbury, Maryland.[29] It remains to be seen whether these cases prove to be exceptional or harbingers of convictions yet to come.

Although financial interests certainly color the decision making of some physicians, they alone do not account for the persistent popularity of prophylactic revascularization despite the advent of the plaque rupture hypothesis. Psychological factors clearly contribute as well. Researchers in San Francisco, curious about why so many heart specialists recommended angioplasty when a survival benefit had not been shown, held focus groups in 2006 with twenty cardiologists. These informants offered many explanations for the decisions they made. Some supported a "zero tolerance" policy for ischemia. Others felt that patients should leave "the lab with an open artery, the best that my interventional partners can deliver." Many doctors experienced "anticipatory regret" and worried that they would feel responsible if any of their nonangioplastied patients suffered a heart attack. Fear of lawsuits heightened the intensity of this feeling. As one cardiologist explained, "The cath lab staff probably wouldn't let us leave the lab unless we did something with the lesion." Another told researchers, "When I see a lesion, the bottom line is that the oculostenotic reflex always wins out." Intervention also assuaged patient anxiety. Imagine how difficult it would be for patients to sleep peacefully knowing they had untreated plaques lurking in their coronary arteries. Cardiac CT scans brought more patients into the screening process, something that culminated for many in a referral to angiography and then angioplasty. Finally, the popularity of drug-eluting stents had "lowered the threshold" for intervention. What was the bottom line? The researchers concluded that the "apparent gulf between evidence and practice appears to be motivated primarily by emotional and psychological factors."[30]

The San Francisco cardiologists were not unique. Two Swiss cardiologists argued that coronary artery stents "have a great future—they give excellent predictive results in angiography, are clinically safe, and, most of all, calm the interventional cardiologist." Opening an artery could be immensely gratifying for cardiologists. This is easy to envision. Standing next to the patient, the cardiologist studies the angiogram for evidence of obstructed flow. Plaques large and small litter the coronaries. What should the doctor do? A few stents, deftly deployed, could leave the patient with a rejuvenated angiogram. If you believe large plaques are time bombs waiting to detonate, the decision is easy. As Boston interventionalist Ashwin Pande told the *New York Times,* "This adrenaline rush is why people like me go into cardiology." Another Massachusetts cardiologist saw this more cynically: "Arguably the financial rush is also sweet, since the median salary for invasive cardiologists is roughly half-a-million dollars."[31]

Attention to such psychological factors opens up a range of other possible contributing factors. From the 1960s into the 1990s, cardiac surgeons and cardiologists in the United States led the world in their enthusiasm for bypass surgery and angioplasty. Their predilection for aggressive therapy may have had deep roots. Although "heroic" therapy has long been controversial, it is woven into the fabric of American medicine. Benjamin Rush, a veteran of the American Revolution and of Philadelphia's yellow fever epidemics, famously pushed his colleagues to take bloodletting, emetics, and cathartics to the limits of tolerance. He believed that dire diseases demanded powerful remedies. In twentieth-century America this tradition of heroic medicine manifested in the aggressive treatments of the "War on Cancer," from radical mastectomy to bone marrow transplants. Pursuit of aggressive surgery was especially pronounced after World War II. Emboldened by their military experiences, surgeons in the United States took the risks required to make transplant surgery and cardiac surgery a reality. This legacy remains alive and well in cardiac surgery and interventional cardiology, not just among physicians but also among their patients. One interventional cardiologist described in 2004 how heroic treatment had become "ingrained in the American psyche": patients think that "the worth of medical care is directly related to how aggressive it is."[32]

One other psychological dynamic may figure into this mix. Throughout the twentieth century, patients and doctors sought technical solutions for what they knew to be social problems, even ones steeped in morality. Should people rein in their sexual appetites and stamp out syphilis and other sexual infections through rigorous self-discipline? Penicillin offered an alternative means of syphilis control that appealed to many people, including health officials. A similar tension has now emerged around coronary revascularization. We inhabit an increasingly moralized discourse about lifestyle and personal responsibility. The burgeoning epidemic of obesity and diabetes has led to increasing calls for individuals to take responsibility for leading healthier lives. Yet gastric bypass surgery has become more and more popular, for those who cannot manage to lose weight through self-control, even for teenagers. Coronary artery disease patients have long had a notoriously difficult time making the lifestyle changes (involving smoking, diet, or exercise) needed to minimize their risk. One of the pioneering plaque rupture researchers, Meyer Friedman, described this reality: "Totally effective prevention can be obtained against coronary artery disease if Western man is prepared or willing to adopt these four habits. Unfortunately he is not. Therefore we are confronted with a constantly increasing incidence of a disease which is almost totally preventable but at a cost few of us are willing to pay."[33] Patients may have sought solace in bypass surgery and angioplasty. It is easier—and may even cost less—to have surgery than to change an entrenched habit. Countless bypass patients, and countless more on statins, hope that treatment will accomplish what discipline cannot.

Surgeons have long recognized the dangers of this strategy. If risk factors remained unchecked after surgery, atherosclerosis would continue its progressive course, not just in the coronary arteries but in the bypass grafts as well. Some surgeons used the teachable moment of open-heart surgery to push for lifestyle reform. Stanford surgeons, for instance, explained in 1972, "Where atherosclerosis risk factors are present preoperatively, an attempt is made to correct them in the period following by appropriate drug therapy or dietary means." They had some success with this: "Most of these patients can be readily persuaded to modify their eating habits or stop cigarette smoking, having been made aware of the gravity of the situation." Bill Clinton's well-publicized case is a notable example. After his bypass surgery, he transformed his habits, lost weight, and adopted a vegan diet. However, commitment to change does not always bring change. As the Stanford surgeons confessed, "Preventive efforts are not always successful." Some recent data bear this out. One study found that patients who had bypass surgery or angioplasty had lower rates of adherence to statin therapy than patients treated with medications alone.[34] The former may have felt that further treatment after revascularization was unnecessary.

Given how pessimistic physicians can be about the prospects of patients "doing the right thing" with lifestyle reform, it is not surprising that they pursue expedient alternatives. They offer prophylactic revascularization to patients with chronic stable angina, despite the many doubts about the logic of this path, because it might help. This attitude, along with the other psychological factors and financial interests, has bolstered elective angioplasty. In a remarkable display of therapeutic autonomy, the procedure persisted and thrived even after the theory that once provided its rationale had been displaced. Therapeutic practice drove plaque rupture to prominence, but the theory could not conform all practice to its dictates.

CHAPTER EIGHT

Fear and Unpredictability

In addition to psychological and financial motivations, another factor may have contributed to the persistence of prophylactic revascularization in the era of plaque rupture consensus: fear. Fear of heart attacks swept popular media and the medical literature in the twentieth century. This anxiety paralleled the growing focus on risk and risk factors in medicine. The idea of a "risk factor" was first formalized by Framingham Heart Study researchers in the 1950s and the 1960s, as they attempted to explain the distribution of heart disease. Their work broadened our sense of who is "at risk" of heart disease, from an initial focus on elite white men, to all men, and now to any person— worldwide—who strays from an ideal of perfect diet and behavior. Risk factors have since become ubiquitous considerations throughout medicine.[1] Fear, however, has a different character than caution based on risk assessment. It has had a more visceral impact on how doctors and patients have thought about and treated heart disease.

Fear of heart attack and sudden death spread quickly as heart disease gained the public's attention in the 1920s and 1930s. Few diseases had a similar ability to strike down previously healthy men and women in their prime. Medical textbooks made this clear. A 1928 edition of *A Text-Book of Medicine by American Authors* observed that "sudden death is always a danger, and cannot be anticipated." A 1950 text noted, "There is perhaps no other disorder in which a patient, apparently progressing favorably, is so likely to die unexpectedly." The frequent lack of prior angina was especially troubling: "A particularly distressing aspect of this disease is the frequency with which it suddenly strikes young men, between 35 and 55 years of age, without prior warning." Even now the first sign that anything is wrong might be the last. Peter Libby's 1998 review of the pathophysiology of atherosclerosis captured this well: a "dramatic acute clinical event, such as myocardial infarction or cerebrovascular accident, may be the first manifestation of atherosclerosis."[2]

Widespread advertisements accentuate the fear. The billions of dollars spent marketing statins and platelet inhibitors fuel the sense that we are all at risk and that we all ought to take the medicine. Something similar happens with bypass surgery and angioplasty, even though they are not marketed as aggressively. Intensive media coverage of celebrities in need of bypass surgery—David Letterman in 2000, Bill Clinton in 2004, Regis Philbin in 2007, Larry King in 2010—contributes to the pervasive fear and the willingness of people to submit to dramatic medical interventions.[3]

Doctors feel this fear with particular acuity. Throughout the twentieth century, they saw themselves as the group most at risk of the ravages of coronary artery disease. In his 1910 account of angina pectoris, William Osler described it as the "morbus medicorum," literally the disease of doctors. James Herrick featured a physician in his early case reports about coronary thrombosis. Experts had little trouble explaining doctors' peculiar risk. When two statisticians from the Metropolitan Life Insurance Company characterized the mortality of physicians in 1947, they were not surprised by the high prevalence of coronary artery disease. They cited the "special activities and burdens" of doctors: "Their call to service at any time, their constant guard against varied infections and their need for mental alertness all may well impose strains that are felt in few other callings." By mid-century the fear was well established. Two New York physicians even called coronary disease "an occupational hazard of the medical profession."[4]

The prevalence, or the perceived prevalence, of heart disease among physicians motivated a particular kind of therapeutic imperative. Physicians desperately wanted better therapies for a disease that so many of them suffered from, but they worried that this desperation colored how they thought about promising new therapies. Eugene Braunwald, perhaps the most influential cardiologist of the late twentieth century, warned his colleagues in 1971 against excessive enthusiasm for bypass surgery, explaining that "members of the medical profession, themselves subject to the ravages of coronary arteriosclerosis in disproportionately high numbers, are often in the vanguard of those having faith in the efficacy of each newly reported therapeutic maneuver."[5] Fear could color judgment.

The traditional model of atherosclerosis as a disease of progressive obstruction had particular appeal in this world of pervasive fear. Yes, some people were struck out of the blue and died from a heart attack without warning. But for most people, the disease was predictable and gradual, offering patients a chance to save themselves. The Framingham Heart Study and other epidemiological studies identified specific risk factors for atherosclerosis, particularly smoking, obesity, high blood pressure, a high-fat diet, and lack of exercise. Doctors could study a patient's lifestyle and predict that person's risk. The disease, once diagnosed, was expected to behave in an orderly, almost fair, manner. When plaques reached a threshold, angina began, a harbinger of bad tidings yet to come: "Angina pectoris is the forerunner of the coronary event; therefore, if something can be done to modify the course of angina, the likelihood of such an event may be reduced." The warning provided by angina, understood within the model of progressive obstruction, gave patients and doctors a chance to respond before it was too late, and it motivated them to do so. The risk factors could be managed.[6] Treatments could be employed before a heart attack. Angina could be a blessing in disguise.

Angiography, bypass surgery, and angioplasty fit into this vision well. Angiography improved physicians' ability to detect and characterize the disease. It could be used to distinguish patients with true coronary artery disease from patients with other diseases that could cause similar symptoms, such as heartburn, muscle strains, or even anxiety. The technique identified dramatic plaques, especially the widow-maker lesion, which could be managed with surgery and angioplasty. When postprocedure angiography demonstrated improved flow, patients and doctors could hope that risk had been reduced. As Donald Effler argued in 1969, this was "prima facie evidence that the needs of that particular individual have been met." More recently, San Francisco cardiologist Grace Lin and colleagues explained that angioplasty could assuage patients' anxiety.[7] Heart disease might be frightening, but the fear was orderly and manageable.

Compare this world with the world of plaque rupture, where uncertainty and fear are not so easy to address. Although the traditional risks factors remain relevant, doctors now view atherosclerosis as a more duplicitous disease. One pathology textbook noted that atherosclerosis "may develop in the absence of any apparent risk factors, so even those who live 'the prudent life' and have no apparent genetic predispositions are not immune to this killer disease." High-risk lesions can be asymptomatic and invisible to angiography. Because rupture is "unpredictable and abrupt," many "asymptomatic adults in the industrial world have a real but unpredictable risk of a catastrophic coronary event." This is an alarming situation: "Regrettably, it is presently impossible to reliably predict plaque disruption or subsequent thrombosis in an individual patient."[8] John Hunter, whose angina attacks were triggered by his short temper, complained that he was at the mercy of rascals. At least he knew his enemy. Those of us with fragile plaques—likely most of the adult population of the United States—are now at the mercy of the macrophages, cytokines, and collagen tendrils that hold our plaque caps in place.

To make matters worse, fragile plaque theory marginalizes the experience and expertise of both patients and doctors. With most illnesses, patients develop symptoms of whatever has gone awry. They can tell that something is wrong, and many develop a sense of its nature and severity. But a person with coronary artery disease and its vulnerable plaques can be asymptomatic until the moment of death, without any awareness at all of the mischief at work within the heart. Meanwhile, physicians cannot be the all-seeing and all-knowing experts they aspire to be. They cannot detect all relevant lesions of atherosclerosis. Of the lesions they can see, they cannot predict which might cause problems someday. For decades physicians had been frustrated because they did not understand the cause of heart attacks. Now they believe that they do, but this knowledge has not taken the fear away. If the plaque rupture hypothesis is right, then coronary atherosclerosis is inherently unpredictable, and nearly everyone is at risk. This is a bitter irony of the plaque rupture hypothesis.

Beyond the number of Americans heart attacks kill and the abruptness with which they strike some people down, heart disease causes fear in subtler

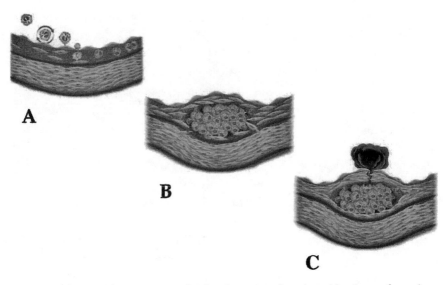

Erupting plaques. Three stages in the development of a vulnerable plaque, from the arrival of macrophages (*A*), through growth (*B*), to catastrophic rupture (*C*). The diagram of a rupture resembles both volcanic eruptions and mushroom clouds. From Peter Libby, Paul M. Ridker, and Attilio Maseri, "Inflammation and Athero-sclerosis," *Circulation* 105 (5 March 2002): 1135–43, fig. 1, p. 1136. With permission from Wolters Kluwer Health.

ways as well. Imagery of erupting volcanoes, exploding bombs, and other catastrophes abound in medical and popular writings about the plaque rupture hypothesis. Progressive obstruction relied on a language of gradual, uniform processes (e.g., slow accumulation of plaque, or steady deposition of thrombus); plaque rupture, in contrast, invokes a language of geological catastrophe. Vulnerable plaques underwent "fissures, breaks, tears, rents, erosions, ulcers, and rupture." The damaged layers of arterial walls bent, folded, and fractured—a microcosm of atherosclerotic tectonics. Even when plaques seemed dormant, they seethed in "smoldering inflammation." The parallels extended into the visual imagery of cardiology. Diagrams of plaque rupture portray plaques as erupting volcanoes, spewing their deadly contents into the lumen.[9]

The parallels are striking. Geologists know well where earthquakes and volcanoes form, and they can predict which places in the earth are most at risk. They can even say how many quakes and eruptions will likely take place globally each year. Yet they cannot say when the San Andreas will rupture or when Mount Rainier will erupt again. Like earthquakes and volcanoes, plaque rupture is predictably unpredictable. Doctors know which patients are most likely to have vulnerable plaques. They know where in those patients' arteries the plaques are most likely to form. They know how many heart attacks occur each year. But they cannot predict when or whom heart attacks will strike next. Instead, they know in considerable detail why so much is, and will likely remain, unknowable.

Doctors have been strategic in their use of the imagery of geological catastrophe. One group of cardiologists and engineers, who founded a company to produce intravascular ultrasound devices to detect vulnerable plaques, deliberately named it Volcano Corporation. When European and American experts met for two days in 2003 to hash out a consensus about the meaning of vulnerable plaques, they chose to meet in Santorini, site of the ancient Minoan civilization that had been obliterated by a volcanic eruption. Dutch cardiologist Ton van der Steen, who helped organize the conference, told me that Santorini was a "deliberate choice": "A volcano is the geological equivalent/analogy to a vulnerable plaque."[10] The conference logo—an irregular red ring—simultaneously evoked the shape of a volcanic crater and a cross-section of an inflamed, atherosclerotic artery.

Plaque rupture science, as it emerged during the Cold War, also adopted the imagery of manmade catastrophes, especially nuclear bombs. The photographs and diagrams of plaque rupture that appear in so many medical publications resemble not just volcanoes, but also mushroom clouds. James Muller, who coined the term "vulnerable plaque" in 1989, was inspired to do so by his work on nuclear disarmament. Muller had helped to establish International Physicians for the Prevention of Nuclear War, a group that won the Nobel Peace Prize in 1985. In the early 1980s Muller and others from this group had debated government officials about nuclear weapons policy, particularly about the money spent to protect the "vulnerable warheads" of land-based missiles. As he began working on fragile plaques, this episode came to mind. His concern with nuclear war and plaque rupture led to the idea of "vulnerable plaque."[11]

Bombs appear often in discussions of heart attacks and plaque rupture in popular media. One cardiologist, writing in *Slate,* described the plaques as "little bombs that blow up suddenly and cause a sudden and devastating blockage." Another doctor, promoting a new drug to treat high cholesterol levels, told the *New York Times* how patients, after their first heart attack, were like "walking time bombs." If they did not receive adequate treatment, their second attack was just a matter of time. Other doctors complained that such imagery was used as a scare tactic to pressure patients to accept angioplasty. One suggested that this bomb language explained why many patients who did not need angioplasty had the procedure done anyway: "I think they have talked to someone along the line who convinced them that this procedure will save their life. They are told if you don't have it done you are, quote, a walking time bomb."[12]

The foreboding, fear, and uncertainty that pervade the literature on fragile plaques and plaque rupture might have slowed the uptake of the theory. Any idea based in part on uncertain and unknowable risk will be a hard sell for patients and doctors, and this dynamic may have been in play in the 1970s and 1980s. The plaque rupture hypothesis only became palatable in the 1990s onward as new diagnostic techniques, such as intravascular ultrasound and near-infrared spectroscopy, held out hope that vulnerable plaques might be detectable, and as new therapeutic agents, such as statins and other plaque-stabilizing agents, met the demands of the therapeutic imperative. These technologies, marketed to patients and doctors through campaigns that accentuated the fear, promised to take some of that fear and powerlessness out of plaque rupture. Pharmaceutical advertising has recrafted individual identity such that an ever-increasing number of patients see themselves as people in need of statins and other preventive medications.[13]

But even when millions of patients take statins and platelet inhibitors, the fear persists. Lifestyle modification and pharmaceutical regimens cannot guarantee that plaques will never rupture. When angiography reveals a worrisome plaque with its volcanic and explosive connotations, a pill and its promise of gradual change hardly seem up to the task. People who are told that their vulnerable plaques make them like a walking time bomb will often rush to revascularization. Amid fears of cardiac catastrophe, patients ask not whether revascularization will help, but whether it could help. The risk and fear inherent in the plaque rupture hypothesis makes any potential benefit feel worth pursuing. Even if patients and doctors know, from a rational, empirical, data-driven perspective, that revascularization might have limited value, they also know that doing something almost always feels better than not doing it when faced with mortality.

Physicians appropriately emphasize the importance of evidence-based medicine. To the extent that they can study the outcomes of medical treatments and use this knowledge to guide treatment decisions, they can contribute both to better patient care and to the continuing refinement of medical science. But physicians draw on a wide range of information when they assess promising treatments and make recommendations to patients.[14] Although outcomes data from clinical trials play a role, physicians also consider whether the treatment is plausible. This judgment arises from their beliefs about the pathophysiology of disease and the mechanisms of therapeutic action. This creates a complex dynamic in which doctors weigh two different kinds of claims of therapeutic efficacy, some based on outcomes, others on plausibility. Physicians also use their knowledge of treatment outcomes to revise their theories about the mechanisms of disease. This creates another dynamic, in which therapy informs theory.

Given the centrality of therapeutics to medical practice, it is no surprise that the treatments doctors use powerfully influence their thinking about disease. This sounds appropriate in principle, but in practice it creates subtle problems. Physicians exhibit a selective permeability to new ideas. The existence of a well-validated disease model, such as plaque rupture in the 1980s, does not guarantee its application in medical practice. Instead, the impact of new disease models depends on the extent to which they are linked to therapeutic innovations. Therapeutic practice can therefore be both a conservative force, stabilizing old ideas as long as old treatments remain popular (as seen with the persistence of prophylactic revascularization alongside intuitions about progressive obstruction), and an innovative force, encouraging new ideas when linked to new treatments (as seen with the advent of the plaque rupture consensus alongside the emergence of statins). The existence of a popular treatment can alter physicians' understanding of pathophysiology such that the new understanding serves to justify the treatment.

This has consequences for how patients, doctors, and policy makers think about medical decision making. We need a healthy skepticism about claims of efficacy, whether they are grounded in clinical research or in the intuitive logic of disease theories. Doctors, and increasingly the general public, have become aware that clinical trials will not be a panacea: comparing two treatments to see which works best is never as straightforward as it ought to be. We should be equally skeptical about our therapeutic intuitions. If a treatment seems to make sense, it is important to wonder about the underlying model of disease. Disease models are subject to the same hopes, fears, and enthusiasms as the treatments associated with them. Paying closer attention to the ambiguities of disease models and claims of therapeutic efficacy could result in more realistic expectations of efficacy, which could improve both the quality of care and the satisfaction that patients and doctors take from it. Medical knowledge must be tempered by wisdom born of awareness of the dynamic interactions between belief and practice, and by humility that arises from our understanding of the limitations of each.

PART II / Complications

CHAPTER NINE

Surgical Ambition and Fear

O n December 13, 1960, Cleveland Clinic surgeon Donald Effler wrote to a Montreal colleague, Arthur Vineberg. Effler and Vineberg, two pioneers in the new field of cardiac surgery, exchanged letters often. They shared stories of their successes and failures as they worked to develop new surgical techniques. This time Effler had bad news. He had recently operated on a fellow physician. The procedure—a valve replacement—had gone well, but twenty-four hours later, the doctor suffered a stroke, which partially paralyzed his left arm. The doctor's clinical situation quickly deteriorated, and by the end of the week he was dead. Effler also described another operation that had gone badly, and Vineberg, always sympathetic, wrote back quickly: "As we both know, these were pretty rough cases and there are bound to be losses such as occurred in this case."[1] Although the surgeons regretted the doctor's death, they knew that such losses were an inevitable part of their campaign to push forward the frontier of cardiac surgery.

When patients agree to try surgery or any treatment, they gamble that the benefits will justify the potential complications. Is this wise? It depends on what their doctors have learned about the intervention. As described in the first part of this book, it can be surprisingly difficult for physicians and patients to develop a realistic sense of a treatment's benefit. Doctors' beliefs about efficacy do not depend simply on the results of randomized clinical trials, the official arbiters of therapeutic success. Instead, doctors also consider their beliefs about what causes a disease and how the treatment might intervene. Along the way, their assessments succumb to other influences—from financial conflicts of interest to the psychological compulsion of the therapeutic imperative.

The situation with treatment safety is just as complex. In these next chapters, I explore why it is often easier for physicians to generate knowledge about the effectiveness of medical treatments than it is for them to characterize a treatment's safety or complications. Doctors end up knowing much more about the benefits of medical treatments than they do about their risks. This asymmetry has profound consequences for medical decision making. The history of cardiac surgery, and particularly the history of knowledge about its cerebral complications, offers a valuable case study.

Between the 1940s and the 1960s, surgeons and engineers transformed cardiac surgery from a forbidden frontier into a showcase of medical science and technology. They knew that this work would be dangerous. If they disrupted the heart's function during surgery, the patient could die or vital organs, especially the brain, could be damaged. To understand the problem posed by cerebral complications of cardiac surgery, it is necessary to understand how cardiac surgery evolved in the first half of the twentieth century. As surgeons developed the techniques and technologies that made cardiac surgery possible, they adopted a particular attitude toward risk that colored their perception of treatment complications. Strokes, seizures, delirium, and even patients' deaths became a macabre testimony to a surgeon's courage.

Cardiac surgery has a short history. Many of its most important procedures were not developed until the 1950s and 1960s, in contrast to appendectomy, Cesarean section, or other familiar operations that had become routine in the late nineteenth century. The lag in cardiac surgery is not difficult to understand. With general anesthesia and increasingly meticulous techniques that reduced the risk of infections, surgeons in the 1870s and 1880s opened the body with greater and greater confidence that their patients would survive and benefit. However, even the boldest surgeons stopped short of the heart. Theodor Billroth, who developed many abdominal procedures that still bear his name, famously told the Vienna Medical Society in 1880 that "any surgeon who wishes to preserve the respect of his colleagues would never attempt to suture a human heart." British surgeon Stephen Paget—another giant of the age—agreed: "Surgery of the heart has probably reached the limits set by Nature to all surgery: no new method, and no new discovery, can overcome the natural difficulties that attend a wound of the heart."[2] Surgeons feared that no patient could survive an operation on the heart.

Initial forays into cardiac surgery involved patients with such wretched prognoses that nearly any amount of surgical risk seemed justified. In September 1896, for instance, Frankfurt surgeon Ludwig Rehn operated on a 22-year-old gardener who had been stabbed in the heart. When Rehn first saw the victim, two days after the stabbing, the gardener was near death. Blood filled half his chest and compressed his left lung. His pulse was barely palpable. Rehn opened the man's chest and saw blood ooze from a wound in the right ventricle with every beat of the heart. Three sutures, nervously placed, closed the wound. The patient survived a postoperative wound infection and was in good health six months later. Other surgeons followed Rehn's lead. With their patients bleeding to death from stab wounds, surgeons had little to lose by trying to operate. The overall toll, however, was grim. When Rehn published the results of his experience with 124 patients, his mortality rate was 60 percent.[3] Only in a setting like a stab wound to the heart could such outcomes be seen as progress.

Another challenge soon captured the interest of surgeons: rheumatic heart disease. Rheumatic fever was once one of the most feared of all diseases. Caused by infection with a particular strain of streptococcus, it was a leading killer of children and young adults in the early twentieth century.[4] Many children survived the initial infection—a routine bout of strep throat—only to suffer chronic inflammation of their cardiac valves. In severe rheumatic heart disease, the valves became so thickened that they could barely open and close, preventing blood from flowing easily through the heart. Men and women in their teens and twenties were "cardiac cripples," bed bound and facing a slow, suffocating death. In the 1920s Henry Souttar, in London, and Elliott Cutler, in Boston, each set out to provide relief to victims of the dreaded disease. They proposed to break open the damaged valves, relieve the obstructed blood flow, and restore a semblance of normal function. Unlike the attempts to repair stab wounds, this would be elective surgery, on patients who, though suffering, did not face imminent death.

Souttar believed that "the heart is as amenable to surgical treatment as any other organ." The basic problem was "purely mechanical." Like all other tissues, the heart would tolerate incision, manipulation, and repair. His approach was daring. He chose a 15-year-old girl with severe rheumatic disease as his first patient, made a small incision in her left atrium, and inserted his index finger into the heart to force open the rigid valve. Although she survived and seemed improved, London physicians refused to refer other patients to him. As Souttar later explained, "the Physicians declared that it was all nonsense and in fact that the operation was unjustifiable." Cutler tried something different. He operated on a 12-year-old girl, inserting a small knife through the wall of her heart and trying blindly to break the valve open. She survived—barely—but without clear benefit. Still, Cutler looked on the bright side. Her survival showed that surgery on the heart "bears no special risk, and should give us further courage and support in our desire to attempt to alleviate a chronic condition, for which there is now not only no treatment, but one which carries a terrible prognosis." His next nine patients, however, all died. Samuel Levine, the cardiologist who collaborated on these "heroic attempts," recognized that the "extremely high mortality in subsequent cases quickly placed this operation in disrepute."[5]

Other surgeons picked up the challenge of cardiac surgery in the 1930s. Instead of attempting to open the heart, however, they mostly worked on the surrounding tissues and blood vessels. For instance, they tried to relieve the constrictions caused by pericarditis or to bring new blood to the heart to treat coronary artery disease. First in Boston, and then in Baltimore, surgeons attempted to repair the hearts of babies born with congenital anomalies. Since most of these children died quickly without surgery, nearly any operative risk could be justified. This work first gained national attention in November 1944 with Alfred Blalock and Vivien Thomas's "blue baby" operation.[6] The child, born with a series of cardiac anomalies, survived the initial operation. Blalock's success inspired surgeons and pediatric cardiologists to take on more ambitious repairs.

Other surgeons gained valuable experience during World War II. Boston surgeon Dwight Harken directed a thoracic surgery unit in England that handled casualties from D-Day. He quickly learned how to treat soldiers with penetrating chest injuries. Removing shrapnel from inside the chest became routine. Harken went further and retrieved shrapnel from inside the cardiac chambers of thirteen patients. Back in Boston after the war, he picked up where Souttar and Cutler had left off and operated on beating hearts to repair mitral stenosis. In 1948 he successfully completed his first valvuloplasty. He made a small incision in the right atrium, inserted his finger into the beating heart, and broke open the mitral valve. When the procedure worked, his patients improved dramatically.[7]

Beating-heart surgery, however, was an imperfect solution. It limited surgeons to quick, blind interventions inside the heart's chambers. The procedures still had horrific mortality rates. Philadelphia surgeon Charles Bailey, who also began performing valve surgery in 1948, lost his first four patients before his fifth survived. Harken's first four patients also died. His subsequent learning curve was grim. Of his initial one hundred patients, thirty-two of the sickest ones died. By the time he had operated on his one thousandth patient, mortality in the direst cases was still high—20 percent. Harken's son remembered the challenge of his father's early work: "It was a very difficult group of patients." Harken was "devastated" whenever patients succumbed.[8]

Surgeons needed a better option. They wanted to be able to stop the heart so they could operate on a quiet and blood-free surgical field. But before they could do this, they had to develop a way to replace the functions of the heart and provide the body with a constant supply of oxygenated blood. Surgeons had long imagined how this might be done. In 1812 a French doctor had suggested that doctors make repeated injections of fresh blood into the arteries for as long as the heart was stopped. In the 1880s German physicians envisioned a device that would use an artificial pump and oxygenator to perfuse the body while surgeons repaired the heart. Significant progress had been made by the 1930s. Physiologists and surgeons had developed devices in their laboratories that could perfuse and preserve the viability of isolated organs. Looking for his next great challenge, aviator Charles Lindbergh teamed up with surgeon Alexis Carrel at the Rockefeller Institute, where they developed pulsatile pumps that Carrel used for research. Meanwhile, Boston surgeon John Gibbon sought a device that would perfuse an entire organism, and not just an isolated organ. By 1937 he had a prototype, which he used successfully to perfuse a cat during surgery. Other surgeons, however, did not yet trust these devices. London surgeon Laurence O'Shaughnessy developed his own perfusion system and tested it in cats and dogs. Although he had some successes, the procedure was not reliable and remained "clearly unfitted for immediate clinical application."[9] World War II put many research programs on hold, but when surgeons returned, emboldened by their war experience, several groups took up the challenge of providing artificial circulation.

The most daring approach, attempted by Minnesota's C. Walton Lillehei, connected the circulatory systems of two people, often a parent and the child needing the cardiac repair. The parent's lungs oxygenated the blood that now flowed through both bodies. Mechanical pumps, adapted from the dairy industry, moved the blood from the donor, through the patient, and back again. The technique required perfect compatibility of blood types and flawless perfusion technique. It also involved unprecedented risk. As one doctor said to Lillehei, "You know, Walt, this is the only operation that carried the risk of a 200 percent mortality rate." Surgeons had gained confidence, however, from decades of animal experiments on cross-circulation. In Lillehei's first attempt, in 1954, the procedure worked for the operation, but the patient, a 13-month-old boy with a heart defect, died within two weeks from postoperative pneumonia. Another case had to be aborted before the surgery even began when the anesthetist accidentally injected an air bubble into the mother's circulation. She suffered "a pretty significant stroke" and required prolonged care in a mental hospital. Of forty-five attempts, twenty-eight patients survived, often with significant cardiac repairs.[10]

The willingness of surgeons and patients to take such risk demonstrates well the desperation that surrounded the severest forms of heart disease in the 1950s. Surgeons had two linked ambitions. They wanted to save the lives of individual patients, and they wanted to develop new surgical techniques. Lillehei's efforts on both fronts earned him fame and notoriety.[11]

The fundamental goal, however, had not yet been met. Surgeons wanted an artificial device that could both pump the blood, without causing clots or damaging its cells and proteins, and oxygenate the blood, without introducing air bubbles or other debris. Heparin, a powerful anticoagulant that had become available in the 1930s, could prevent the blood from clotting. Pumps were solved next. Michael DeBakey, while a medical student at Tulane, developed a roller pump that squeezed blood through flexible tubing. Gibbon adapted this for his devices. Lillehei, meanwhile, continued to use his dairy pumps. But all of these raised another problem. They required new plastics that could endure the pump's repeated compressions without shedding plastic fragments into the blood. Lillehei found a solution at a Minneapolis plastics manufacturer that made polyvinyl tubing used in the dairy and beer industries.[12] By adapting these industrial components in the early 1950s, surgeons produced reliable pumps that could safely move blood through a circuit.

What about oxygen? Lungs oxygenate blood by bringing red blood cells nearly into contact with air, separated only by the thinnest membranes of capillaries and lung cells. Oxygenators attempted to replicate this process in different ways. Some engineers used wire mesh screens or rotating disks to create thin films of blood across which oxygen diffused. Others bubbled oxygen directly through blood, with exchange occurring across the surface of the bubbles. But bubbling turned blood plasma into an unmanageable froth. As Boston surgeon Roe Wells explained, "Without defoaming, the operating room would soon be submerged in plasma foam."[13] Everyone knew how difficult it would be to create a viable heart-lung machine. And everyone knew they would have to take risks.

With the initial pump and oxygenator innovations in hand, surgeons refined their devices in dogs and other laboratory animals before beginning human trials. John Gibbon, for instance, resumed his work in Philadelphia after the war and captured the imagination of IBM's founder, Thomas Watson. Watson provided funding and assigned skilled engineers to the project. They produced an elegant 2000-pound device that combined a DeBakey roller pump with stainless steel screen oxygenators and sophisticated control systems. Gibbon took the device through experiments in dogs and then, in 1952, began human trials. His first patient, a 15-month-old girl, had been misdiagnosed: expecting one problem, the surgeons encountered another, and she did not survive the surgery. A Detroit group, meanwhile, worked with engineers from General Motors to develop a pump that allowed partial bypass, replacing the pumping functions of either the left or right ventricle while still relying on the patient's own lungs for oxygenation. John Kirklin brought Gibbon's design to the Mayo Clinic and modified it to suit his purposes. Most other hospitals, however, lacked the expertise and resources needed to manage the precise Mayo device.[14]

Lillehei's group at the University of Minnesota wanted something simpler. They tried a rotating disk oxygenator, but their first two patients died. Lillehei asked one of his residents, Richard DeWall, to find a better solution. DeWall knew that chemists at Dow Corning had developed special agents to remove bubbles from the fuel lines of new jet engines. These defoaming agents had also been adopted by the dairy industry. DeWall realized that defoamers would make it possible to develop a simpler device based on a bubble oxygenator.

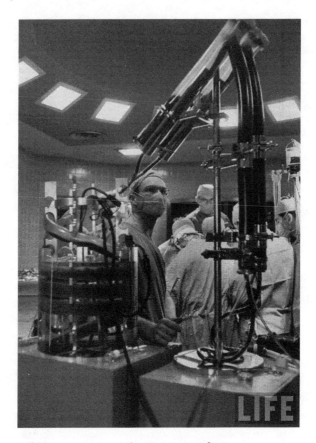

Bubble oxygenator at the University of Minnesota.
Photograph by Al Fenn. From *Life* magazine (1957).
With permission from Getty Images.

Their first patient, in 1955, survived. By 1957 the Travenol Corporation had
begun to market a preassembled, disposable pump oxygenator.[15]

Early heart-lung machines were not easy to use. When René Favaloro first
arrived at Cleveland in 1962, he had to clean their heart-lung machine—
which staff called "the monster"—after each use. This task took at least two
hours and initially limited the Cleveland team to a single open-heart procedure each day. Early heart-lung machines also consumed vast amounts of
blood. Gibbon's initial device had to be primed with blood from twenty-five
volunteers.[16]

In the end, however, the work proved to be a productive collaboration of surgeons, engineers, and the burgeoning medical device industry. Surgeons benefited from the expertise of automotive and computer engineers and from the practical solutions of the dairy and chemical industries. Once they had workable prototypes, the physicians collaborated with emerging medical device companies to move the devices from laboratories to the marketplace. The Minnesota team led the way. Surgeons came to Minnesota from all over the country and the world to see Lillehei's successful and simple device. As Denton Cooley later recalled, "Walt Lillehei brought the can opener to the cardiac surgery picnic." The visitors could watch an operation in the morning, learn how to use the device in the engineering labs in the afternoon, order the materials, and then build their own device for just $500. Cardiac surgery became a prime example of postwar American scientific and engineering preeminence, a marker of the extent to which the center of innovation in medicine had moved from Europe to the United States.[17] American surgeons trained surgeons from Europe and elsewhere and sent them forth to colonize surgical centers worldwide.

The advent of "safe enough" pump oxygenators in the early 1950s transformed cardiac surgery. An article published in *Reader's Digest*, then the most widely read magazine in America, dramatized the work of Cleveland surgeon Donald Effler. Before heart-lung machines, surgeons were "required to work blindly in a pool of blood, attempting to repair a writhing, slippery object." The heart-lung machine allowed Effler to stop the heart completely. The drama was palpable. "There are a few fluttery last beats. At 11:25 all motion ceases; the sick heart comes to a standstill. Always before such an event has been tantamount to death. This time it offers the hope of life." Effler explained the machine's significance: "With a dry, quiet, clearly visible field in which to work, we believe that a new era of heart surgery has opened." Denton Cooley, who brought the Minnesota device to Houston, described his own excitement in *Time*: "This is the ultimate in heart surgery—the achievement we have been waiting for . . . We can now repair some of the most serious damage there can be to the adult human heart."[18] The future seemed full of promise.

Surgical ingenuity, however, had its price. Surgeons and engineers tested their prototypes in hundreds—probably thousands—of cats, dogs, and other animals. Some people protested the price these animals paid in the name of surgical progress. As one surgeon later recalled, "I was accused at one point of trying to destroy all the dogs in Pittsburgh." And despite this preparatory work with animals, surgeons still experienced appalling mortality rates when they tried the devices in humans, just as they had with the early beating-heart valve operations. Gibbon lost three of his first four patients and never performed an open-heart procedure again. Kirklin lost four of his first eight patients at Mayo. John Ochsner, who trained with DeBakey in Houston, described how they lost five patients before having their first success with a mitral valve replacement.[19]

It is difficult for us now to imagine a world in which such outcomes could be tolerated by surgeons, patients, or their families. Danger, however, had always been part of surgery and medicine.[20] Whether in ancient Greece, medieval Europe, or early nineteenth-century America, surgeons knew that every time they cut into a patient's body, they would cause pain and bleeding. Any experienced physician had seen many patients die during surgery or soon thereafter. When battlefield surgeons amputated limbs before anesthesia was available, they did so only if death from bleeding or gangrene seemed imminent. When surgeons attempted Cesarean sections, they did so only as a last-ditch effort to save the mother or the infant. If a patient died, the surgeon had done no harm: the patient would have died anyway. If the patient survived, however, then the surgeon had snatched someone from the jaws of death.

The balance of risk and benefit shifted in the mid-nineteenth century. When anesthesia became available in the 1840s, it presented surgeons with a new and difficult choice. They knew that surgery was painful, but they also knew that patients could endure brief pain. Anesthesia prevented suffering but also introduced unknown dangers. What if the patient never woke up? Surgeons had to perform a "calculus of suffering" each time they determined whether pain relief justified the risk of anesthesia. Lobotomy, though now notorious, also reflected a sincere effort to weigh the risks of disease and treatment in the 1930s and 1940s. Yes, lobotomy often left patients with constrained personalities, without initiative or spontaneity, but surgeons, psychiatrists, and even patients' families understood that if the procedure freed some patients from confinement on the back wards of a desolate asylum, then perhaps the risks were worth it.[21]

Cardiac surgeons felt these pressures acutely. Their patients faced the direst prognoses. Many were infants born with severe congenital anomalies or young adults crippled by rheumatic heart disease. Since these people would die or suffer terribly without surgery, nearly any surgical risk seemed acceptable. Surgeons' language here was harsh but revealing: they hoped to salvage the lives of "cardiac cripples" and "cardiac derelicts." English surgeon Geoffrey Wooler remembered how "in those early days the operative mortality was at least 50 percent, but the patients were all high risk and the mortality of those on the waiting list was almost as high." Effler lost five of his first eleven patients, but *Reader's Digest* forgave him, explaining to its readers, "The score may seem poor, but any new heart surgery is attended by risk, which decreases as experience is gained." Furthermore, Effler operated on "the most desperate cases he could find." Faced with such daunting odds, Effler told the reporter, "We feel rewarded that so many have lived.' "[22]

This environment fostered the emergence of a particular kind of therapeutic bravery in cardiac surgery in the 1960s, something that has been called "the courage to fail." Although surgeons hoped to improve their patients' lives, it was good enough if the patients merely survived. Even if they did not, their operations could still contribute to the advancement of surgical knowledge and technique. Within this mindset, there was no truly bad outcome. Operative mortality rates of 15 to 35 percent remained common and tolerated during this early work on rheumatic valves and congenital malformations.[23]

The public acclaim surgeons received for their work reinforced their sense of purpose. For instance, in 1969 a dispute between DeBakey and his protégé, Denton Cooley, over Cooley's unauthorized use of an experimental artificial heart, received national attention. Readers did not condemn the surgeons' egos and bickering. Instead, when *Life* magazine described the dramatic feud in 1970, adoring readers sent letters that defended the surgeons. One Texas man dismissed their foibles: "Without the skill developed by both of these surgeons in their years together, hundreds, and perhaps thousands, of us would now be dead." A New Yorker whose wife had been saved by DeBakey wrote that the surgeon "is a god, or at least a saint."[24]

Surgeons persevered in their efforts, and by the early 1960s, their operative mortality rates had fallen to under 5 percent. Effler completed 408 operations in 1965 with only 10 deaths in the operating room (though another 45 died before discharge from the hospital). The routinization of operative mortality is startling. Effler complained to the chair of the clinic's surgical committee that the requisite mortality reports were "becoming an increasing burden in our Department." To spare surgical residents this effort, Effler told them "to peel the summary information to barest minimum." Stanford surgeons were pleased enough with their skill that, despite a 6.4 percent mortality rate, they recommended surgery enthusiastically: "Every person with a cardiovascular defect should be seriously considered for surgical correction."[25] Death, woven into the fabric of the field, stopped no surgeon. Cardiac surgery boomed.

CHAPTER TEN

Suffering Cerebrums

Pioneering cardiac surgeons faced other problems in addition to high mortality. Surgeons' initial wariness about operating on the heart had arisen from their fear that any manipulation of the organ would threaten the rest of the body. All tissues depend on the constant flow of oxygenated blood that the heart and lungs provide. A brief disruption that did not kill a patient might still leave other organs injured. Attention was focused on the brain, the most important and sensitive organ of all. Surgeons soon realized that they were right to be worried.

Concern about the brain appeared early. Henry Souttar laid out the basic problem in 1925. Although it was possible to stop the heart and the circulation of blood in a cat or a dog for up to two minutes, this "would never be justifiable in a human being, in view of the extreme danger to the brain from even the shortest check to its blood supply." This fear loomed over would-be cardiac surgeons in the 1930s. They knew that the field could never develop until they found a way to protect the brain during surgery. As Laurence O'Shaughnessy noted in 1939, "so long as the surgeon is faced with the certainty of irreparable cerebral damage if he interrupts the circulation for any appreciable period, he has little incentive to devise operative procedures which can never be carried out."[1] O'Shaughnessy hoped that his efforts to develop a method of artificial perfusion would make cardiac surgery possible.

When Dwight Harken, Charles Bailey, and other surgeons worked to resuscitate beating-heart surgery in the late 1940s and 1950s, the old fears of cerebral complications were realized. Many patients survived the finger-fracture valve surgery only to suffer devastating brain injuries. In Bailey's initial series of 235 patients, 12 suffered a substantial stroke. Of these, 9 experienced hemiplegia and 3 partial quadriplegia; half never regained full consciousness. Eight died as a result of their strokes. Of the four who recovered, "none became completely well." A team of Chicago surgeons had patients who suffered a "stormy postoperative course," and were sometimes left paralyzed, blind, or unable to speak. Psychiatric complications occurred as well. Some of Harken's patients became psychotic and experienced hallucinations, amnesia, psychogenic paralysis, and other "hysterical" reactions. Other changes were subtler. One math teacher "with a very high total I.Q." showed a "marked regression to infantile concepts" after surgery. For six months she lost interest in teaching and wanted only to read mystery novels.[2] Although this last case is open to many interpretations, the strokes and other complications were not ambiguous. The toll of neurological and psychiatric complications was devastating to the patients, their families, and their physicians.

Surgeons had little difficulty deciphering what had gone wrong. As their fingers and knives forced open the valves, they dislodged blood clots that had formed in the heart's chambers or calcifications that had grown on the valves. This material entered the bloodstream and traveled throughout the body. Some of these traveling fragments, called emboli, lodged in the brain and caused strokes. When emboli occurred spontaneously in patients with rheumatic heart disease, Bailey's team wrote, "it is considered a part of the natural course of events, and accepted by the physician, patient, and relatives as an unavoidable complication." However, when it happened after surgery, "the complication is attributed directly to the manipulations, and justly so, but it is not well accepted by either the physician or the patient if he recovers,—or by the relatives if he does not."[3] Anger, regret, and recriminations can all be seen beneath the veil of dry clinical prose.

Surgeons struggled to protect the brain from these dangers. Harken tried to make only "slow, gentle motions" with his finger inside the heart to minimize the number of thrombi knocked loose. He also had his anesthesiologist compress the carotid arteries for a few moments as he manipulated the heart, hoping that the emboli would end up someplace in the body other than the brain and cause less mischief. Bailey wanted to filter blood as it left the heart, but this was not possible. Nor could he stop flow to the brain for the length of the operation. Doing this for the three minutes required would be "fraught with dire consequences." His team tried Harken's method of temporary carotid compression and then refined it. They looped a piece of surgical tape under the carotid arteries and used it to kink the vessel and stop blood flow as they manipulated the heart. Limiting the carotid obstruction to 30 to 60 seconds, they performed the maneuver of "protective carotid obstruction" in eighty patients without complications. Harken had less success. He placed rubber loops around the arteries and pulled them tight, watching the clock and waiting for 60 or even 90 seconds at a stretch. He would then release the tension to restore blood flow to the brain for two minutes. The occlusion could be performed repeatedly as needed to allow work on the heart. Unfortunately the technique failed doubly: it "has not prevented embolization and the anoxia that it produces may have caused more harm than the theoretical benefit it confers."[4]

Harken, Bailey, and others in this first wave of beating-heart surgery saw that the needs of the heart and those of the brain existed in tension. Patients needed relief from their debilitating, chronic valve disease, but providing it with the surgical techniques then available inevitably put the brain at risk. A trade-off was unavoidable. As Harken wrote, "a compromise will obviously have to be made between the protection of the brain and the maintenance of the cardiac mechanism."[5]

The desire to avoid this compromise, by finding a way to protect the brain during cardiac surgery, had motivated the dream of developing a heart-lung machine. As O'Shaughnessy explained in 1939, "the real key to further advance in the surgical treatment of established cardiac defects will only be provided by the provision of some simple and efficient method of maintaining the cerebral circulation while the heart is temporarily out of commission." However, surgeons knew that the heart-lung machines would also introduce risks. C. Walton Lillehei, who led the effort at the University of Minnesota, saw Richard DeWall's bubble oxygenator as a dangerous gamble: "In 1954 you could have quizzed all of the physiologists, surgeons and cardiologists who were working on this, and you would have found 100 percent agreement that the one way not to oxygenate the blood was by bubbling, because, of course, bubbles in the circulation would be a terrible danger to the brain."[6] They knew that the devices would expose patients' brains to periods of low blood pressure, changes in the permeability of the blood-brain barrier, changes in blood chemistry and acidity, degraded blood plasma proteins, air bubbles left behind in cardiac chambers, and debris knocked loose from scarred and calcific valves. Surgeons also suctioned and recycled blood that accumulated in the chest, a maneuver that introduced air, fat, and other particulate debris.

Worried about the many dangers of open-heart surgery, surgeons and engineers tested their pump and oxygenator prototypes in animals before trying them on their patients. Since they were especially concerned about the vulnerability of the brain, surgical researchers watched their lab animals closely for signs of neurological damage. One of O'Shaughnessy's dogs "made an uneventful recovery and is now in every way a normal animal." A cat, observed for two years after surgery, "behaves in a perfectly normal manner." George Clowes, at Case Western, paid attention to whether the animals had "the ability to walk, drink, respond to commands or attention, to see, and to urinate." Surgeons at the University of Pennsylvania even performed thorough neurological assessments on their dogs. They were reassured by how well the animals recovered. Even when dogs were left blind or paralyzed by surgery, "these signs would usually recede slowly after the lapse of several days to a week." When DeWall finally thought he had a working device at the University of Minnesota, he invited Lillehei to examine his postoperative dogs. "I remember walking them out in the field in front of the grassy meadow by the research lab," DeWall described years later, "We were just walking the dogs around and analyzing them and checking them over. There was no apparent damage to them. It was then that Walt decided that we would try this in the operating room."[7] Surgeons' success with protecting dog brains provided the reassurance they needed to proceed with human trials.

Some doctors did worry that their animals provided a poor proxy for the cognitive abilities of humans. O'Shaughnessy warned that brain injuries would be "of greater importance in man than in the lower animals." But reassured by how well animals recovered from neurological injuries, and desperate to save patients, surgeons hoped that their devices would be safe enough. Some early experiences were reassuring. Although three of John Gibbon's first four patients died during or soon after surgery, his one survivor—an 18-year-old woman with an atrial septal defect—did well neurologically. As he described in his postoperative notes, "At no time during her convalescence from her operation did she show any signs of cerebral anoxia or central nervous system injury." John Kirklin's team at Mayo downplayed the risk altogether, writing in 1958 that "properly conducted open intracardiac surgery utilizing extracorporeal circulation has no discernible adverse effect on the brain."[8] They believed that safe open-heart surgery could be achieved.

Experience quickly dashed such hopes. Detailed clinical narratives from the 1950s and 1960s paint a grim picture of what could go wrong. Patients suffered massive brain injuries and never woke up. Others survived but developed psychosis, seizures, or strokes. Still others seemed intact initially only to exhibit subtle changes in memory, concentration and personality later. The experience of surgeons at the Cleveland Clinic was typical. Like all cardiac surgeons, Effler had his share of bad outcomes. Over 20 percent of his mitral valve patients experienced strokes. Others became delirious. One patient received a revascularization procedure for coronary artery disease, and the operation was a technical success: his angina disappeared. A postoperative stroke, however, left him aphasic and hemiplegic. Effler wrote to his mentor, Arthur Vineberg, "he is lost to us for all practical purposes."[9]

Complications followed heart-lung machines wherever they went. When surgeons brought open-heart surgery to Iowa City, they faced the increasingly familiar problems. A 12-year-old boy, initially irrational and paralyzed on his left side, recovered. A 4-year-old girl never awoke, suffering seizures until her death on postoperative day three. A 30-year-old woman "had right hemiparesis and disported herself in an irrational and rather wanton manner. She completely recovered in 1 week." Even in patients spared catastrophic complications, skilled observers could find subtler findings. Boston neurologist Sid Gilman studied Harken's patients and documented problems with speech, memory, visual fields, naming objects, reading, drawing, and calculation. One patient, who had worked in a credit bureau, could no longer do math: "Subtracting 7 from 100 serially, he gave the following sequence before giving up in exasperation: 93-91-73-89-87-86." As New York Hospital neurologist Fred Plum commented in 1960, "neurological complications of cardiac surgery are perhaps the single most worrisome aspect of this surgical problem."[10]

No one doubted, at least initially, that the problems resulted directly from surgery, as one English neurologist observed in 1964: "One of the most interesting features of these disorders is that they are truly new illnesses, pathological conditions which essentially result from therapy and were never encountered when these patients were left untreated." Surgical practice had created a novel category of disease, yet surgeons offered many reasons for proceeding with their work despite knowing what could go wrong. Sometimes a risk of neurological catastrophe existed prior to the surgery. Strokes were common in patients with rheumatic heart disease. Emboli from atrial thrombi or valve calcifications broke off spontaneously, even without surgery. Ellis and Harken, and their patients, were all too aware of possible complications: "The knowledge of this hazard, with the possibility of death or permanent paralysis occurring at any time, is a sword of Damocles over the heads of patients." They hoped that the surgery would make up for its immediate neurological risks by reducing the long-term danger of spontaneous embolization.[11]

Other surgeons argued that the surgery's potential benefits to patients, especially the cardiac rehabilitation, made the risk of neurological complications a reasonable gamble. These doctors hoped to restore "cardiac derelicts" and "cardiac cripples" to productive function. Any effort against such serious disease entailed certain risks. The hope of salvaging these wrecked lives justified this risk for surgeons and families. In 1964 a London surgeon admitted, "There is a small but appreciable neuropsychiatric morbidity to take into account when advising such operations." However, he added, "This morbidity is, of course, to be set against the undoubted and dramatic physical, social, and psychological improvement which results from these operations and with which one could not fail to be impressed."[12] Every decision to operate had to weigh such considerations.

A related argument explained away the risk as a passing problem. Physicians knew that any new procedure, when first developed, would have its difficulties. Over time, however, physicians would improve the technique and reduce its risks. A 1964 *Lancet* editorialist explained, in a column titled "Cerebral Injury," that the complications accompanying any innovation were "an unavoidable concomitant to progress."[13] Expectation of a better future justified the risk and bad outcomes. As will be seen in coming chapters, however, this faith that cerebral complications are but a passing phase in the history of progressive innovation still endures among cardiac surgeons and interventional cardiologists.

As surgeons grappled with the danger and complications of surgery in the 1950s and 1960s, the ways in which doctors thought about risk started to change. Informal notions of risk had long been part of surgeons' language, with discussions of the risk of disease or of operative risk. Much of the new postwar thinking about risk involved ideas about what it meant to be at risk of developing a disease, as is evident in the Framingham Heart Study's concept of risk factors for coronary artery disease. As disease risk became something measurable and calculable, surgeons began to measure and calculate the risk of complications. Physicians' ideas about risk management developed in parallel. When internists discussed risk factors for heart disease, they emphasized the ways in which these factors could be modified and reduced through changes in lifestyle or medications. Surgeons took steps as well. Harken and Bailey tried to lessen operative risk by using carotid occlusion to reduce the likelihood of cerebral emboli. When surgeons began to use heart-lung machines, they did what they could to minimize the risks introduced by those machines. For instance, whenever they opened the cardiac chambers, they tried to remove all the air and make sure the chambers had filled with blood before restarting the heart. They often placed a needle vent at the highest point in the aorta to allow residual bubbles to escape.[14] Surgeons were conscious of risk as an entity that had to be characterized and managed.

Hoping to find more ways to reduce the dangers of open-heart surgery, surgeons and their colleagues studied the cerebral complications with impressive energy and imagination. Their efforts provide a treasure trove for anyone interested in how researchers visualize, quantify, or otherwise make knowable clinical phenomena. In the first part of this book, I described pathologists' and cardiologists' efforts to understand the causes of heart attacks. Much of this research involved a literal kind of visualization. Pathologists took an organ—the heart—and tried to find the most reliable way to reveal its atherosclerotic plaques and coronary thromboses. They looked, touched, sectioned, and stained the tissues to understand the disease. Surgeons could also see and feel coronary plaques. Cardiologists used x-rays, intravenous contrast, and elaborate angiography machines to visualize the coronary arteries of living people. When clinicians and researchers set out to understand the cerebral complications of cardiac surgery, they faced a form of this visualization challenge, but one more elusive than the challenge of coronary plaques and clots. They had to reveal, measure, and analyze the functions and dysfunctions of a human brain.

Research began with the familiar and well-established techniques of the medical interview. Doctors have long spoken to patients to learn their symptoms and to "take" the history of their illness.[15] Neurologists could sometimes learn much from what a patient said—or could not say—about their ability to move, feel, or see. Psychiatrists used the interview to probe other aspects of brain function. This verbal exchange, heavily influenced by the popularity of psychoanalysis in American psychiatry in the 1950s and 1960s, formed the basis of pre- and postoperative assessments of the psychological state of cardiac patients. Psychiatrists believed that they could use these techniques to reveal the hidden structures of the unconscious mind.

Doctors had many other ways of seeing their patients. Starting in the nineteenth century, physical examination—a mix of seeing, hearing, touching, and smelling—gradually became a routine part of medical care. In contrast to an interview, where doctors tried to learn what patients knew and could express about their illness (as well as what patients knew but might not have felt relevant), physical exams often sought to extract information from a patient's body that the patient was not aware of, such as pulse rate, skin hue, or subtler signs of illness. Called the clinical gaze, this approach to bodies treats them as seats of disease and objects of medical scrutiny.[16] This way of interrogating patients' bodies takes its most elaborate form in the complex and ritualized examination that neurologists use to probe patients' strength, sensation, reflexes, and mental functions. When they tested patients in this disciplined way, neurologists could make visible, audible, or palpable the consequences of neurological injury.

Doctors did not stop at interviews and physical examinations as ways to gather information. In the nineteenth century, they learned to use stethoscopes, thermometers, blood pressure cuffs, and x-rays to detect and measure bodily phenomena that are not directly accessible to the senses. These devices convert hidden structures or processes into forms that can be perceived by physicians, often into a tracing or inscription they can then read. When the tracings were transduced from the patient by a device without intervention or interpretation by the observer, researchers and clinicians believed that the device provided objective images of the phenomena. However, using a new technology to generate objective data about a vital process is never that simple. Instead, physicians must learn how to use and trust the devices and assign significance to the findings the machines reveal. It took many years for doctors to be convinced that blood pressure cuffs or x-rays produced reliable information that improved on what they could learn from their own senses. Physicians often acknowledge that visualization techniques provide imperfect representations of the studied phenomena. Despite this, doctors set these subtleties aside in pursuit of any knowledge that might be clinically useful.[17] The devices used to visualize or otherwise represent the cerebral complications of cardiac surgery demonstrate this well.

Certain technologies could be used during surgery, when the deep state of anesthesia prevented any traditional assessment of brain function. Neurologists were quick to use electroencephalograms (EEGs) to monitor the brain waves of animals and then patients during surgery. These devices, developed in the 1920s and 1930s, used electrodes placed on the scalp to record the electrical activity of the brain. EEG pioneers hoped that the tracings would allow them to "read" brain activity. From the 1930s into the 1950s, researchers attempted to read these "brainscripts" to find recognizable markers of personality, intelligence, and even specific thoughts. *Life* magazine even featured one team recording the brain waves of Albert Einstein as he thought about relativity. As open-heart surgery programs began in the 1950s, neurologists adapted the technology for a less glamorous task: revealing evidence of disrupted brain activity during surgery. EEGs were routine, and often alarming, when the Mayo Clinic began its open-heart surgery program: "The manifestations of failing circulation accompanied by an electro-encephalographic change indicate a decrease in cerebral blood flow below a safe level, and require immediate and vigorous treatment." Effler used EEGs to monitor the adequacy of cerebral flow as he transitioned his patients from the heart-lung machine back to their own beating hearts. A Montreal neurologist interpreted EEG changes as "an expression of the cerebral suffering." When an EEG did not return to normal after surgery, it was "a grave sign," "a sign of serious cerebral disturbance which may lead to vital brain stem damage."[18] By the 1950s neurologists had refined EEG interpretation into an elaborate science, one that could detect subtle seizures, localize brain tumors, or reveal occult signs of disordered function. In the setting of cardiac surgery, however, the gravest tracing was one that even a nonexpert could read, a flat line indicative of no brain activity.

Doctors also employed diagnostic devices to probe the causes, and not just the consequences, of brain injury. Radiologists and neurologists used Doppler ultrasound, adapted from military sonar technology in the 1960s, to detect emboli and other debris passing through carotid arteries. Air bubbles "appeared on the Doppler ultrasonic signal in the form of whistles, chirps, and snaps." Their significance, however, remained uncertain. Early experience showed that "that overt central nervous system complications may not develop even

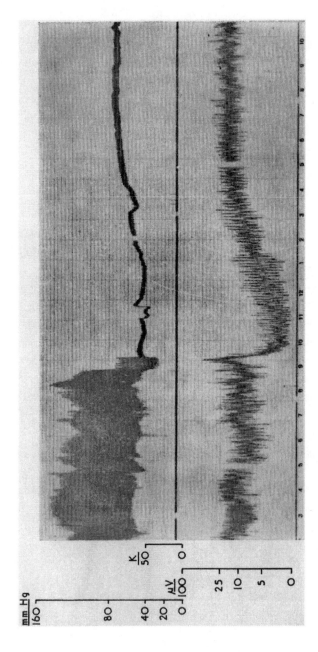

EEG as cerebral function monitor. Blood pressure (*top*) at the onset of perfusion with the heart-lung machine; note the transition from normal pulsatile flow to the uniform pressure of the perfusion pump. Electrical activity of the brain (*bottom*), with a dramatic drop after the pump is turned on. The patient suffered a stroke and was hemiplegic postoperatively. From M. A. Branthwaite, "Detection of Neurological Damage during Open-Heart Surgery," *Thorax* 28 (July 1973): 464–72, fig. 5, p. 467. With permission from BMJ Publishing Group, Ltd.

when thousands of micro-emboli are accumulatively directed to the cerebral circulation." Despite this ambiguity, the Seattle team that pioneered Doppler technology recommended that all heart surgeons use it: "Even though it may shatter some of your complacency . . . it may help to explain that occasional patient who does not wake up, or wakes up slowly, and often has a focal neurologic change."[19] The evidence revealed by Doppler could help doctors troubleshoot what had gone wrong.

In addition to hearing emboli, doctors realized that they could see them directly—in the eyes of their patients. Ophthalmologists learned in the 1850s how to examine the retinas of their patients and use special lenses to inspect the blood vessels for evidence of high blood pressure, diabetes, or other diseases. Since the retina is essentially an extension of brain tissue, what doctors see there can reveal what might be happening deeper in the brain. Concerned about the problem of cerebral emboli, ophthalmologists began to examine the retina of patients during surgery. They saw white plugs and even "glinting specks and larger sparkling plaques" moving through the retinal arteries. Careful study suggested that the plugs were made of platelets and that the sparkling debris could be either cholesterol or silicone. Ophthalmologists watched as the emboli wound their way through vessels, paused, broke in two, and caused small infarcts.[20] Since there was no reason to suspect that emboli from the heart preferentially ended up in the retinal vasculature, doctors assumed that what they saw there also took place throughout the brain.

Once patients awoke after surgery, neurologists and psychiatrists could use their familiar methods to test for injuries. Some findings were obvious, as with delirium or massive strokes. Others were subtler, such as the deficits in reading, writing, and calculation described by Gilman. Careful testing could

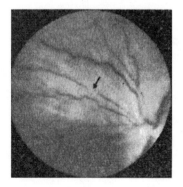

Watching retinal emboli. An embolus (marked with the arrow) displacing blood from an artery. The ophthalmologist could watch as it moved through the artery during the operation. From Isla M. Williams, "Intravascular Changes in the Retina during Open-Heart Surgery," *Lancet* 298 (25 Sept. 1971): 688–91, fig. 2, p. 689. With permission from Elsevier.

reveal deficits that had not been noticed by either patients or their surgeons. This was difficult work: there is no limit to the ways in which brain functions can be probed. As researchers studied patients' cognition, memory, reasoning, and even personality, they realized that they needed techniques to standardize and, ideally, to quantify their assessments. In the 1950s neurologists, psychiatrists, and psychologists began to deploy extensive batteries of cognitive tests on patients undergoing cardiac surgery. One South Carolina group used the Wechsler Adult Intelligence Scale, the Trails Test, the Tapping Test, the Graham-Kendall Memory-for-Designs Test, the Minnesota Percepto-Diagnostic Test, the Minnesota Multiphasic Personality Inventory, and the Cardiac Adjustment Scale.[21]

As the tests proliferated, everyone from psychologists to surgeons debated which tests to use, how to interpret them, and the significance of the results. Neuropsychological tests converted subtle aspects of cognitive function into scores that could be averaged, compared, and graphed to reveal differences between groups of patients and changes within individuals and groups over time. Surgeons valued these assessments because they offered a concrete measure of cognitive function. Postoperative scores were either worse or not. Harken, for instance, praised the efforts of the South Carolina group: "Your objectivity, Dr. Lee and colleagues, is sorely needed in psychiatry. You have brought to view things that have concerned us." The tests functioned as inscription devices, transforming cognitive functions into legible traces. In some of the tests, the patients themselves were the inscription device. The Draw-a-Person Test, for instance, required the patient, before and after surgery, to draw a picture of a person. Changes between the pre- and postoperative drawing provided a picture of changed cognitive abilities.[22] The existence of the tests, however, did not settle the question of their significance. Debate about the role and interpretation of psychological testing in cardiac surgery continues today.

When patients died, pathologists offered postmortem analyses. Dissection of the brains of patients who suffered cerebral complications revealed an astonishing range of problems, with evidence of focal and diffuse hypoxic damage, infracts, and swelling. Debris introduced into the bloodstream by the operation often ended up in the brain. A fragment of a sponge caused a stroke in one patient. Many patients suffered emboli of the antifoam used with bubble oxygenators. Fat emboli could enter the circulation when the sternum was sawed in half or when surgeons suctioned and recycled blood that accumu-

Draw-a-Person Test. Three drawings of a person made by a patient before surgery (*left*), soon after surgery (*middle*), and one year later (*right*). The patient, a 39-year-old factory worker, suffered a stroke after surgery to open a rheumatic mitral valve. The researchers emphasized the clumsiness in the middle drawing and the "general deterioration in the ability to produce an integrated picture of a person." The patient eventually recovered most function. From Misha S. Zaks et al., "The Neuropsychiatric and Psychologic Significance of Cerebrovascular Damage (Strokes) Following Rheumatic Heart Surgery," *American Journal of Cardiology* 5 (June 1960): 768–76, fig. 2, p. 772. With permission from Elsevier.

lated in the chest cavity during surgery. One study found fat emboli in the brains of 82 percent of patients who died soon after surgery. This did not surprise surgeons, who had become "familiar with the appearance of glassy globules of fat floating on the surface of blood in and around the cardiotomy wound."[23] In some cases the pathologists found evidence of more extensive injury than had been suspected clinically, leaving everyone wondering about the significance of the different insults.

Taken together, these techniques and technologies allowed clinicians and researchers to see the unseen and to capture transient phenomena so that they could be studied and understood. Surgeons could see retinal emboli with ophthalmoscopes and hear carotid bubbles with Doppler. Microscopes revealed the minute details of pathological specimens while EEGs converted brain activity into legible tracings. Psychologists could quantify cognitive function with standardized tests and convert subtle phenomena into a single numerical score. These techniques transformed the elusive or imperceptible into the measurable and knowable. This work required collaboration across many disciplines. And the collaboration, in turn, distributed knowledge across many fields. Surgeons, anesthetists, internists, cardiologists, neurologists, psychiatrists, and neuropsychologists all studied the cerebral complications of cardiac surgery. The varied techniques produced inscriptions of the phenomena, converting brain activity and brain injury into charts, graphs, or other tracings that could be printed, published, and reprinted. Readers of a wide range of journals, from *Nature* or the *New England Journal of Medicine* to specialty journals in surgery, cardiac surgery, anesthesia, psychiatry, neurology, and cardiology, could access the results of this research.

The evidence of injury existed in tension with two kinds of ambiguity. First, when neurological and psychiatric complications did occur, they were often transient. In a large case series from the University of Michigan, all patients, even those who had suffered dramatically, recovered completely: "Neurological manifestations range from hyperexcitability and mild hallucinations immediately after operation to serious cerebral damage with focal neurological symptoms. The mild symptoms clear in a day or two; the severe symptoms persist for a week or two, but clearing has been gratifyingly complete and no patient has been left with recognized permanent damage."[24] Many other reports relayed similar narratives of injury followed by recovery, allowing doctors to hope that the problem might not be as serious as it had first seemed.

Second, studies estimated the incidence of cerebral complications across an astonishingly wide range—from zero to 100 percent of cardiac surgery patients. As researchers tried to sort out what contributed to the divergent estimates, they came to realize that the harder they looked for complications, the more often they found them. Using retrospective chart reviews, a common and relatively easy clinical research technique, researchers could ascertain only the most serious complications, the ones that had merited mention in the surgeons' progress notes or discharge summaries. However, when neurologists or psychiatrists studied patients before, during, and after surgery, they detected subtler and more prevalent deficits. Many questions arose. Was one approach better than the other? Did diligent search reveal the true burden of disability, or did it simply produce overly sensitive estimates of inconsequential problems? If the surgeon or patient had not noticed anything wrong, was the problem revealed by careful testing even relevant? Some researchers dismissed deficits if they only mattered "in a rare case in which an individual was in a situation which demanded intellectual functioning at the very peak of capacity."[25]

The quest to quantify brain function, such as with x-rays or other diagnostic technologies, demonstrated that visualization was not a simple process of rendering a truth visible. Instead, it involved debates about effort, thresholds, and interpretation. However, even when such debates went unresolved, their existence had a clear message. Demonstrable brain damage could occur without being relevant to most patients most of the time.

Even with these caveats, most researchers were convinced that the problem was substantial. Surgeons saw air bubbles passing through pump tubing, the fatty sheen on the surface of recycled blood, cholesterol emboli sparkling in the ophthalmoscope's light, and the flat line of an ominous EEG tracing. They heard the crunch of instruments fracturing calcified valves, the chirps of Doppler probes, and the rants of delirious patients. They touched large emboli in cerebral arteries, softened patches of infracted brain, and the insensate limbs of stroke patients. The techniques revealed both the mechanisms of injury (e.g., air emboli) and the outcomes of injury (e.g., patients' symptoms). The myriad tools and techniques made both the complications and the underlying cerebral damage a concrete, polysensory reality for surgical teams. The various findings demonstrated, as one leading researcher warned in 1977, that "what was clinically observed was but the tip of an iceberg."[26]

The language used in physicians' published reports conveyed how seriously they took the problem. They described "cerebral suffering" and "serious cerebral disturbance." They wrote articles about "cerebral injury" and "damage to the brain." They described "gross neurological deficit," "gross neurological dysfunction," and "overt cerebral dysfunction." Often they used the starkest and most familiar label: "brain damage."[27] The most likely suspect to cause these complications was obvious. Surgeons and anesthesiologists knew that the heart-lung machines produced bubbles, damage to blood proteins, and variations in blood pressure. Pathologists found evidence in patient's brains— including air emboli, fat emboli, shards of plastic, and other debris—directly linking complications to the machines. Many doctors concluded that the heart-lung devices created the problem. One British team later explained that the many complications likely resulted "from hazards inherent in the ECC [extra-corporeal circulation] process, which remains an imperfect replacement for normal functions of the heart and lungs." At some point doctors began referring to the patients' suffering as "pump head."[28] Colloquial use of pump head to describe the complications reflected not just the morbid humor of physicians but also a specific etiological claim: heart-lung machines caused most of the injuries.

CHAPTER ELEVEN

Deliriogenic Personalities

Surgeons quickly learned that their concerns about the brain's vulnerability during open-heart surgery were justified. First with the early finger-fracture techniques on rheumatic valve disease, and then with their efforts to bring heart-lung machines to fruition, physicians witnessed a devastating toll of cerebral complications. Committed to overcoming this problem, they worked to visualize the complications, gauge their severity, and eliminate their causes. By the late 1960s it had become clear that many aspects of surgery, especially the disruptions caused by the heart-lung machines, could cause brain damage.

Yet even amid stark evidence of the damage caused by heart-lung machines and surgical manipulations, the question of responsibility remained open to interpretation. This can be seen most clearly in how physicians wrote about delirium. Common after surgery, delirium can be subtle, with patients confused about time or place, or describing small shadows flitting around their hospital rooms. But it can also be dramatic and frightening, with delusions, hallucinations, and agitation that disrupt efforts of doctors and staff to provide medical care. Surgeons and psychiatrists believed that open-heart surgery produced the most dramatic deliriums of all. As one New York team described in 1971, "no other form of surgery, no matter how severe or psychologically traumatizing, produces psychiatric complications of the reported kind and magnitude." The varied symptoms provided "a kaleidoscopic view of mental illness," from catatonia and stupor to excited and "beautiful psychedelic experiences." One patient thought armed secret agents lurked outside his room, while another thought he was in the circus, complete with the smell of horses and the songs of circus bands.[1]

The psychiatric literature from the 1960s and 1970s overflows with detailed clinical narratives that provide a vivid sense of patients' experiences. On post-op day five, one 56-year-old lab technician began to hear rock 'n' roll music. Sometimes she heard laughter, as if at a party. She grew paranoid and believed that her friends had hidden a record player under her bed. At times she heard voices in the music, as if people were laughing at her. She became convinced that this was all part of a plot to torture her. According to her doctors, she developed other bizarre ideas, even suspecting that one of her nurses was having an affair with one of the married doctors. After several days, the symptoms cleared as suddenly as they had come on. Rachel MacKenzie, a fiction editor at the *New Yorker*, underwent open-heart surgery to repair an aneurysm of her left ventricle. Her surgeon estimated that she had a 35 percent risk of dying during the operation. She survived, but even though her doctors and nurses had prepped her about how difficult recovery would be, it was far worse than she had imagined. In and out of intensive care, she endured weeks of pain and confusion, hallucinations of people coming to her bedside, and overhearing terrifying conversations without knowing whether they were real or imagined. Once she finally recovered, she published a long account in the *New Yorker* of those "nightmare days," an account that was then republished as a book.[2] The problem of post-op delirium was dramatic, frightening, and well known.

Psychiatrists now see delirium as a disease of the brain, a response to toxins, infections, inadequate oxygen, and countless other insults. It was not difficult for doctors to understand why delirium so often followed cardiac surgery. By the late 1960s, they had documented the many kinds of physical damage that the brains of cardiac surgery patients endured. The prevalence of delirium could easily be explained as a direct consequence of cardiac surgery and injury. Such mechanistic accounts of the etiology of delirium, however, existed in parallel with quite different explanatory frameworks.

One prominent theory attributed postoperative delirium to the trauma and disorientation of intensive care units (ICUs). ICUs today are one of the most dramatic and technologically sophisticated sites of medical practice. In the 1960s, however, the units were still new and only partially developed. They had emerged out of a series of crises, including the Cocoanut Grove fire in Boston, the epidemics of paralytic polio, and the demands of trauma care during World War II. They combined intensive monitoring of patients with round-the-clock nursing and medical care. Specialized teams worked together to keep patients alive who, only years before, would have died quickly. As ICUs began to take shape in large hospitals in the 1950s, cardiac surgery was one of the first specialties to develop them.[3] Only with such aggressive, skilled care could patients be brought through the rigors of recovery from open-heart surgery.

In the early years of this new approach, physicians and nurses had learned how to keep some patients alive, but they had not yet learned how to make the experience tolerable for patients. With postoperative pain, constantly beeping monitors, lights on all night, and near continuous interruptions for monitoring and treatment, patients suffered terribly. As one psychiatrist observed, "It is sufficient to observe here only that the oxygen tent, cooling blanket, monitoring devices, monotonous noise background, painful and weakened physical state, relative immobilization, and frequent interruption of sleep— all superimposed on the usual clouded, postanesthetic mental state—combine to form a unique variety of psychological torture, rivaled only by the brainwashing techniques portrayed in recent science fiction thrillers."[4] It was no surprise that such conditions provoked psychotic reactions.

Other analyses focused attention on patients themselves as the source of the delirium. This was, after all, the heyday of Freudian psychodynamic theory in American psychiatry. Working in Vienna in the late nineteenth and early twentieth century, Sigmund Freud had developed a theory that traced psychopathology to repressed conflicts that people inevitably accumulated over their lifetime. Unacceptable urges, often related to sexuality or violence, were quashed into the unconscious mind only to leak out later, expressed as neurotic or psychotic symptoms. Freud developed techniques of free association and psychoanalysis that helped him identify the repressed impulses that gave rise to psychiatric symptoms. His ideas only slowly took hold in the United States, but by the 1950s they had considerable influence, especially among psychiatrists at the elite medical centers where cardiac surgery had begun.[5]

Motivated by Freudian theory, leading psychiatrists interpreted delirium as an extreme form of postoperative maladjustment, brought on by marital conflict, oedipal relationships, domineering parents, or patients' unresolved guilt over their illness. Those with rheumatic heart disease, who required valve surgery to rescue them from heart failure, had been chronically ill for decades and had used their illnesses "for neurotic purposes," often as an excuse to avoid adult responsibilities. When surgery cured their physical disease and took away this excuse, they broke down. One woman developed psychotic rage toward other people, especially her husband. The psychiatrist was not surprised: "It was apparent that the patient was very disturbed by the prospect of giving up heart disease, which had provided her a somewhat acceptable means for satisfying her dependency needs and, in addition, a way of avoiding the more mature but anxiety-provoking sexual role in life." Another woman became delirious five days after surgery. She was convinced that she was dead, that her veins had collapsed and her "flesh was cold." At night God and the devil struggled to control her life. When called for consultation, psychiatrists worked to explain the nature of her delirium. With assistance from her family, they constructed a narrative in which "the patient's mental symptoms were the expression of unrecognized feelings of guilt." She had lived a life of "indulgent self-centeredness," which she had justified as compensation for her crippled state. When surgeons repaired her heart, her "psychic equilibrium" failed. Specific symptoms also made sense. The sensation "that her flesh was cold and her blood stagnant conveyed her sexual frigidity which had made a mockery of her marriage for twenty years."[6]

Men had their own special psychological susceptibilities. Some could not tolerate the dependency that surgery and hospitalization entailed: "The cultural demands on men to control themselves, their families and their environment as well as the castration-injury-death threats of the operation itself may well have made these men more susceptible to the break." Masculine vulnerability became even more striking with the shift toward bypass surgery in the 1970s. Some of these patients—the typical type A overachievers then seen at risk for coronary artery disease—were "characterized by dominance, aggressivity, and self-assuredness, reflecting a characteristically active orientation in coping with life's problems." These traits, "ordinarily assets, may deprive these persons of the opportunity for pre- and postoperative ventilation of emotions and anticipatory anxiety."[7] Men, like women, could seemingly drive themselves mad.

In the most severe cases, psychological maladjustment could cause not just delirium but even death, an extreme pathological wish fulfillment by patients who could not give up the dependency of chronic illness. A psychiatrist and an internist from Montefiore Hospital described one 36-year-old mother who suffered the "ultimate complication." When she learned that her mitral stenosis required surgery, she became consumed by dread. Convinced that she would die during the diagnostic catheterization, she postponed the test as long as possible. Her family physician prescribed a powerful sedative so that she appeared calm on admission to the hospital. Nonetheless, she panicked in the catheterization lab and the cardiologist had to abort the procedure. She began to vomit blood, became acutely psychotic, and died the next day. Only then did the family learn from an undertaker that she had made "complete arrangements for her own funeral" before entering the hospital. Columbia University's Donald Kornfeld noted in his commentary on this case that "the direct effect of psychic stimuli on the cardiovascular system creates a situation in which emotional factors can literally spell the difference between life and death."[8]

With the benefit of hindsight, we could borrow a phrase from physician and anthropologist Paul Farmer and suggest that these explanations represent "immodest claims of causality." Farmer coined this phrase to critique social scientists who focused on superficial causes of patient behavior and health outcomes instead of acknowledging the systematic inequalities that inflict the worst health outcomes on the poorest, most marginalized patients.[9] In the case of cardiac surgery, psychiatrists turned to patient factors to explain delirium despite having good reason to believe that delirium was a direct result of damage done to patients' brains by cardiac surgery.

How else could responsibility for the delirium be assessed? Responsibility is an elusive and contentious topic in medicine, law, and philosophy. When can you say that one act caused another to happen, that the first act is responsible for the second happening? Much of the effort in medical science in the twentieth century focused on these questions. Researchers have been especially interested in sorting out the causes of disease (e.g., of lung cancer, AIDS, or autism), or in determining what caused some change in the prevailing

burden of disease (e.g., why did tuberculosis decline, or lung cancer and autism increase?). Claims about causation become especially difficult when multiple factors interact, which was certainly true with postoperative delirium.
Was it caused by sexual maladjustment, by ICU-induced sleep deprivation,
or by intra-operative emboli? All of these might be involved. Reviewing the
problem of postoperative delirium, Kornfeld traced the phenomenon to "a
combination of factors: organic changes present in the brain before surgery
or produced by the surgery itself; sleep deprivation and sensory monotony
produced by the environment of the open-heart recovery room; and the overwhelming anxiety produced by the event and the environment."[10] A single
outcome could have multiple causes.

Physicians' suspicions about multiple interacting causes complicated their
attempts to assign responsibility for cerebral complications. What does it
mean to be responsible for something? Some meanings are volitional: I chose
to do something, so I am responsible for what happened. Some meanings are
contextual: poverty puts patients at risk for disease and is responsible for the
diseases they suffer. The ambiguities of responsibility have been demonstrated well by lawsuits over cigarettes. When smokers sue the tobacco industry, they argue that the companies, by marketing an addictive, lethal product,
are responsible for any deaths that ensue. The companies, in turn, argue that
smokers are responsible for the consequences of the informed choices they
make.[11] The debates have been contentious because they involve both pragmatic questions of financial liability and existential questions of meaning.

Similar ambiguities about responsibility cropped up in cardiac surgery.
Patients consented to the surgery and its risks. Surgeons had encouraged the
patients to submit to surgery and had performed the operation on the heart.
Technicians operated the heart-lung machines. Engineers had designed the
machines, which, by the 1960s, were manufactured by device companies.
Patients needed surgery to begin with because of illnesses acquired through
infection, lifestyle choices, genetics, or bad luck. Many people could be held
responsible whenever bad outcomes arose. As a result, even amid the wealth
of evidence that researchers produced about the ways in which cardiac surgery damaged patients' brains, it was still possible for psychiatrists to examine these patients and place blame on the patients' unresolved emotional
conflicts.

Another aspect of responsibility becomes important here as well. Who was
responsible for doing anything about the problem of cerebral complications?

Who was responsible for studying the problem, funding research studies, or ensuring that surgeons' articles about their successes included adequate discussions of the complications? Who was responsible for making sure that everything feasible had been done to optimize the safety of surgical procedures? In 1962 two Pittsburgh doctors had beseeched their colleagues and journal editors to pay more attention to complications, not just of new medications but also of surgery: "Individually or collectively, the members of the medical profession must accept their responsibility for therapy."[12] During this era of valve surgery, cardiac surgeons took this responsibility seriously. From Bailey and Harken's efforts at "protective carotid obstruction" to the efforts by many others to describe and understand the causes of the complications, surgeons set out to minimize cerebral complications and make the surgery safer. Could they have done more? Did their willingness to place responsibility in part on the patients allow them to take the pressure off themselves? These questions do not have easy answers.

Questions of responsibility are part of a more general question at the core of the medical enterprise. How far must physicians go in their efforts to provide the best possible care for patients? Medicine is unfortunately full of outcomes that are not as good as they could be. Doctors and patients can blame each other or the dysfunctional health care systems in which they work, but they dodge the crucial question. If something more could be done, does someone have an obligation to do it? Physicians could provide tireless care, they could devote themselves to laboratory research, or they could advocate for health care reform. Patients could accept similar responsibility to do everything possible to maximize their health and adhere to medical regimens. But no one does everything they can do. Patients and doctors act as though responsibility extends only so far. This, regrettably, introduces a touch of fatalism into medical practice, in which suboptimal outcomes are expected.

Throughout the 1960s and into the 1970s, as mortality rates of open-heart surgery continued to decline, complication rates decreased as well. Surgeons and anesthesiologists focused their efforts on improving the safety of heart-lung machines. They questioned their initial enthusiasm for bubble oxygenators and looked again at membrane oxygenators. Though the latter were initially less reliable, they offered a potentially safer solution. Surgeons also experimented with using blood filters to catch emboli and medications to control intra-operative blood pressure. One study found that such efforts reduced the incidence of neurological damage from 19.2 percent in 1972 to 7.4 percent in 1975. But 7.4 percent (roughly 1 in 14) was still substantial. Did the problem persist because surgeons had not yet done everything possible? Was the problem that open-heart surgery carries inevitable risks? A 1975 *Lancet* editorial set the bar high: "There can be no room for complacency until unexpected cerebral damage after cardiac surgery has been eliminated completely."[13] But then something remarkable happened. As coronary artery bypass surgery became the most common cardiac operation in the 1970s, the problem of cerebral complications was nearly forgotten.

CHAPTER TWELVE

The Case of the Missing Complications

After spending fifteen years refining the techniques of cardiac surgery on patients with congenital or rheumatic heart disease, cardiac surgeons in the 1970s took on a new problem: coronary artery disease. Coronary artery surgery had existed on the fringe of cardiac surgery since the 1930s, but it was overshadowed by valve surgery in the 1950s and 1960s. A few surgeons, however, had recognized the potential of a bypass procedure. With coronary artery disease killing one-quarter of all Americans, the market for this promising operation would be vast.

When Cleveland Clinic surgeons published their first bypass surgery case series in 1968, surgeons from all over the nation flocked to do the procedure. By 1977, more than one hundred thousand bypass surgeries were being performed each year.[1] The growth, however, occurred amid a striking absence. Even though physicians had assiduously studied the cerebral complications of valve surgery, they at first paid little attention to these same complications with coronary artery bypass grafting. How and why this happened illuminates the forces in medicine that make complications so difficult to study.

Early descriptions of bypass surgery focused on its positive features. Surgeons at the Cleveland Clinic, for instance, emphasized how well the operation restored blood flow to the heart, how well it relieved angina, and how many patients survived the procedure. Of their first ten published articles about bypass surgery, only one mentioned neurological outcomes at all: two patients had suffered strokes in a series of eighty patients who'd had bypass surgery combined with another surgical procedure. Although surgeons paid close attention to other sorts of complications of bypass surgery, including heart attacks, arrhythmias, bleeding, heart failure, and clotting problems, they did not seem to consider cerebral outcomes (or, at least, they did not document them). Surgeons in Milwaukee and New York also published early and influential bypass surgery case series, and only one of the first five reports from each city mentioned cerebral complications. As surgeons gained more experience with the procedure, they published ever larger series to demonstrate their superior outcomes. Denton Cooley's team at the Texas Heart Institute described 4,522 consecutive cases, and they noted that some of their patients eventually died of strokes (or cancer, ulcers, murder, and suicide), but they made no mention of any postoperative neurological or psychiatric problems.[2]

The inattention to cerebral complications can be quantified. Of the first two hundred articles published about bypass surgery between 1968 and 1973, nearly all described the direct cardiac effects of the procedure: graft patency (97% of the articles), relief of angina (84%), and impact on cardiac function. Discussions of complications, when present, focused on operative mortality (86%), arrhythmias, and myocardial infarction. Neurological damage was noted in some (19%), usually as part of a listing of causes of death. Only four made more than a passing mention of neurological or psychiatric outcomes. Textbooks of the era either did not address complications at all or discussed only operative mortality, heart attacks, and graft failure.[3]

In 1971 the American College of Chest Physicians surveyed surgeons about their experience with bypass surgery. It requested data about angiography, graft patency, angina, and in-hospital fatalities, but not about complications in general or cerebral complications in particular.[4]

When a collaborative group of Veterans Administration hospitals conducted the first major randomized trial of bypass surgery, they initially reported operative mortality, angiographic outcomes, ventricular function data, and long-term survival rates. Subsequent reports described quality of life, including outcomes such as nonfatal heart attacks, angina, exercise capacity, and work status, but not neurological or psychiatric status. Researchers cast their net wide, with publications on the characteristics of noncompliant patients and on the cerebral complications of angiography (4 strokes in 3,044 angiograms). Only one of the forty-six publications that emerged from this trial mentioned a cerebral outcome related to bypass surgery, a single patient who did not regain consciousness after the operation.[5]

The Coronary Artery Surgery Study (CASS), orchestrated by the National Institutes of Health, followed a similar trajectory. Although the researchers collected data about patients' neurological histories and any observed neurological or psychiatric complications, the study's primary reports focused on operative mortality, long-term survival, and quality of life. Some of the scores of reports published in the 1980s did mention neurological outcomes, but only for particular subgroups, such as elderly patients, women, or African Americans. Finally, in 1992, in an obscure journal, researchers published specifically about the incidence of stroke, but when they did, they admitted that the study protocol, which lacked formal neurological assessments, had likely missed most cerebral complications: "The observed incidence of stroke represents only the most obvious persistent neurologic deficits. This results in an underestimation of the true stroke rate and other more subtle neurologic deficits."[6]

Given what was known about the cerebral complications of cardiac surgery in the 1950s and 1960s, it would have been natural for surgeons to study the cerebral complications of bypass surgery. Such research would have been especially important as the procedure quickly became the most common type of cardiac surgery performed. Nonetheless, the problem received scant attention in the procedure's first decade.

The lack of attention persisted even as some prominent physicians, skeptical of the benefits of bypass surgery and concerned by its rapid spread, subjected the procedure to increasing critical scrutiny. The debate, however, focused on questions of efficacy and the applicability of randomized clinical trials to surgical interventions. Even amid the fierce controversy, the technique's most strident critics ignored the potential impact of the operation on patients' brains. The scores of editorials and letters instead focused on the inadequacies of surgical research and the financial conflicts of interest in the bypass surgery industry. Eugene Braunwald, one of the nation's leading cardiologists, challenged surgeons to answer a series of questions about the operation (e.g., about survival rates, future risk of heart attack, fate of the vein graft, comparison with medical management), but none of the questions involved complications, cerebral or otherwise. Seattle cardiologist Thomas Preston condemned surgeons' financial conflicts of interest but said nothing about complications. Henry McIntosh, chief of medicine at Houston's Methodist Hospital (where DeBakey was chief of surgery), did worry about complications, especially perioperative myocardial infarction, but did not mention cerebral ones.[7] These vociferous critics had many pressing concerns about bypass surgery. Cerebral complications were just not among them.

By the end of the first decade of bypass surgery, physicians were in a remarkable situation. Even though surgeons performed over one hundred thousand bypass operations each year, no one had a clear sense of what risk the surgery posed to their patient's minds and brains. The experience of neurologist Anthony Breuer at the Cleveland Clinic was especially revealing. In 1980, recently hired by the clinic, Breuer spoke at a conference on cardiac anesthesia. He commented on how interesting it was that even though much had been written about the neurological complications of valve surgery, and some had been written about the neurological effects of heart transplants, "there is virtually no information in the literature whatsoever on the subject of neurologic difficulties encountered in the patient population undergoing coronary artery bypass graft procedures."[8] This statement is remarkable. Thirteen years into the bypass surgery era, a neurologist at the nation's leading bypass center could find no significant publications about the procedure's cerebral complications.

What were patients told about the potential risks of the procedure before they agreed to surgery? Discussions about safety and expected benefit have a long history in surgery. When the implementation of anesthesia and aseptic techniques allowed the rapid proliferation of elective surgery in the 1890s, surgeons often had to convince prospective patients that the operation was worth doing, that the expected benefits outweighed the risks and costs. In at least some cases, surgeons carefully informed patients, and patients chose to proceed or not. Practice certainly varied from surgeon to surgeon, and we do not know how common or how thorough these discussions were in the early twentieth century. A series of developments in mid-century, from the Nuremberg war crimes trials to the rise of patients' rights movements, increased attention on the problem of informed consent in medical research and practice.[9] By the 1960s, informed consent was well established in principle. Patients wanted to know what options they had for medical treatment and what each option involved. Physicians understood that patients had the right to make the final decision about what would be done or not. Legal precedents clearly supported patients' rights. But when it came to day-to-day medical care, the practice of informed consent varied considerably. When bypass surgery began to spread in 1968, surgeons remained ambivalent about the role of formalized consent procedures.

At the Cleveland Clinic, the surgical committee, with advice from the clinic's attorneys, actively debated the best policy. Standard practice through the 1960s had been to require specific consent forms only for a few specific operations and situations: sterilizations, abortions, kidney donations, and all surgery on children. These decisions had particularly high stakes or involved vulnerable populations. For routine surgical procedures, however, procedures were not standardized. Surgeons presumably had a conversation about risks and benefits, and might have scribbled a note about the conversation in the chart, but practices varied from patient to patient and surgeon to surgeon. At a 1969 meeting the committee saw no reason to change this policy. Surgeons could use consent forms if they chose, but this was merely a formality. The form itself did not "alter one way or another the legal responsibility of the physician to his patient"—which was presumably to inform the patient adequately regardless of the use of a form. Moreover, a signed consent form would not protect surgeons from a lawsuit if something went wrong.[10] The implicit understanding was that Cleveland Clinic surgeons would continue to do the right thing, which was to inform and gain consent from their patients, and that the paperwork generally did not matter.

Surgeons in Houston took a more cautious approach. In November 1968 a clinical research committee at Methodist Hospital clarified its definition of human experimentation. The committee specifically included the "performance of any new operation or technical procedure which has not received acceptance by the general field of medicine or specialty employing the procedure." Any procedure that fit this definition required approval from the committee on human experimentation, and the surgeon had to use specific informed consent paperwork. In March 1970, neighboring St. Luke's Episcopal Hospital also acted to formalize its consent procedures, going much further than Methodist had. It mandated that consent forms be used for all nonemergency operations. These forms specified that risks and complications had been explained to the patient, with the qualification that "no guarantee or assurance has been made as to the result that may be obtained." A subsequent legal brief enforced the idea that it was the substance of the consent conversation, not the form itself, that mattered, especially for new procedures.[11]

The existence of such policies and expectations, however, provides only an indirect measure of how carefully surgeons informed their patients. Anecdotal evidence suggests that actual practice varied widely. John Ochsner, who had trained under DeBakey, listed stroke along with bleeding and infection on the consent forms he used for bypass surgery. But not all surgeons were so careful. Denton Cooley's first bypass patient at the Texas Heart Institute described how quickly consent could happen: "Dr. Cooley walked in the room, pulled out a scratch pad, drew me a picture and said 'This is what I'm going to do,' like he did them every day. I had no idea I was his first bypass patient." Another patient described having a brief conversation about the risk of death and the prospects for good outcomes, but the conversation seemed a mere formality: "Dr. Brown never really asked if I wanted the operation. Obviously, I needed it desperately." Cardiologist Thomas Preston, who took his critique of bypass surgery to the *Atlantic Monthly,* doubted that patients were adequately informed about stroke and the many other possible complications.[12]

Two groups were curious enough about the value and impact of informed consent to undertake studies of how patients responded. When radiologists at the Cleveland Clinic began performing angiography (on organs other than the heart—cardiologists perform all cardiac and coronary angiography), they debated the value and need for a formalized consent process. Most feared that if they provided a comprehensive list of possible complications, they would scare patients away from the procedure. One radiologist set out to test this fear. He gave a detailed consent form to 132 patients, and then, after his colleagues expressed concern, he gave a less frightening (i.e., less thorough) consent form to the next 100. Much to his surprise, only 4 of the 232 patients declined the procedure after being informed of its risks.[13] Eager to learn the diagnostic information promised by the procedure, patients remained willing to consent despite possible complications. This was, after all, in the heyday of the theory of progressive obstruction and faith in prophylactic revascularization.

Another team, from Montefiore Hospital in New York, reached a more pessimistic conclusion about informed consent. Conversations about procedures often took place at a difficult time for patients. They might be experiencing an acute exacerbation of their illness, or they might be stressed about hospitalization and pending surgery. Do they actually understand and remember what they are told? The Montefiore team decided to find out. The physicians audiotaped detailed conversations they had with patients, before cardiac operations, about the diagnosis, proposed operation, risk of death and complications, and possible alternative treatments. Four to six months later, they selected twenty patients and asked them what they remembered from the original conversation. Only two of them recalled that complications had been discussed and sixteen actively denied that major portions of the conversation had taken place: "While these patients were well informed and comprehended their situations prior to operation, they subsequently forgot most of what they had understood and made other qualitative errors in their attempts to recall the consent interview."[14] This finding was a major challenge to the implementation of informed consent. Doctors could respond in many ways. The authors of the Montefiore report simply advised their colleagues to document their consent conversations carefully. That way they would have evidence of their efforts to provide informed consent when their patients forgot.

All this effort, from the debates about using specific consent forms to studies of how patients responded to informed consent, was premised on a false faith. Physicians can only inform patients about what they know, and they only know what they have studied. If they do not have good knowledge about particular complications, then they cannot provide useful guidance to their patients. This is especially true in the setting of significant innovations. Anecdotes are again revealing here. When Michael DeBakey, earlier in his career, developed new techniques to repair abdominal and thoracic aortic aneurysms, he tried to be candid with his patients. When he attempted his first repair of a thoracic aneurysm in 1952, he warned his patient, "I have done these in the abdomen, but I haven't done these in the chest and nobody else has, but I think the same technique will apply." However, he did not realize at the time that the thoracic repair carried a risk of cutting off the blood supply to the spinal cord. As a result, he could not and did not describe this risk to the patient. As he later explained, "We didn't have any experience at that time with the possibility of ischemic damage to the spinal cord, so I never said anything to him about that."[15] The problem was not DeBakey. The problem was, and is, the impossibility of informed consent in the absence of perfect knowledge.

The situation arose again in the early years of bypass surgery. Even when physicians did provide diligent informed consent, they presumably discussed only those complications they knew and worried about. As documented above, surgeons knew next to nothing about the cerebral complications of bypass surgery in the early 1970s. Could they have provided adequate warnings about cerebral complications on the basis of their prior experience with valve surgery? Probably. But if they were as worried about these complications with bypass surgery as they had been with valve surgery, then presumably they would have studied the problem. The fact that they had not studied the problem carefully in the procedure's early years suggests that they were not worried. And if they were not worried, then would they have mentioned the problem in their consent process? Ochsner did, but others did not.

This inconsistent attention demands explanation. Surgeons in 1968 knew well that cardiac surgery posed a serious threat to the brain. Years of experience with beating-heart surgery and then heart-lung machines had made this perfectly clear. Yet during the early years of bypass surgery, as surgeons worked to establish and extend their innovation, their attention was elsewhere. Fierce debates focused on whether the procedure worked well enough to justify the risk—meaning operative mortality—and the cost. Few physicians studied the complications. Most of the hundreds of thousands of patients who had bypass surgery in the 1970s were probably inadequately informed about its risk of cerebral complications, if only because the risks had not yet been well characterized.

CHAPTER THIRTEEN

Selective Inattention

On April 4, 1962, Johns Hopkins surgeon David Sabiston used a piece of saphenous vein to bypass an obstruction in the right coronary artery of a 41-year-old man. This feat might have earned him international fame: he was the first surgeon to perform what would become one of the most important operations known to surgery. Instead, the patient suffered a stroke and died on the third day after the procedure. A clot had formed where the bypass graft was stitched into the aorta. A piece from that clot had evidently broken off and been swept into the patient's cerebral circulation, where it plugged a vital artery. Discouraged by this outcome, Sabiston abandoned the procedure for six years—until René Favaloro reported his successful case series. Sabiston did not publish an account of the technique or its neurological complication until 1974, long after bypass surgery had been established by other surgeons.[1]

In April 1975 almost exactly the same thing happened to a patient at the Mayo Clinic. A 45-year-old man suffered a fatal embolic stroke arising from a clot at the aortic end of a bypass graft. This time, however, Mayo neurologists thought that the outcome was significant enough to justify a case report in a prominent journal.[2] The juxtaposition of the responses to these two events demonstrates an important aspect of medical practice: complications are a fact of medicine, but their relevance is not. The same adverse outcome can have different meanings and consequences depending on the setting in which it occurs.

Early cardiac surgeons had become dreadfully familiar with dramatic psychiatric and neurological symptoms in their patients after surgery. The complications had been studied by a range of specialists, who had documented many possible mechanisms of injury. When bypass surgery began to spread in the late 1960s, surgeons could easily have expected that it too would cause neurological and psychiatric complications. But as we have seen, these complications were nearly ignored during the first five years after the innovation, and as late as 1983—fifteen years into the bypass surgery era—physicians at the Cleveland Clinic, one of the nation's preeminent centers for cardiac surgery, could assert that little was known about the "neurologic difficulties" in their bypass surgery patients.[3]

The poverty of the published literature on the cerebral complications of bypass surgery compared with the abundant literature on the these complications of valve surgery provides an opportunity to examine how a particular phenomenon becomes an object of scientific scrutiny or medical concern. As scientists work to understand the natural world, they cannot study everything they observe. They often pay most attention to results that surprise or puzzle them. Although unexpected results sometimes prompt scientists to abandon long-held ideas, at other times, the results are explained away as inconsequential anomalies. The decisions scientists make about what is relevant have enormous significance. A complex mix of interests and contingencies influence how an observation becomes constituted as an object of study, making it not just a matter of fact but a relevant matter of concern. Understanding these processes requires recognizing what is at stake for the people involved—physically, socially, economically, and morally.[4] The ability to visualize or hear or feel a phenomenon is only part of the story. Something else must happen to direct researchers' attention to the problem. With cerebral complications of bypass surgery, that something did not happen, at least not initially.

Explaining why something did not happen is difficult—it is hard enough to explain something that did. But this has not stopped scholars from trying. In the 1980s one sociologist described how professionals, faced with increasingly complex problems, narrowed their focus to what they defined as their professional purview. This made it easier for them to develop and deploy their expertise effectively. However, it also left them "selectively inattentive of data that fall outside their categories." One historian has recently offered a more elaborate account. Interested in understanding why certain kinds of facts or ideas were not known, he proposed a threefold taxonomy of ignorance: "ignorance as native state (or resource), ignorance as lost realm (or selective choice), and ignorance as a deliberately engineered and strategic ploy (or active construct)." Had some fact not yet been discovered, had it been discovered but forgotten, or—more nefariously—had it been actively suppressed? These dynamics have been studied in many situations, most famously with the "controversies" about global warming and the dangers of cigarettes.[5] The history of a decade in which clinicians paid little attention to the cerebral complications of bypass surgery demonstrates how even the most mundane concerns of physicians' daily lives could erect barriers to recognizing treatment complications.

Certain hypotheses about the long disregard of bypass-related cerebral complications can be ruled out easily. Surgeons did not come to bypass surgery as a blank slate. Every cardiac surgeon who performed bypass surgeries in the late 1960s and early 1970s had trained at a time when valve surgery dominated the field. All knew well the serious neuropsychiatric complications that valve patients experienced. At some centers, including the Cleveland Clinic and the Massachusetts General Hospital, surgeons were intimately involved in postoperative care. Patients who experienced strokes or delirium remained hospitalized for weeks and provided a prolonged reminder of the dangers of open-heart surgery. Moreover, the inattention to the operation's complications was not limited to surgeons. Neurologists and psychiatrists, who had written so much about the complications of valve surgery, wrote little in the 1970s about bypass surgery. In terms of the threefold taxonomy of ignorance, the inattention to cerebral complications could not have been a native state. Physicians knew that the risk existed; they just did not study or write specifically about it.

Could this silence have been a deliberate strategy? It is easy to imagine why surgeons might not have wanted to publicize the complications of their work. Sabiston, for instance, did not rush to publish an account of his failed first bypass operation. But no surgeon publishes case reports every time something goes wrong, and no evidence suggests that Sabiston deliberately concealed the outcome. However, there is at least on case in which a surgeon—or perhaps the surgeon's reputation—did complicate an effort to publish a report about complications.

In September 1969 two medical students at Baylor College of Medicine, Ottis Layne and Stuart Yudofsky, completed a review of psychiatric complications of surgical patients at Methodist Hospital (Michael DeBakey's surgical service) and St. Luke's Episcopal Hospital (Denton Cooley's surgical service). They found psychotic reactions in eight of the fifty-seven cardiac surgery patients but none in the twenty vascular surgery patients. The difference was revealing. The two sets of patients had similar risk factors, but the vascular surgery had not required the use of a heart-lung machine. When the students wrote up the results with their mentor, psychiatry professor Hilde Bruch, she showed "the outstanding findings" to DeBakey. This was a delicate move. DeBakey, the best-known surgeon in the United States and the president of Baylor, was a force of nature. Named the "Texas Tornado" in a 1965 *Time* magazine cover story, DeBakey that September had forced the resignation of Cooley, his protégé, after a dispute over research ethics. DeBakey's anger had received national media attention.[6]

DeBakey saw a draft of Layne and Yudofsky's paper in March 1970, but he refused to "approve publication of this manuscript in its present form." He spoke with Bruch about his concerns and was eager to discuss them with the students. What was he worried about? Yudofsky recalls that DeBakey doubted their results because he had never noticed an unusually high rate of psychosis in his cardiac patients. Bruch and the students, however, had taken pains to make the study as rigorous as possible. They made revisions and submitted the manuscript to the *New England Journal of Medicine* in August. In her cover letter, Bruch acknowledged some of the tensions at Baylor. As she explained to the editor, there was an "anomaly of the relationship between the two professors of surgery"—presumably DeBakey and Cooley—"who will not permit to have their names mentioned in the same paper, but do not object to the publication without reference to surgery."[7] But this was not the whole story.

When the journal accepted the paper, Bruch warned Yudofsky that there was "a slight damper concerning the manuscript." One of her senior colleagues, Alex Pokornoy, had "some misgivings about publishing it without Dr. DeBakey's knowledge or consent." Bruch resolved the tension by removing her name from the paper before it was published. She told the students that since the paper was largely their work, she did not merit authorship. As she explained to one of the journal's editors, the initial inclusion of her name "was a courtesy that the two young doctors wanted to show me." Her fear of De-Bakey, however, clearly played a role in her decision. She told Yudofsky, by then an intern in New York, "Since Ottis and you are away from Baylor the Mighty Arm of the Superior Power can no longer reach you."[8]

Bruch's concerns proved unwarranted in the end. Just one week after she asked the journal to remove her name, she received word that DeBakey had "okayed the paper as is." But she left well enough alone, and the paper appeared without her name on it. DeBakey's role remains ambiguous. The students had one perception: he was always a supporter and an advocate of medical students; he welcomed their proposal to study his patients, and although he respectfully disagreed with their findings, he never intervened with them to block the paper. Bruch must have had another. Although she told the journal that she did not deserve to be listed as an author, she had originally submitted the paper with her name on it. Had she really reassessed the scope of her contribution? According to Yudofsky, Bruch had played a central role in designing the research and revising the paper. It is extremely unusual for researchers—and thoroughly against their own interests—to remove their names from a paper after it has been accepted for publication, especially when the paper had been accepted by one of the world's most prestigious journals.[9] She must have been worried about what DeBakey might think or do.

Similar concerns might have led other doctors at other institutions to keep quiet about their own observations and data concerning the cerebral complications of bypass surgery. After all, many cardiac surgeons had considerable institutional power. The financial stakes of bypass surgery—with millions of Americans suffering from coronary artery disease—were vastly higher than for valve surgery. Entrepreneurial surgeons, such as Donald Effler and Denton Cooley, billed millions of dollars each year for the operation. I can only speculate here, but recent scandals over the complications of widely used medications suggest that commercial interests can lead researchers to conceal bad outcomes.[10]

However, there are also reasons to be skeptical about a dominant role for financial interests and fears of reprisal in the early 1970s. The most famous and powerful surgeons did nothing to hinder study of their patients. When Sid Gilman approached Dwight Harken, or when Layne and Yudofsky approached DeBakey, the surgeons allowed these neurologists and psychiatrists to study their patients, including their complications. Not even the most strident critics of cardiac surgeons, including Eugene Braunwald, Thomas Preston, or Henry McIntosh, suggested that cardiac surgeons deliberately suppressed complication data in order to profit from the procedure. Furthermore, even if surgeons showed little initiative in studying the cerebral complications of bypass surgery in the early 1970s, they cooperated—as will be seen—when approached by neurologists in Cleveland, London, and elsewhere in the late 1970s.

If the neglect was neither a native state nor an active strategy, what might it have been? When I have spoken with surgeons and anesthesiologists who cared for patients in the early 1970s, some have said that at the time, they believed that bypass surgery would be spared the complications that plagued valve surgery, for various reasons. The operation, as initially performed by René Favaloro, did not require cardiopulmonary bypass. Since surgeons did not open the cardiac chambers (coronary artery bypass involved only the arteries that ran along the heart's surface), bypass was less likely than valve surgery to introduce air emboli into the patient's circulation. Moreover, it was done on a different set of patients. Most valve surgery had been done on patients with rheumatic heart disease. That disease, the consequence of a prior episode of rheumatic fever, was often associated with brain injury even in the absence of surgery. Rheumatic fever itself could cause neuropsychiatric complications, as could years of chronic heart failure and cardiac insufficiency. Already damaged in some ways, these patients seemed uniquely susceptible to the cerebral complications of cardiac surgery.[11] These explanations would suggest that the inattention of surgeons, anesthesiologists, and neurologists in the early 1970s was reasoned: they did not expect that there would be a problem, so they did not study it.

However, the published literature from the early 1970s complicates these recollections. Surgeons, even Favaloro, quickly began to use heart-lung machines for bypass surgery because they could not operate on the dominant left coronary artery without it. By 1975, 398 of 401 surveyed coronary artery surgeons reported using heart-lung machines for most of their operations. Once this change had taken place, bypass surgery introduced nearly all the same risk of valve surgery. As Mayo Clinic neurologists observed in 1976, the "pathogenetic mechanisms implicated in the production of cerebral symptoms during open-heart surgery (i.e., hypotension, hypothermia, hypoxia, microembolization of particulate matter, etc.) are equally operational during ACBG [bypass] surgery."[12] Moreover, surgeons quickly realized that bypass surgery patients, who often had extensive atherosclerosis in their aortas, carotid arteries, and cerebral arteries, had their own risk factors for stroke before having the surgery.

Some physicians at the time did comment on how little attention had been paid to the problem considering the known cerebral complications of cardiac surgery, the similar risk profile of bypass and valve surgery, and the particular risk of operating on patients with severe vascular disease. The Mayo neurologists noted in their case report that it was "surprising" that the complications, even in 1976, remained "unknown, overlooked, or disregarded." William DeVries, who later became famous for implanting the first artificial heart in a patient, was frank in his reminiscences: "We were surprised when we had the first stroke in a patient. It does not make sense now that we could have been that stupid, but we really were a little stupid about it."[13] If doctors then, and now, think that they ought to have been aware of and interested in the problem, then it is worth continuing to investigate what might have contributed to their inattention at the time. In the mid-1970s two sets of factors certainly played a role: the many demands made on the time and attention of the relevant clinicians, and clinicians' conscious choices about the relevance of heart and brain.

The mundane realities of time and money distracted physicians in the 1970s. Good research on cerebral complications of any procedure was difficult and expensive to do. Neurologists could invest substantial time and resources assessing one hundred preoperative bypass surgery patients only to find five with major postoperative strokes, too low a yield for useful study. Neuropsychological testing was undermined by debates about which tests to use and which ways to analyze the data. Even if doctors had wanted to study the complications, they might have balked because of concerns about the feasibility of the research.

The experience of the Cleveland Clinic in the early 1970s is suggestive. As cardiac surgery boomed in the late 1960s and early 1970s, the surgeons faced a constant struggle to manage the increased volume. Nursing shortages prevented them from using all their hospital rooms. They had to negotiate with the operating room scheduler to find additional slots for their surgeons. The increased work spilled over to the cardiologists, who performed angiography and clinical assessments on constantly increasing numbers of patients. Inadequate space and nursing shortages hindered their work too. Long wait lists for both angiography and surgery added to the constant stress of daily operations. Outside forces caused headaches as well. At the Cleveland Clinic, after decades of growth, cardiology and cardiac surgery volumes leveled off and then declined in 1977, the result of winter blizzards that closed operating rooms for a week in January, increasing competition from community hospitals, a sluggish economy, and negative publicity that followed the publication of the Veterans Administration study of bypass surgery in September. Angiography and surgery rates did not recover until the following spring.[14]

Even as Cleveland Clinic surgeons did thousands of bypass procedures annually, only four neurologists were on staff in the early 1970s, and all of them were already busy with their own patients. Moreover, the clinic's surgeons and neurologists did not have a history of collaborative research on neurological complications of cardiac surgery. The neurologists' research focused on different projects—studies of narcolepsy, seizures, brain recordings, and cerebrovascular disease. None of them rushed to study the neurological complications of the new procedure.[15] This is not just a generic situation of people being too busy to do everything they could possibly do—that is often true in any high-powered medical center. No matter how busy the physicians were at baseline, things got dramatically worse in the heady, early days of bypass surgery. This is one irony of a boom. Even as the expansion of bypass surgery provided new opportunities for studying its complications, the participants in the boom lacked the time to step back and study them.

When research did happen, whether the earlier studies of valve surgery complications or the later studies of bypass surgery, it often had to happen in whatever spare time busy clinicians had, and at their own expense. Neurologists sometimes had to see patients for research purposes in their free time, on nights and weekends. Such challenges made it difficult to begin a study, much less bring one to fruition. A 1969 collaborative project between the Peter Bent Brigham Hospital and the Massachusetts Institute of Technology demonstrated this. Researchers set out to study the effects of open-heart surgery on cognitive function, but even with funding from the National Institute of Mental Health and the hospital's Department of Surgery, as well as a research assistant who tested patients for two years, the analyses were never completed or published.[16] Moreover, at a time when neurology and psychiatry rewarded those who deciphered novel mechanisms of disease or developed new treatments, neurologists could gain little academic prestige by cataloging the complications of a specific surgical procedure.

Logistical challenges and competing demands on the time of surgeons, neurologists, and other clinicians made it possible for studies of complications to fall between the disciplinary cracks. No single group had the necessary expertise, interest, and sense of responsibility to ensure that the studies were done quickly. The problem was not inevitable. After all, in the 1960s these same groups of specialists had come together to conduct extensive studies of the complications of valve surgery (and would eventually come together again, in the late 1970s, to study bypass surgery). The different specialists could have found ways to negotiate and discover enough common interests to make the research possible and productive, but in the early 1970s, something hindered these collaborations.[17] One obstacle might have been simple happenstance. If the leading bypass center, Cleveland Clinic, did not have neurologists able to do the research, similar misalignments must have occurred at other medical centers.

Additional obstacles to productive research might have been conceptual. The long-standing need to balance the interests of heart and brain may have influenced the urgency with which clinicians confronted the problem. In the early 1970s, surgeons saw bypass surgery as a matter of life and death. Narratives of coronary artery disease portrayed death by myocardial infarction as an inevitable, even imminent, reality for anyone with angina. A *Cleveland Magazine* cover story in 1973 exhorted its readers to have bypass surgery before it was too late.[18] With surgeons focused on what they viewed as necessary life-saving repair of the heart, other concerns might have seemed secondary.

Sometimes the conflicts between competing demands were explicit. Bypass surgery required surgeons to sew the vein or artery grafts into position with great skill and precision. Surgeons knew that the long-term prospects for the graft were best when they took the time to construct meticulous suture lines. But they had also learned that the more time patients spent on heart-lung machines, the more likely they were to experience complications. Neurological deficits "rose precipitously" in patients on bypass for more than two hours. Effler put it bluntly: "A patient's tolerance for the heart-lung machine is limited and you just can't stand around there wondering what you're going to do next." Cooley, who became famous as the most dextrous and efficient of all cardiac surgeons, set the bar high: "The important thing in cardiopulmonary bypass was to get the patient the hell off the pump as soon as possible. If

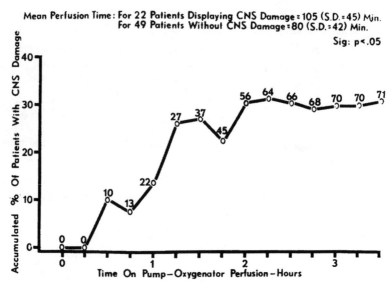

Mean Perfusion Time: For 22 Patients Displaying CNS Damage = 105 (S.D. = 45) Min.
For 49 Patients Without CNS Damage = 80 (S.D. = 42) Min.

Sig: p<.05

Relation between pump time and cerebral damage. Note the dramatic increase in the incidence of brain injury between one and two hours on the heart-lung machine. From William H. Lee et al., "Effects of Extracorporeal Circulation upon Behavior, Personality, and Brain Function: Part II, Hemodynamic, Metabolic, and Psychometric Correlations," *Annals of Surgery* 173 (June 1971): 1013–23, fig. 3, p. 1020. With permission from Wolters Kluwer Health.

you wanted to have a living patient, perfusion periods of fifteen to twenty minutes were ideal."[19] Time was of the essence.

Surgeons faced a difficult choice. They could operate as quickly as possible to minimize cerebral complications, or they could work more slowly and methodically to optimize cardiac outcomes. They had been making such compromises for decades. In the 1840s and 1850s, with the introduction of anesthesia, surgeons could for the first time perform meticulous and careful operations. But longer operations increased the patient's risk of blood loss and infection. Physicians had to learn to balance their desire for optimal surgical technique with the safety of the patient.[20] Familiarity with such trade-offs, however, did not make them any easier to make, especially in the early years of bypass surgery, when the predictors of good outcomes and the risks of complications both remained ill defined.

Surgeons and anesthesiologists faced a similar choice with oxygenators.

Bubble oxygenators—simple, reliable, and affordable—dominated cardiac surgery in the 1960s. However, surgical teams grew increasingly dissatisfied. Citing a 50 percent incidence of "subtle or severe central neurological deficits," one team argued that the spread of bypass surgery increased the need to achieve "decreased morbidity from heart-lung perfusions." Some teams reintroduced membrane oxygenators. Though more complex and less reliable, membrane oxygenators posed less risk to the brain. Comparative studies— focused on the complications of different oxygenators (for valve or bypass surgery) and not on the complications of bypass surgery itself—soon showed that membrane oxygenators did indeed produce fewer neurological complications. Each team had to weigh the risks of neurological complications and the benefits of increased reliability. Some centers, including the Cleveland Clinic, switched over to membrane oxygenators for most of their open-heart surgeries (87% in one series). At other sites, however, bubble oxygenators remained in regular use into the 1990s.[21]

Surgeons also had to choose their field of view. Some bypass surgery pioneers argued that the procedure could only be done properly on the tiny coronary arteries with the aid of an operating microscope.[22] This provided the magnification needed to optimize operative technique. Many others used special eyeglasses with magnifying lenses built in. But these devices literally focused the surgeon's view on the coronary artery and not much more. With so much effort and attention devoted to the technical quality of the surgical intervention, they might have had less attention for problems beyond their coronary focus.

Pragmatic realities made this problem worse. With so much demand for their services, prominent surgeons performed many bypass operations each day. They had little time for thorough postoperative assessments of their patients. Layne and Yudofsky, the two Baylor students, speculated that the high prevalence of psychosis in postcardiotomy patients was underappreciated by surgeons of that era because the symptoms were often not revealed during surgical follow-up rounds. Surgeons conducted rapid, highly structured postoperative evaluations.[23] They focused on basic questions: Was the patient stable? Were there signs of infection? There was little opportunity for neurological or psychiatric complications to come to their attention.

The problem, in fact, was broader than just the neurological and psychiatric complications of bypass surgery. With any major therapeutic intervention, a host of outcomes might matter to patients, even if these outcomes seem less immediate and relevant to the surgeons. David Eddy, a physician and decisions analyst, captured this problem in a 1984 discussion of the challenges of medical decision making: "Most procedures have multiple outcomes and it is not sufficient to examine just one of them. For example, a coronary artery bypass may change the life expectancy of a sixty-year-old man with triple vessel disease, but it will also change his joy of life for several weeks after the operation, the degree and severity of his chest pain, his ability to work and make love, his relationship with his son, the physical appearance of his chest, and his pocketbook. Pain, disability, anxiety, family relations, and any number of other outcomes are all important consequences of a procedure that deserve consideration. But the list is too long for practical experiments and many of the items on it are invisible or not measurable at all."[24] It simply did not seem possible for doctors to study every imaginable outcome of their therapies.

What might explain what happened instead? Eager to demonstrate the value of their new operation, overwhelmed by the flood of patients seeking bypass surgery, and working with colleagues in other specialties who faced comparable challenges in their own work, cardiac surgeons in the early years of bypass surgery left many relevant outcomes unstudied. This history reveals several obstacles to the production and use of knowledge about treatment complications. Medical institutions and professional relationships create barriers to optimal collaboration between different specialists. Doctors sometimes see complications as inevitable, especially in the sickest patients, and assume that the incidence of complications in patients will inevitably decrease as their therapies continue to progress. The reward systems in medicine create incentives for some kinds of work but not others. All these factors, and countless others, conspire to direct physicians' attention away from the problem of treatment complications.

With the benefit of hindsight, we can see how the history might have unfolded differently had someone set his or her mind to the task. Surgeons and anesthesiologists could have anticipated and measured the most likely complications of bypass surgery that required a heart-lung machine. Neurologists and psychiatrists could have recognized a pattern of frequent consultations to cardiac surgery intensive care units. Hospitals could have monitored complication rates and taken action to optimize patient outcomes. The National Institutes of Health and other funders could have allocated research funding specifically to this question. Journal editors, accustomed to publishing reports about the cerebral complications of valve surgery, could have demanded that such complications be reported in publications about bypass surgery. Regulators could have reviewed the safety and efficacy of new operations and surgical devices, just as they did for medications. In the early years of bypass surgery, however, none of these groups sought systematic evidence. Surgeons did not need formal studies or approvals to launch bypass surgery. They needed only authorization from their hospitals and referrals from cardiologists. Peer reviewers and journal editors did not reject surgical case series if they lacked data on neurological outcomes. The problem fell through many cracks.

CHAPTER FOURTEEN

The Cerebral Complications of Coronary Artery Bypass Surgery

As surgeons succeeded in making cardiac surgery the most prominent specialty in American medicine in the 1960s, they witnessed the complications their patients suffered, from strokes and seizures to cognitive dysfunction and personality change. But despite being aware that cardiac surgery put their patients' brains at risk, they paid little attention to cerebral complications during the early years of coronary artery bypass grafting. As bypass surgery entered its second decade, however, the situation changed. Starting slowly in the 1970s and then more quickly in the 1980s, data about its cerebral complications emerged. The shifting dynamic provides a test case of sorts. If one set of factors, described in the previous chapter, had diverted physicians' attention away from the problem in the 1970s, did these same factors somehow change in the 1980s and make the complications a matter of concern? Much depended on the increasing scale of coronary artery bypass surgery in the 1970s and 1980s and the new opportunities and incentives for doctors to study its complications.

Reports about the neurological and psychiatric complications of bypass surgery began as a trickle. David Sabiston, for instance, finally described in 1974 his patient with the fatal stroke from 1962. However, he did so only in passing during a general lecture about coronary circulation, which was later published in the *Johns Hopkins Medical Journal*. Other reports followed, including the one from Mayo describing a similar case in 1976. The first report dedicated to the psychiatric complications of bypass surgery appeared in 1975. Groups of bypass surgery patients also began to appear as parts of larger case series. In 1974, for instance, a pair of Swedish researchers—a surgeon and a psychologist—described neuropsychological test results in 144 cardiac surgery patients, including 18 bypass surgery patients. Cerebral outcomes also appeared as part of broader studies of surgical outcomes. When Cleveland Clinic cardiac surgeons reviewed their first 9,000 bypass surgery patients in 1979, they found that stroke rates varied from year to year between 0.9 and 2.1 percent.[1]

The slow emergence of literature on the cerebral complications of bypass surgery reflected a linked set of developments. The emergence was slow, in part, because some research touched on bypass surgery only passively. For instance, some researchers interested in cerebral complications framed their research around the complications of heart-lung machines. They examined whatever patients happened to be available. In the 1960s they studied patients with valve disease. As the focus of cardiac surgery shifted from valve surgery to bypass surgery in the early to mid-1970s, the composition of the heart-lung machine case series shifted as well, and more data about bypass surgery emerged.

Other researchers had specific interest in bypass surgery patients. Physicians knew that the two procedures, valve replacement and bypass surgery, involved distinct pathological conditions: rheumatic heart disease, requiring valve surgery, and coronary artery disease, requiring bypass surgery. According to a team of New York psychiatrists, the situations differed "in terms of etiological agent, duration of illness, and type of surgery performed." As a result, there could be "significant differences" in the rates of complications, something that justified specific studies of bypass surgery patients. Some doctors, such as the Mayo Clinic neurologists, were skeptical of the low rates of complications mentioned in passing in the existing surgical literature about bypass surgery and sought "a more critical evaluation," specifically a careful prospective study "to ascertain the true incidence of cerebral complications."[2]

The rapid increase in the scale of bypass surgery also played a motivating role. In a 1980 abstract, Cleveland Clinic neurologists noted that even though

72,000 patients had received bypass surgery in 1978, the cerebral complications had not yet "been widely appreciated or systematically studied." By the time they published a full report in 1983, the information deficit had only grown larger. Surgeons performed one hundred thousand to two hundred thousand bypass operations each year in the early 1980s, still without good information about its cerebral consequences. The increasing scale of bypass surgery might have drawn neurologists to study its complications for several reasons. More surgery meant more complications, more consults for neurologists and psychiatrists, and more material for their research. Neurologists, as diligent, responsible clinicians, would have wanted to characterize the complications of the important new procedure. They might also have wanted to ride on the coattails of cardiac surgeons and draw attention to their own work by showing its relevance to the burgeoning bypass surgery industry. Some neurologists and psychiatrists might have set out to study the operation's complications in a deliberate effort to dampen enthusiasm for an operation that many physicians felt was overused.[3]

These are speculative motivations, but one motivation was clearly in play. Neurologists and psychiatrists quickly became interested not in what they could learn about the hazards of bypass surgery, but in what bypass surgery could teach them about the pathophysiology and treatment of neurological and psychiatric disease. They realized that bypass surgery offered a valuable opportunity. Open-heart surgery, with its attendant risk of brain injury, was a scheduled event. Researchers could study patients before and after surgery to explore the causes and course of cerebral complications. As one New York team recognized, "the cardiac-surgical patient provides the unique opportunity to study the development of psychopathology under optimal observational circumstances." Cleveland Clinic neurologists agreed: "Stroke in the setting of open-heart surgery provides a unique opportunity to study the pathogenesis and modification of both focal and global brain ischemia." Researchers had high hopes for what they might learn: "The 'CABG-stroke model' offers a unique opportunity to study antistroke therapy."[4] Complications, no matter how regrettable, could be put to good use.

The first large study to take advantage of the opportunity afforded by bypass surgery examined the possible causes of post-op psychiatric complications. A team of New York psychiatrists studied fifty-one patients before and after the procedure. They found a range of syndromes, including delirium, depression, hallucinations, and paranoia, in 16 percent of the patients. This was lower than the 41 percent incidence they had seen in an earlier series of valve operations. They speculated about many possible reasons for the difference, including the brain damage caused by rheumatic fever in the valve patients, the shorter average time on the heart-lung machine in the bypass surgery patients, and differences in the personalities of the two sets of patients. Specifically, they wondered whether the "hard, driving, aggressive coronary personality" of the bypass surgery patients protected against delirium: "The coronary bypass patients' lifelong aggressive competitive striving may facilitate reality-grasping activities." Other studies, however, found rates of psychiatric complications to be twice as high in bypass surgery patients as they were in patients who'd undergone other cardiac procedures.[5] Such inconsistent results have plagued the study of cerebral complications for decades.

Neurologists undertook similar studies. One of the first large prospective studies took place at the Cleveland Clinic. By 1979, the clinic had expanded its neurology program from the four neurologists of the late 1960s and early 1970s to a staff of thirteen. Two of the new hires, Anthony Breuer and Anthony Furlan, were stroke experts. Looking for possible research projects, the pair found a hospital full of bypass surgery patients, with over 2,000 operations performed annually. As they explained, "advantage was taken of the opportunity." With some help from anesthesiology and with the blessing of cardiac surgeons (including two future chief executive officers of the clinic), Breuer and Furlan orchestrated a prospective study of 421 bypass surgery patients. They developed an elaborate set of templates to collect and organize data on hundreds of potentially relevant variables: patient demographics, preoperative neurological assessment, details of the surgery and anesthesia (bubble in 52, membrane in 369), and postoperative course.[6] They published two abstracts in 1980; three major reports soon followed.

Breuer and Furlan found delirium in 11.6 percent of patients and strokes in 5.2 percent — these were severe in 2 percent. Another 13 percent suffered injuries to peripheral nerves, usually from positioning during surgery. The researchers could not include cognitive testing because the clinic did not have a neuropsychologist on staff. The absence was sorely felt. The department chairman highlighted this limitation in his annual report to the board of governors: "We desperately need to work this area of deficiency." Even before the neurologists had compiled their results, they began to see the impact of their study. The surgeons, for instance, had become familiar with the "audible

TABLE 3 *Clinical Features of 24 CNS Infarcts in 22 Patients*

Patient	Age	Sex	Infarct location	Severity
1	46	M	R cerebral hemisphere	mild
2	55	M	R cerebral hemisphere	mild
3	56	M	R cerebral hemisphere	mild
4	59	M	R cerebral hemisphere	moderate
5	67	M	R cerebral hemisphere	moderate
6	59	M	R cerebral hemisphere	severe
7	64	M	R cerebral hemisphere	severe
8	44	M	L cerebral hemisphere	mild
9	52	M	L cerebral hemisphere	mild
10	64	M	L cerebral hemisphere	severe
11	64	F	L cerebral hemisphere	severe
12	65	M	L cerebral hemisphere	severe
13	55	M	brainstem	mild
14	55	M	brainstem	mild
15	49	M	brainstem	moderate
16	70	M	brainstem	moderate
17	60	M	brainstem	severe
18	39	M	retina	mild
19	53	M	retina	mild
4	59	M	retina	mild
20	64	M	retina	mild
10	64	M	retina	moderate
21	61	M	optic nerve	severe
22	63	M	optic nerve	severe

Features of strokes in the Cleveland Clinic study. The correlation between age and severity can be seen, with all severe strokes occurring in patients 59 years old or older. From A. C. Breuer et al., "Central Nervous System Complications of Coronary Artery Bypass Graft Surgery: Prospective Analysis of 421 Patients," *Stroke* 14 (Sept.–Oct. 1983): 682–87, table 3, p. 684. With permission from Wolters Kluwer Health.

crunch" heard when they put clamps on the aorta so that they could suture the saphenous vein graft into place. Knowing that the neurologists would examine their patients in detail and would detect any neurological anomalies, surgeons became more careful in their handling of the aorta. [7] This was an illustration of the Heisenberg uncertainty principle of sorts: the fact of observing a phenomenon changed the phenomenon itself.

Research on the complications of bypass surgery launched an enduring collaboration between neurologists and cardiac surgeons at the Cleveland Clinic. They published many papers about the neurological outcomes of other cardiac procedures and diseases. One clinic magazine issue featured a picture of Tinman and Scarecrow from the *Wizard of Oz* on its cover, to celebrate the clinic's doctors for "exploring uncharted paths as they search for the relationship between the heart and brain."[8]

Other groups undertook similar long-term prospective studies in the 1980s. For instance, neurologists in Newcastle, England, had become concerned, like the Cleveland group, about the rapid rise in bypass surgery and the absence of adequate knowledge about its complications. Calls to increase the funding for bypass surgery substantially in England added a sense of urgency. Teaming up with a cardiologist and a cardiac anesthetist, a team of four neurologists followed 312 patients through bypass surgery. They used more thorough screening and a more inclusive definition of injury than the Cleveland team had used and found neurological deficits in 61 percent of patients: stroke in 15 patients, altered consciousness in 10, visual problems in 78, psychosis in 4, aberrant reflexes in 123, and peripheral nerve problems in 37. At 6 month follow-up, 85 patients still had "detectable neurological signs." Unlike the Cleveland team, the Newcastle group was able to conduct neuropsychological tests on 298 of the patients. These tests revealed postoperative cognitive dysfunction in 79 percent.[9]

The English researchers hoped that their findings would draw attention to "hazards inherent in the extracorporeal circulation process, which remains an imperfect replacement for normal functions of the heart and lungs." To do so, they compared these outcomes to outcomes in 50 patients who underwent peripheral vascular surgery (without a heart-lung machine). Neurological findings (61% vs. 17%) and cognitive deficits (79% vs. 38%) were more common in the bypass surgery group. The conclusion seemed clear: "The difference in frequency and severity of central nervous system complications between the 2 groups is likely to reflect cerebral injury resulting from cardiopulmonary bypass."[10] Although it had taken researchers nearly twenty years, they had finally generated substantial data that confirmed what they had feared since the 1920s and known since the 1960s: cardiac surgery threatened the brain.

THE CLEVELAND CLINIC FOUNDATION

Challenge

Homage to the collaboration between neurologists and cardiac
surgeons at the Cleveland Clinic. The familiar image from the
Wizard of Oz used on the cover of a Cleveland Clinic magazine
celebrates the efforts of clinic neurologists and cardiac surgeons
to understand the relationships between the brain (Scarecrow)
and the heart (Tinman). From *Challenge: The Cleveland Clinic
Foundation* (Fall 1986), front cover. With permission from the
Cleveland Clinic Foundation and from Warner Bros. Entertainment Inc. *The Wizard of Oz* (1939) © Turner Entertainment Co.
A Warner Bros. Entertainment Company. All Rights Reserved.

Once researchers began looking for evidence of bypass surgery complications, reports accumulated quickly. The steady maturation of the field is demonstrated by an ambitious multicenter study published in the *New England Journal of Medicine* in 1996. The project team had familiar motivations: despite the ever-increasing number of revascularization procedures, researchers still worried that surgeons, and therefore patients, had inadequate knowledge of the risk of significant cerebral complications. Starting in September 1991, the investigators enrolled and then followed 2,108 patients at 24 surgical centers across the United States. This was the largest prospective study of the complications of bypass surgery that had yet been done. Once the results were in, researchers reported stroke, stupor, or coma in 3.1 percent of the patients and seizures or cognitive deterioration in another 3 percent. Furthermore, the risk increased steeply in older patients, from less than 2 percent in patients younger than 60, to 8 percent in patients older than 80. The study authors were deeply concerned by these results: "Adverse cerebral outcomes after coronary bypass surgery are relatively common and serious; they are associated with substantial increases in mortality, length of hospitalization, and use of intermediate- or long-term care facilities. New diagnostic and therapeutic strategies must be developed to lessen such injury."[11] They hoped that efforts to reduce the frequency of complications would reduce suffering and deaths as well as save health care resources.

By the 1990s the story of the cerebral complications of bypass surgery resembled that of the complications of valve surgery in the 1960s. Widespread awareness of the seriousness of the problem existed across the relevant specialties, from neurology and psychiatry to surgery, anesthesiology, and pathology. It generated enough interest to support international conferences, books, and special issues of relevant journals. Researchers also had more potential sources of funding. While the Cleveland team received no external funding for its initial study, both the Newcastle group and the multicenter collaboration received support from foundations interested in improving the treatment of heart disease. The National Institutes of Health began funding projects as well.[12] Increased interest, increased funding, and more substantial datasets all helped place the cerebral complications of bypass surgery firmly on the agenda of physicians and researchers from many disciplines.

As had happened with studies of valve surgery, researchers used a range of techniques—now even more sophisticated—to make the complications a polysensory reality. Refinements in Doppler technology had transformed its

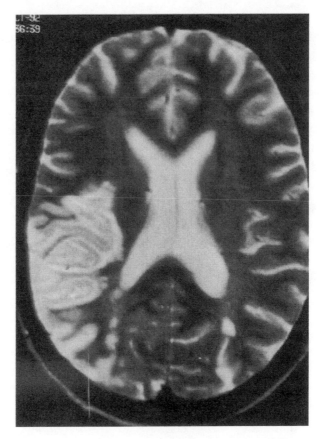

MRI evidence of brain injury after bypass surgery. A large
infarct on the left side of the brain image indicates that
the patient suffered a stroke. From Roxanne Deslauriers,
John K. Saunders, and Michael C. McIntyre, "Magnetic
Resonance Studies of the Effects of Cardiovascular
Surgery on Brain Metabolism and Function," *Journal of
Cardiothoracic and Vascular Anesthesia* 10 (Jan. 1996):
127–38, fig. 1, p. 128. With permission from Elsevier.

familiar chirping into visual traces: "Flurries of emboli are frequently detected
at the inception of bypass and cannulation of the aorta." Researchers could
watch in real time as debris flowed through the carotid arteries. As one neurologist described, "On video it looks like a snowstorm." Ophthalmologists
continued to use the retina as "a 'window' for the study of cerebral micro-
circulation." One study of sixty patients found retinal occlusions in all of them.

With ever-improving imaging techniques, radiologists could use CT scans and then MRI to visualize brain injuries. MRI findings correlated with decreased cognitive test scores, suggesting that the evidence of brain injury had functional significance. Abnormalities could also be found on MRI even in patients "without overt neurological deficits." A parallel search for chemical markers of brain injury in either blood or cerebral spinal fluid yielded some useful results, with markers of inflammation increased in patients with postoperative cognitive dysfunction. Pathologists also refined their methods. They developed special stains to detect microvascular injuries called small capillary and arteriolar dilatations, or SCADs. Careful analysis found scads of SCADs, an estimated 13.5 million in a single postoperative brain.[13]

History had repeated itself. As cardiac surgeons embarked on new procedures—valve surgery in the 1950s and 1960s, coronary artery bypass in the 1970s and 1980s—they confronted an alarming toll of neurological and psychiatric complications in their patients. They developed many techniques to visualize the brain injuries, both literally and figuratively. They could hear and see emboli, they could detect damaged tissue with MRI and blood tests, and they could examine brains under microscopes and find evidence of extensive disease even in patients who otherwise seemed well. And, of course, they could speak with and examine their patients. All the resulting evidence made cerebral complications a concrete reality for clinicians. The complications were a matter of fact. At first pass they also seemed to be a matter of concern, drawing attention and effort from a range of researchers. Despite the growing evidence base, however, inattention became common once again in the 1990s.

CHAPTER FIFTEEN

A Taxonomy of Inattention

In 1991 two of the most influential organizations devoted to heart disease, the American Heart Association and the American College of Cardiology, teamed up and released their first official guidelines about coronary artery bypass surgery. The guidelines reviewed the most accurate information then available so that physicians and patients could make the best decision possible about the procedure. They had much information to work with. Over nearly twenty five years, bypass surgery had "become the most completely studied operation in the history of surgery." Surgeons had learned which patients could expect relief of angina or even improved life expectancy. They had begun to ferret out the relative merits of bypass surgery and angioplasty. And they had finally generated substantial data about the cerebral complications of bypass surgery. However, the ACC/AHA consensus guidelines downplayed the significance of these complications: "The damaging effects of the cardiopulmonary bypass usually required for the operation result in neurobehavioral disturbances in some patients. These are sufficiently mild that they may not be apparent unless patients are tested specifically for them." What about strokes and other dire complications? The guidelines offered reassurance as well. "Gross neurological defects," such as strokes, were "considerably less common," at least in patients younger than seventy years old.[1]

The ACC/AHA consensus guidelines were not unique in this respect. Despite the efforts of many groups to characterize cerebral complications, they often remained on the periphery of the science of bypass surgery. Even as textbooks of medicine and surgery devoted increasing space to the operation from the 1970s into the 1980s, many leading texts did not mention the cerebral complications at all in the 1990s. Others mentioned them in passing.[2] The mere fact that knowledge has been produced does not ensure that the knowledge will affect medical thought and practice.

In the early 1970s a wide range of factors hindered research on the cerebral complications of bypass surgery. Focused on fixing the heart to save patients' lives, surgeons saw cerebral outcomes as secondary considerations. Meanwhile, surgeons, neurologists, and psychiatrists all struggled to adapt as the rapid expansion of bypass surgery reshaped the medical landscape. With many demands on their time, doctors did not focus on the complications in these early years. Only in the procedure's second decade did they produce substantial studies. Then in the 1990s a new dynamic came into play. Knowledge of the complications existed, but physicians often wrote about complications in a way that downplayed their significance.

Examining the circumstances in which complications did and did not receive close scrutiny can help to identify the many professional, institutional, methodological, psychological, and economic factors that complicate medical knowledge. A close reading of the cardiac surgery literature reveals the different arguments used to explain away or minimize the problem of cerebral complications. The arguments generally fell into one of five categories: physicians described cerebral complications as a recently recognized problem, they emphasized their continuing uncertainty about the complications' significance, they described the resilience of the brain in the face of injury, they minimized the consequences of the damage, and they continued to argue that the benefits of the surgery justified its risks.

First, many cardiac researchers in the 1990s and into the next decade foreshortened the history of the problem they studied. They described the cerebral complications of bypass surgery as only recently emerging as a problem. Specifically, they argued that cerebral complications became relevant only after operative mortality and cardiac complications decreased dramatically.

As one expert wrote in 1993, "As mortality falls, attention inevitably turns to morbidity." Even if the bypass grafts worked, an operation "cannot nowadays truly be described as 'successful' if the patient sustains cerebral injury during the operation, sufficient to modify the patient's personality or intellectual performance postoperatively." Another researcher felt that it was a luxury that surgeons could even worry about subtle problems like cognitive dysfunction: "It is a testimony to the dedication of the cardiovascular research and clinical community that death and stroke are so uncommon, tests of subtle higher cortical function are necessary to evaluate new interventions." Many writers in the 1990s asserted that since the cerebral complications had only been recognized recently, "their incidence and effects have not been rigorously investigated."[3] These narratives were grounded in real developments. Operative and anesthetic techniques had improved, and mortality had dropped. But they obscured the long history of awareness of the problem: the risks of open-heart surgery had been characterized in the 1960s, and the specific risks of bypass surgery had been rigorously investigated in the 1980s.

The mischaracterizations of this history may have offered solace to physicians. It might have been easier for clinicians and researchers to forget the extensive prior study of the complications than to acknowledge that they had been aware of them for decades and had been unable to resolve them. The few physicians who did acknowledge the long history of complication research noted an intriguing puzzle. From the 1970s into the 1990s, researchers had used different diagnostic techniques and produced different estimates of the prevalence of the complications. Recall the low estimates of the initial case series, the higher estimates of the Cleveland prospective studies, and the highest estimates of the Newcastle group. One number, however, often resurfaced across the literature: a remarkably stable estimate of a 2 to 3 percent risk of stroke. Some experts interpreted this stability as a "just-so" story of perfect balance. Surgical techniques had improved, which would have ordinarily lowered complication rates. But the increasing safety of the procedure had enabled surgeons to operate on older patients, and these patients carried an increased risk of adverse outcomes. According to this narrative, the two trends combined to produce stable rates over time.[4] The stability, however, could also reflect another kind of equilibrium effect. Researchers had known since the 1960s that their estimates of the incidence of complications depended on how assiduously they examined their patients. Given this, the stable stroke rate might mean that researchers looked just hard enough to find the expected rate.

Second, despite decades of accumulated data, physicians often emphasized their continuing uncertainty about the incidence and severity of the complications. Researchers throughout the 1990s complained, as if repeating a mantra, that the incidence of complications remained poorly defined: "Even after 30 years and more than 50 studies addressing in some way the issues of incidence of and risk factors for delirium after cardiac surgery, these questions remain unanswered." Another ten years did not help: "Despite >40 years of clinical experience, it remains unclear whether CABG [bypass surgery] has an impact on basic mental faculties such as problem solving and memory." Researchers often traced this continuing uncertainty to problems with their sources of information. One group found that patients were unreliable witnesses to their own cognitive function: "Patients who considered that aspects of their cognitive function had deteriorated, were not found to have reduced functions as assessed on appropriate neuropsychological tests." Neuropsychological testing was also imperfect. With estimates of cognitive dysfunction ranging between zero and 79 percent, experts in 1995 complained that "even today there is no consistency in the methods, the statistical analysis, and the results of NP [neuropsychological] test batteries used by various research groups."[5]

Had such protestations come only from surgeons, a suspicious mind might wonder whether they emphasized the uncertainty as part of an effort to reassure patients and convince them to accept surgery. But anxieties about the uncertainty and the research methods that generated it were also pervasive among neurologists and psychologists, who had no obvious motive for minimizing the problem.[6] Quite the contrary, they had a motivation to increase attention paid to the problem, because this would increase the relevance of their expertise. Given this, the rhetoric of uncertainty may have appropriately reflected the complexity of the problem of cerebral complications. The brain and its functions could not be reduced to simple laboratory tests or imaging studies. Instead, complicated studies, especially the neuropsychological tests, were required, and these inevitably left room for interpretation and re-interpretation. Regardless of its appropriateness, the continuing refrain of uncertainty could have contributed to an impression among nonexperts, patients, and the media that the complications were less real or worrisome than they would otherwise have seemed.

Third, even when researchers did acknowledge the concrete reality of heart surgery's risk to the brain, they often noted the remarkable ability of the brain to withstand and recover from damage. The team that developed ultrasound was surprised by how many air bubbles they could detect during surgery as well as by the resilience of the brain to this assault: "The tolerance to gas embolization in the various tissues is unknown, but from our experiences it is clear that overt central nervous system complications may not develop even when thousands of micro-emboli are accumulatively directed to the cerebral circulation." Cleveland Clinic surgeons described how patients could have good neurological outcomes even when Doppler detected thousands of intra-operative emboli: "Apparently the human body has a tolerance for microparticles and in most instances can withstand a persistent bombardment." One London anesthesiologist was reassured by the complete recovery of most patients, which made him wonder, "Does this mean that there is little or no residue of dead neurons—or do we have more brain tissue than we need for normal function?" Brain damage became evident only if it occurred in an "eloquent brain area" that controlled some vital function.[7] The implication here was clear: yes, the brain might face injury, but it was a remarkably resilient organ, and we don't need all of it anyway.

Fourth, when reports did document complications, they often emphasized the transience of the deficits or downplayed their effect on actual activities of daily life. This can be seen in the writing of the Newcastle researchers, who had documented some of the highest complication rates. Although they found neurological deficits in 191 of their 312 patients, only 48 had trouble managing daily activities. At six-month follow-up, 85 patients still had "detectable neurological signs, but these were often minor and of little functional importance." The researchers determined that only 10 had actual neurological disabilities. Similarly, although cognitive testing revealed dysfunction in 235 patients, only 89 complained about their impairment, and only 23 were "overtly disabled by their intellectual dysfunction." The neuropsychological tests, which yielded the highest rates of postoperative deficits, were easiest to dismiss. As a prominent Oxford surgeon argued, "The functional significance of cognitive impairment is uncertain, because many activities of daily life do not require the level of performance called for during neuropsychological testing."[8] If a tree falls in the forest, but no one hears, did it make a sound? Here the clinicians did not deny that detectable brain damage had occurred. Instead, they argued that most of the damage did not matter.

Finally, even when they acknowledged the risk as significant and the damage as real, surgeons and referring physicians continued to argue that the benefits of surgery justified the risks. Replaying the claims made about valve surgery in the 1960s, they argued that brain damage was a fair price to pay for the cardiac benefit of bypass surgery. As one psychiatrist wrote in 1993, "Cardiac surgery may sometimes compromise brain functioning even though it is necessary to preserve life and enhance cardiac functioning." This faith in the benefits of the surgery spilled over into media discussions of bypass. In 2001 *Time* magazine described a Duke University study that found cognitive impairment in 42 percent of bypass patients at five-year follow-up. The authors of the study hoped that the results would not "scare anybody away from getting necessary surgery." Since some of the patients likely would not have survived five years without the operation, "a little trouble with memory or attention might seem worth it." A sidebar question-and-answer was even starker. Why would anyone choose bypass surgery and its dangers? As *Time* explained, "It's better than dying."[9] The rhetoric of a compromise between heart and brain remained alive and well fifty years into the history of heart-lung machines.

As doctors moved back and forth between documenting the complications and downplaying their significance, an old predilection reemerged. Just as psychoanalysts in the 1960s and 1970s had traced delirium to their patients' unresolved psychic conflicts, researchers in the 1990s and beyond often highlighted the importance and culpability of patients' intrinsic characteristics.

The basic puzzle was well known. Cerebral complications affected some but not all patients. Serious disability occurred in only a minority. What was the source of variation in the outcomes? Doctors wanted to know whether the disparate outcomes arose from inconsistencies in surgical and anesthetic techniques or from differences between patients. Techniques of multivariate analysis, adapted for medicine in the 1970s, made it possible to answer this question. Sophisticated statistical algorithms allowed researchers to examine dozens, even hundreds, of variables in a research study and determine which subset of them made the most significant contributions to the outcomes. These methods provided yet another way of visualizing the problem of cerebral complications, revealing not the consequences of the complications but their possible causes.

For instance, as part of its prospective study of bypass surgery, the Cleveland

Clinic team recorded and analyzed 451 characteristics of each patient and bypass procedure. It found that several characteristics were significant predictors of cerebral complications: length of operation, oxygenator type, patient age, and comorbid disease. Each of these made sense given what doctors knew about the interaction between the heart-lung machines and patients' brains. Hopkins researchers determined that 74 percent of the risk could be attributed to patient characteristics, specifically previous stroke, hypertension, carotid artery disease, age, and diabetes. A Pennsylvania group made an even bolder prediction. Patients with four risk factors—mural thrombus, diabetes, tobacco use, and aortic calcifications—had a 93 percent risk of sustaining a permanent neurological deficit.[10] Most of the risk resided in specific attributes of each individual patient.

Genetic studies soon reported something similar. It should come as no surprise that in the 1990s—the decade of the Human Genome Project—surgeons and neurologists turned to genetics to understand the distribution of risk in their patients. This decade saw unprecedented interest and faith in genetics in American science and society. Researchers in nearly every field searched for genetic correlates of whatever they studied. Cardiac surgeons and neuroscientists were no different. They turned to genetics to explain why only some patients suffered cerebral complications of cardiac surgery. A Duke study looked at a gene associated with risk of Alzheimer's disease (Apo E-ε4) and found a significant association with postoperative cognitive dysfunction, especially in poorly educated patients. They speculated about the possible basis of this association: "Cardiac surgical patients may be susceptible to deterioration after physiologic stress as a result of impaired genetically determined neuronal mechanisms of maintenance and repair." Another study used a state-of-the-art gene chip to test forty thousand genetic markers in forty-two patients. Finding differences in gene expression between patients who did and did not have cognitive decline, they concluded, "Patients with neurocognitive decline have inherently different genetic responses to cardiopulmonary bypass compared with those of patients without neurocognitive decline."[11]

The implication of both sorts of studies seems clear. Although most patients tolerated open-heart surgery without any trouble, some vulnerable few suffered complications. The problem, in this light, was not the surgery but the patients who needed it. A group of neurologists and surgeons at Johns Hopkins, who published many important studies of the problem, captured this sentiment: "From a neurological standpoint, cardiac surgery as currently practiced is a safe and effective procedure for the great majority of patients. Nonetheless, a subset of patients with preexisting risk factors for cerebrovascular disease are at high risk for stroke or encephalopathy."[12] The problem could be an inherent genetic liability of patients, it could be their age, or it could be whatever comorbid conditions patients had happened to acquire. Some of the factors, such as aortic calcifications, increased the risk of brain injury. Other factors, such as genetic markers of inflammation and injury response, influenced how an individual patient's brain coped with injury. In either case, the problem was intrinsic to the patient. In the absence of these individual liabilities, the surgery would have been safe.

The efforts by physicians to pinpoint the source of surgical risk raise once again the question of responsibility. In the 1960s, amid abundant evidence of the brain pathology produced by heart-lung machines, psychiatrists had found space for their discussions of deliriogenic personalities. In the 1990s the surgical teams could see embolic debris flowing through carotid arteries en route to the brain and then see the resultant infarct on a postoperative MRI. But this did not mean that the surgery (and its surgeon, anesthetist, and perfusionist) was solely responsible for the complications. Clinicians can look in several directions when they set out to explain treatment complications. A procedure could be inherently flawed: any attempt to use a bubble oxygenator without a defoaming agent, for example, would inevitably ravage the brain and other organs. Or a procedure could be sound in principle but happen to go wrong in a particular case: a mistake, or even just bad luck, can undermine even the best designed procedures. Or a procedure could be sound and flawlessly executed, only to be foiled by some peculiarity of a high-risk patient.

The pursuit of therapeutic progress requires that physicians ask these questions. Risk factor analysis enables doctors to understand the pathophysiology of treatment complications and design strategies for improving outcomes. It reflects the recognition of idiosyncrasy in medical practice: all patients are different. Multivariate analysis helps doctors to ferret out which differences might be the most important determinants of treatment response. It enables them to provide patients with a nuanced prognosis and with the individualized risk assessment needed for informed consent. The intent is not to blame but to understand. But blame may be an inevitable and unfortunate consequence. Multivariate analyses and genetic studies each suggest that the problem is not what surgery does to patients, but what liabilities patients bring into the operating theater. Physicians must consciously ensure that such displacement of risk and responsibility onto patients never undermines the assiduity with which they work to optimize the safety of their therapies.

CHAPTER SIXTEEN

Competition's Complications

Imagine a conversation that takes place a thousand times each day in this country. A cardiac surgeon recommends that a patient undergo coronary artery bypass surgery. The patient asks if there will be any complications. The clinical evidence supports a wide range of possible answers. One surgeon could say, "Yes, we have to put you on a heart-lung machine; this is an imperfect substitute for the heart and introduces the risk of stroke, cognitive decline, and personality change." Another surgeon could say, "Long-term cognitive outcomes are just as good in patients who have surgery as in those who do not, and sometimes even better." The fact that each claim is well supported by recent research on the cerebral complications of bypass surgery poses a dilemma for patients and doctors. It also presents a key puzzle for scholars interested in medical knowledge and a critical challenge for those who stake their hopes in evidence-based medicine.

In previous chapters, I have explored why treatment complications sometimes do not come to the attention of clinicians and researchers. Here, the problem is different. Beginning in the mid-1990s, the cerebral complications of bypass surgery received unprecedented attention in certain arenas. Motivation for this effort came from therapeutic competition. As angioplasty, especially with stents, surged past bypass surgery in the 1990s, one of its selling points was its lower rate of cerebral complications. Surgeons, working to overcome this liability, introduced several new forms of minimally invasive bypass surgery and claimed that each reduced the risk of stroke. Yet even as competing researchers shined a spotlight on cerebral complications, they could not produce decisive evidence of the prevalence or severity of the complications. Their efforts, thus far, have simply demonstrated how difficult it can be to produce definitive knowledge. Even after fifty years of bypass surgery, much remains uncertain about its impact on the brain. Who is responsible for remedying this situation? That is a vital question for medical knowledge and practice.

From the 1960s into the 1990s, the cerebral complications of bypass surgery often ended up on the sidelines of the medical literature. There was one setting, however, in which they came to play a prominent role: competition between different techniques of coronary revascularization. As bypass surgery proliferated in the 1970s, it opened up a vast niche in medicine—surgical treatment of coronary artery disease. The procedure grew from its origins in the late 1960s to 100,000 cases per year by 1977, 205,000 by 1984, 407,000 by 1991, and finally peaking at 607,000 in 1997.[1] Even amid this dramatic growth, physicians maintained expansive ideas about the promise of this niche. They believed that bypass surgery still reached only a small share of the potential market for revascularization. After all, millions of patients received new diagnoses of coronary artery disease each year. Competition—coronary angioplasty—soon appeared.

When angioplasty first emerged, cardiologists' ambitions were modest. Andreas Grüntzig anticipated in 1979 that it would be suitable for only 10 to 15 percent of the candidates for bypass surgery. This modesty soon disappeared. As cardiologists gained confidence, they expanded their conception of whom angioplasty could help, from patients with stable single vessel disease to those with multiple lesions and acute syndromes, especially patients in the throes of a heart attack. The growth of angioplasty was even steeper than that of bypass surgery: 133,000 procedures in 1986, 227,000 in 1988, 428,000 in 1994, and finally exceeding bypass surgery in 1996 with 666,000 procedures.[2]

Angioplasty fueled fierce debate about whether medications, angioplasty, or bypass surgery provided the best treatment for people with coronary artery disease. Comparative trials, which always focused on mortality and long-term survival, typically showed comparable survival rates for angioplasty and bypass surgery in most patients, especially those with mild to moderate disease. This allowed patients and doctors to consider other factors as they made their decisions. Cost, convenience, and the likelihood of requiring a subsequent procedure became especially important. Here the advantages of angioplasty seemed self-evident. Cardiologists did not have to saw open the patient's rib cage to reach the heart. No heart-lung machine was required. Cardiologists and patients assumed, with good reason, that risk was lower and recovery would be easier. When Grüntzig reported his initial outcomes in 1979, he described excellent results with few complications: "In the 50 attempts and three repeated dilatations, there were no deaths, no evidence of embolization, no central-nervous-system deficits and one femoral hematoma requiring evacuation."[3]

The rise of angioplasty in the 1980s coincided with the emergence of the large prospective trials of the cerebral complications of bypass surgery. Each development was certainly a response to the dramatic growth of coronary artery surgery. As described earlier, neurologists and psychiatrists responded to the rise of bypass surgery by subjecting the popular operation to close scrutiny. Cardiologists responded by developing a competing therapy. Cardiologists likely benefited from the neurologists' work as well. Whether neurologists reported a stroke rate of 5.2 percent (the Cleveland studies) or neurological deficits in 61 percent of their patients (the Newcastle studies), angioplasty had a competitive advantage as long as people assumed it was safer.

The rise of angioplasty, like bypass surgery before it, quickly triggered calls for close scrutiny of its risks and benefits. As early as 1979, some of the same skeptics who had demanded better data about bypass surgery in the early 1970s demanded better data about angioplasty. They called for more careful study of the procedure to learn "its range of mishaps and side effects." As larger studies—initially just case series—steadily appeared, they usually affirmed the safety of the procedure. The National Heart, Lung, and Blood Institute, for instance, established a registry to monitor angioplasty outcomes. One report on 1,500 patients found complications in 314, but only five (0.3%) of these were neurological: one patient with a stroke, three with transient ischemic attacks, and one with delirium. Cleveland Clinic neurologists, noting that the prevalence of "neurologic events during [angioplasty] has been virtually ignored," offered an analysis of the clinic's 1968 angioplasty patients. They found only four strokes (0.2%): two from plaque emboli after the inside of the aorta was scraped with the catheter tip, one from an air bubble inadvertently injected through the catheter, and one from a period of postprocedure low blood pressure.[4] Such studies reassured cardiologists. However, all the studies relied on retrospective analysis of hospital charts, exactly the kind of data long known to generate underestimates of the incidence of cerebral complications after cardiac surgery.

In the 1990s the results of large randomized trials comparing angioplasty and bypass surgery began to appear. The reports, like the original trials of bypass surgery in the 1970s and 1980s, focused on survival, relief of angina, and exercise capacity. Most did mention stroke, but only in passing. None looked at subtler neurological or psychiatric changes. The results were surprising. Even though strokes occurred more frequently in the surgical patients, the differences were not large enough to be statistically significant: 1.5 versus 0.5 percent in one trial, 1.2 versus 0 percent in another, and 0.8 versus 0.2 percent in a third. Only when researchers compiled a meta-analysis of 23 trials with 5,019 patients did the difference—1.2 versus 0.6 percent—achieve statistical significance. These studies were prospective, which meant they overcame one limit of the first generation of reports. However, the fact that the stroke rates were lower in the surgical patients in these studies than they were in studies dedicated to the cerebral complications of bypass surgery (e.g., 3.0% in the 1996 multicenter study) suggests that the researchers in the surgery-angioplasty comparisons were detecting only the most serious injuries or were enrolling lower risk (e.g., younger) patients.[5]

Even though the results of these studies were mixed—low stroke rates after bypass surgery, and only slightly lower rates after angioplasty—the intuitive benefit of angioplasty won out. Surgeons and patients saw the convenience and lower complications as a major advantage. As a cardiac surgeon and anesthetist wrote in 2007, cognitive decline after bypass surgery "has been perceived as a key factor that has contributed to the shift to percutaneous intervention for coronary disease."[6] When it came to competition, complications mattered.

Surgeons did not go down without a fight. Denton Cooley pushed surgeons to innovate. Specifically, he encouraged them to reconsider their long-standing reliance on heart-lung machines: "We have gotten so comfortable now in wanting to do all of our coronary bypasses in a nice quiet operating field, overlooking the fact that the patient has to go through some trauma so that we can do our operation in a quiet operating field. We need to improve our skills so that we can do some of these operations with the heart beating."[7] Concerned that heart-lung machines introduced most of the risk in bypass surgery, surgeons revived the old technique of beating-heart bypass or, as it has come to be called, off-pump coronary artery bypass.

René Favaloro had done his initial bypass surgeries in 1967 and 1968 without a heart-lung machine. This had been possible because he had initially targeted the right coronary artery, which is accessible to surgeons along the top surface of the heart. Use of cardiopulmonary bypass became routine when surgeons began operating on left coronary arteries, which run along the posterior surface of the heart and are much harder for surgeons to reach. Some groups did continue to dabble with beating-heart surgery. Surgeons at Cooley's Texas Heart Institute performed 191 off-pump procedures between 1969 and 1983, but only achieved satisfying results on the right coronary. A Sao Paolo team, however, made concerted efforts to develop the technique in the 1980s. The surgeons cited both their desire to reduce the cerebral complications and the competition from angioplasty as motivation. In a series of 593 patients, with most of the grafts done on the left coronary artery, they found neurological complications in only 1.1 percent of the patients, compared with 3.8 percent in patients with traditional bypass surgery. When they presented their results at a European cardiac surgery meeting in Munich, one English surgeon—an expert on the cerebral complications of bypass surgery—initially responded with "amazement and incredulity." However, careful study of the group's meticulous write-up convinced him that the excellent results really "can be achieved."[8]

Under increasing pressure from angioplasty, more and more American surgeons turned to off-pump coronary artery surgery or another variant, minimally invasive coronary artery bypass. In 1997 the Food and Drug Administration approved the Octopus, a device that used suction cups to immobilize a small area of the beating heart's wall. This provided the stabilization needed for careful suturing. Surgeons adopted off-pump surgery enthusiastically. They especially hoped that this approach would solve bypass surgery's stroke problem. As one Ontario team suggested, "With beating heart surgery, avoidance of the embolic potential associated with aortic cannulation and decanulation, and of the generation of microgaseous and microparticulate emboli from the pump circuitry, would be expected to significantly decrease cerebral embolic load and improve outcomes." Early results supported these hopes. One study looked closely at Doppler emboli counts (3954.5 vs. 11), markers of neuronal injury (3.76 vs. 0.13), and incidence of cognitive impairment (90% vs. 0%) to make the case for off-pump surgery: "This supports the hypothesis that both microemboli and cognitive impairment are strongly related to CPB [cardio-pulmonary bypass]. With this in mind, the advantages of off-pump operations on neurocogntive outcome appear to be promising."[9]

Surgeons' hopes reached into popular periodicals. The *New Yorker*, for instance, published a glowing review of the technique's promise—"Heart Surgery, Unplugged." In this celebration of surgical bravado, Jerome Groopman described how surgeons now did "something that until recently would have been inconceivable—operate on a beating heart." The essay's narrative created a sense of daring, radical innovation. Heart surgeon Billy Cohn, who had spent most of his college years playing bass guitar in a punk rock band, brought the procedure to Groopman's hospital in Boston, using a stabilizer that he had first designed in his basement out of soup spoons purchased from Stop & Shop. Why attempt beating-heart bypass surgery? Cohn emphasized pump head, the cognitive complications caused by the heart-lung machine. A colleague explained, "These guys come up after beating-heart surgery as clear as a bell." Full of hope for both the new era of bypass surgery and his Cohn Cardiac Stabilizer (manufactured by Genzyme Surgical Products), Cohn described traditional bypass surgery with its heart-lung machine as "simply barbaric."[10] It is surely no coincidence that this early occurrence of the phrase "pump head" came as part of an effort to distinguish off-pump coronary artery bypass from its predecessor.

A Turkish team took innovation even further and performed beating-heart surgery on awake patients to remove the risk of general anesthesia. The need to compete was explicit: "Such a strategy may ultimately combine the patient comfort of percutaneous revascularization procedures with the advantages of coronary artery bypass grafting with arterial grafts." They were pleasantly surprised by how well their conscious patients tolerated the rigor of open-heart surgery, especially the noise and sensations of having their rib cage sawed open: "No patient in our series was converted to general anesthesia because of intolerance to median sternotomy in the awake setting." Most American and European surgeons, however, thought this went too far. One Stanford anesthesiologist was pointed in her dismissal: "There is no place for this trick in the cardiac anesthesiologist's armamentarium."[11]

Cardiologists and cardiac surgeons were certainly not the first physicians to use complication rates as a means of distinguishing new treatments in the marketplace. Physicians, pharmaceutical companies, and medical devices companies do this often, especially when the new treatments are more expensive than existing options and offer little benefit in terms of effectiveness. Advocates argued that rofecoxib (Vioxx) would cause less stomach irritation than aspirin and ibuprofen, that atypical antipsychotics would avoid the movement disorders seen with haloperidol (Haldol) and chlorpromazine (Thorazine), and that designer drugs in oncology would provide all the benefits of conventional chemotherapy but without the terrible side effects. The problem with such logic is that results often do not turn out as expected. Rofecoxib, which unexpectedly increased patients' risk of heart attacks, is one notorious example. Atypical antipsychotics, although safer in terms of movement disorders, may cause diabetes instead.[12] This pattern of hope and foiled expectations is a common problem with therapeutic innovation.

Similar surprises turned up in the competition between bypass surgery, off-pump bypass, and angioplasty. When researchers looked at six-year outcomes in one of the randomized trials of bypass surgery and angioplasty (the BARI trial), they found no difference in cognitive function: "Although there are significant perioperative effects of CABG on cognitive function, the long-term effects appear to be either relatively small or relatively uncommon, or both." A comparison of on- and off-pump bypass surgery found similar patterns of cognitive decline and recovery. The researchers concluded, "Cardiopulmonary bypass is not the major cause of postoperative cognitive impairment."[13] Even though early case series had favored both angioplasty and off-pump over traditional bypass surgery, longer studies did not confirm differences in cognitive function. Expectations of relative safety, both of angioplasty versus surgery and of off- versus on-pump surgery, went unfulfilled.

The situation has been equally complex for stroke. A long-awaited trial of bypass surgery and angioplasty published in 2009 favored the less-invasive technique, with a stroke rate of 2.2 percent for bypass and only 0.6 percent for angioplasty. Results with off-pump surgery remain less clear. One large study in 2009 actually found higher stroke rates with off-pump surgery compared with traditional bypass surgery—1.3 versus 0.7 percent—but the difference was not statistically significant. The researchers did not offer a clear explanation for why this had happened. They did, of course, call for further research: "Future studies with neuropsychological control groups are warranted to evaluate the frequency of cognitive impairment or improvement over time." This presents a real puzzle. Heart-lung machines introduce all sorts of neurological risks. Decades of research have documented this well. Avoiding them should reduce the incidence of brain damage. But clinical trials have not shown this consistently. Is there some flaw in the studies? Is there an error in our understanding of the surgical procedures and the possible mechanisms of injury? The Hopkins researchers offered another possibility: the lack of difference between on- and off-pump surgery suggests that "patient-related risk factors, such as the extent of preexisting cerebrovascular and systemic vascular disease, have a greater effect on both short- and long-term neurologic sequelae than do procedural variables."[14] After decades of evidence of the dangers of cardiac surgery and heart-lung machines, responsibility has once again shifted back onto the patients.

Despite fifty years of information about cerebral complications of cardiac surgery, surgeons and patients face lingering uncertainty. One set of recent studies underscores the seriousness of the problem. In 2001, for instance, a Duke study reported a 42 percent incidence of long-term cognitive decline five years after bypass surgery. The authors condemned those who dismissed the deficits as subtle, transient, or subclinical, because "such descriptions minimize the importance of these changes to clinicians, patients, and their families." Other studies found a correlation between emboli counts and cognitive outcomes, which suggests a causal link between the two. Magnetic resonance imaging revealed subclinical brain injury. If obvious signs are the "tip of the iceberg," then MRI could show "the lesion load below the waterline." This was true not only in bypass surgery patients, in whom imaging techniques had long demonstrated evidence of postoperative injury, but in coronary angiography patients as well. MRI revealed lesions in as many as 15 percent of patients after the procedure. Studies from elite institutions, meanwhile, continued to find high rates of stroke and delirium after bypass surgery. Hopkins researchers reported rates of 4.5 percent (stroke) and 13.8 percent (delirium) in 2004, higher than those of other studies conducted since 2000. Patients who suffered stroke or delirium had lower long-term survival rates than bypass surgery patients who had been spared those complications. Consensus guidelines from the American College of Cardiology and the American Heart

Association devote increased attention to the problem, as do textbooks of medicine and surgery.[15]

One former president of the American College of Surgery, C. Rollins Hanlon, captured the current state of concern well in a 2000 interview: "When you put somebody my age or even younger on bypass for forty-five minutes, all sorts of things happen to the neurological system. It's not anything gross, such as an inability to do serial sevens backwards. We don't know if it's fat or bubbles, but whatever it is, people sometimes come out of that operation not exactly the same person that they were before. It's becoming increasingly apparent that the idea that cardiopulmonary bypass is a benign procedure, either to the heart or the head, is false. It is not benign." This phenomenon, of course, had been known for decades. Testimonials by cardiac surgery patients also painted a grim portrait of what could go wrong. Bruce Stutz, onetime editor of *Natural History* magazine, faced a "long, dumbfounding struggle against what seemed to be the sudden onset, at 51 years of age, of attention-deficit disorder or incipient senility," the result of valve surgery on a heart-lung machine.[16] When he published an account of his struggles, in *Scientific American*, he drew considerable attention to the problem.

Other good evidence, however, supported an opposite conclusion. One study examined fifty-two patients who carefully controlled their vascular risk factors after bypass surgery. Many quit smoking, took statins, and fastidiously managed their high blood pressure and diabetes. The patients showed no decline in cognitive function after five years. A study of twins, from the National Academy of Sciences, National Research Council Twin Registry of World War II veterans, found that men who had undergone bypass surgery had lower rates of dementia than their twins, who had not, possibly because improved postoperative cardiac function led to improved blood flow to the brain. One New York surgeon even described postoperative cognitive improvements in his patients. As he told the *New York Times*, "Many will tell you they're much more creative and prolific, that they haven't suffered any cognitive decline whatsoever."[17] I have personally seen this in one of my colleagues. He had experienced several years of cognitive decline, which he had attributed to aging (he was in his early 80s). When a routine checkup found evidence of severe coronary artery disease, he had bypass surgery. Once he had recovered, he found to his delight that his energy and cognitive function had both improved to where they had been years previously.

The most rigorous data to date have come from a study started at Johns

Hopkins in 1997. Researchers followed two groups of people with coronary artery disease: one group had bypass surgery and another did not. They found no differences in cognitive function at three months, twelve months, three years, or six years. Patients in both groups had declined to a similar extent over the years, presumably because all shared many of the risk factors for cerebrovascular disease. The conclusion was clear: "Late cognitive decline after CABG is not specific to the use of cardiopulmonary bypass." Instead, the sorts of people who need bypass surgery—people with advanced vascular disease—are likely to experience cognitive decline over time whether they have surgery or not. Moreover, cognitive outcomes were not different in these patients whether they were treated with medications or surgery. The Hopkins group has even sought to rebrand the complications not as "brain damage" but simply as "cognitive change."[18]

Uncertainty over the magnitude of the problem occasionally receives widespread attention. When Bill Clinton had bypass surgery in 2004, newspapers described the confusing state of affairs, with some surgeons worrying about pump head while others "still pooh-pooh it." When Clinton behaved erratically on the campaign trail for Hillary Clinton in 2008, journalists freely speculated that bypass surgery had caused his deterioration. Clinton's office offered a fierce rebuttal.[19] Has Clinton experienced cognitive decline since 2004? If he has, could it be traced to his bypass surgery? Both questions should have clear answers. The presence of cognitive decline, and its causes, should both be simple matters of fact. But for many reasons they are not. Even if Clinton submitted to testing now, it would be impossible to demonstrate decline unless he had been tested before surgery. Careful surveillance of brain function is still not a routine part of cardiac surgery, despite knowledge of the surgery's dangers. Even if pre- and postoperative tests existed, controversy would persist about the best way of interpreting them. And even if decline were detected, it would be difficult to pinpoint its cause.

The existence of conflicting, credible data leaves patients and physicians in a difficult situation. When patients, needing surgery, ask about their prospects for future cognitive function, physicians can offer widely different estimates, each evidence based. The risk of subtler problems, such as personality change, remains even less clear. Most of the research has focused on stroke, but even with stroke many questions remain. Consensus exists for three reasonable claims: risk of intra- or postoperative stroke is inherent to major surgery, it may be increased by heart-lung machines, and it is a liability of atherosclerotic brains. But beyond those points, the consensus breaks down. Although logic suggests that the risk of stroke should vary substantially between angioplasty, bypass surgery, and off-pump operations, research findings remain inconclusive. Certainly the risk seems low: a 2 to 3 percent chance of a significant stroke may be a fair price to pay for cardiac rehabilitation. But the wisdom of that trade-off depends on many factors, especially on a patient-specific estimation of the risk of brain injury and on a realistic prediction of the individual's likely benefit from the surgery. Patients might be willing to risk some cognitive function to extend their lives, but what if the procedure only relieves angina? The likelihood of these desired outcomes, as described in the first part of this book, often remains uncertain, undermining the ability of patients and doctors to make thorough assessments of the benefits and risks.

Fortunately, there are many ways to improve the situation. For example, the Society of Thoracic Surgeons, as part of its effort to improve the quality of cardiac surgery, monitors surgeons' outcomes, including their rate of peri- or postoperative strokes. Despite the lingering uncertainty about the frequency and severity of cerebral complications, some surgical teams have decided that enough is known to justify designing interventions that might reduce them. One collaborative group in northern New England recommended careful study of the links "between processes of clinical care and sources of emboli," and then "redesign of clinical care to reduce and prevent their occurrence." Simple measures, however, are often omitted. Louis Caplan, a prominent Boston neurologist, worries that basic precautions, such as taking careful neurological histories or screening for carotid and aortic atherosclerosis, are not done consistently. Doing so would "save brain tissue and lives by reducing the frequency and severity of adverse neurological outcomes, and may even save some money." But careful preoperative screening of neurological risk has not yet become the standard of care.[20] Why haven't all possible measures already been implemented? The reason must reflect some mix of cost, complexity, and trade-off between benefit and risk. It could also be that physicians remain unconvinced about the significance of the problem.

The challenges involve not just clinical care but knowledge production itself. Whenever medical innovations appear, there is much to learn about safety and efficacy. This effort faces substantial epistemological challenges. When researchers undertake a study, they decide in advance what outcomes to record. These decisions are based on their expectations of how the treatment will interact with their patients. According to the highly refined rules of clinical trials, analysis of such predefined endpoints yields the most reliable answers. In addition, researchers are taught to be skeptical of post hoc analyses, of efforts to mine the data for outcomes that had not been defined at the outset. These policies exist for good reason, but they create a problem. Can researchers anticipate and study every possible complication? This is often not feasible.

It is easy for doctors to define in advance what they hope a treatment will accomplish. They can then look carefully at clinical outcomes or more proximate measures (so-called surrogate markers) to see if the desired outcome was achieved. No similar strategy exists for generating comparable knowledge about treatment complications. Medical interventions, whether pharmaceutical or surgical, are notorious for producing unforeseen side effects. Some potential complications can be predicted based on knowledge of the mechanism of therapeutic action, but many others are unanticipated. Sometimes unexpected complications are recognized within research studies, but not always. Even systems of postmarketing surveillance can miss evidence of substantial complications.

Even when complications are expected, methodological, economic, and social obstacles intervene. Sometimes complications continue to receive inadequate attention. A particularly telling illustration of this emerged recently. In the 1990s intensive care units began to use extracorporeal membrane oxygenation (ECMO) to save the lives of patients suffering from reversible, but dire, injuries to their hearts and lungs. The basic idea is similar to heart-lung machines: the patient is connected to an external pump and oxygenator so that a constant flow of oxygenated blood is maintained until the patient's lungs recover. Even as the use of ECMO increased, researchers paid little attention to its neurological complications. As recently as 2011, and despite rates of brain death as high as 21 percent in some case series, Mayo neurologists wrote, "Neurological consequences of ECMO in adults are likely common but uncharacterized." Their own study of eighty-seven patients found neurological events in half of them, with evidence of hypoxic or hemorrhagic injury in nine of ten brains examined at autopsy.[21] The close parallels of this history to that of open-heart surgery raise many questions. Were the complications unexpected? Given the history of heart-lung machines, this seems unlikely. Were the complications seen as unimportant, since ECMO was used only as a last-ditch effort to save dying patients? Possibly. Careful study is needed to understand the full dynamics of this case.

There is now intriguing evidence of how inconsistently researchers ascertain the complications of medical treatments. One group at Harvard compared many different drug trials, looking specifically at the side effects and complications experienced by patients who had been given placebos. Assuming that one study's placebo is no more toxic than another study's placebo, complication rates in the placebo groups of different studies should be roughly similar. The researchers found something different. The rates of side effects in patients randomly assigned to placebos vary substantially between different studies. The conclusion seemed inescapable: the differences must reflect differences between the research studies in how patients and doctors discerned and reported the side effects.[22]

The methodological obstacles to studying treatment complications and the variable attention that researchers give to the problem create a fundamental challenge for medical decision making. Clinicians, researchers, and drug and device companies all pursue evidence of therapeutic efficacy with great enthusiasm. They are not nearly as tenacious in their efforts to document treatment complications. This creates a marked asymmetry in medical knowledge in which there is more evidence about efficacy than about risk. When patients and doctors make decisions based on unbalanced knowledge, they inevitably lean toward intervention.

Who or what is responsible for treatment complications? Who has responsibility for ensuring that adequate knowledge about them is produced and shared? Who has the obligation to work to eliminate them, and how? These questions must be answered so that physicians can take on appropriate obligations and patients can have realistic expectations. Many actions could be taken toward this goal. It is certainly not possible for researchers to measure every possible outcome, but it might be possible for researchers to reorient their studies to capture more thoroughly the effects of treatments on patients' lives and bodies. Instead of focusing attention on a few predefined endpoints and surrogate markers, they could cast a broader net and ask patients specifically about a wide range of adverse outcomes. This approach would increase the work for clinical researchers, but the payoff could be huge: better knowledge of the safety of medical treatments.

The quality of medical knowledge cannot solely be the responsibility of researchers. Journals have responsibilities as well. For instance, in 1996 the *New England Journal of Medicine* published the influential multicenter study of cerebral complications of bypass surgery; in 2012 it published a major review of the complications. In the intervening years, it published sixteen major clinical trials of bypass surgery. Of these, six make no mention of cerebral complications, two mention stroke only in passing, seven include them as secondary endpoints or as part of a composite primary endpoint, and only one focuses on the problem.[23] None of them provides thorough data about cognitive function, personality change, or the subtler phenomena that have been described for decades. Editors could adopt a higher standard and refuse to publish studies that do not include important outcomes. Similar expectations could be set by the National Institutes of Health and the Food and Drug Administration. Adequate study of complications could be a precondition for research funding and approval of new therapies. Finally, government agencies, professional societies, and medical centers could create incentives that reward clinicians and researchers who make the effort to study complications seriously. If clearer expectations existed for clinicians and researchers about monitoring, analyzing, and preventing complications, then physicians would be more likely to produce the data needed to overcome the current asymmetry in the knowledge of therapeutic benefits and risks.

Without good evidence of both the risks and benefits of treatment, the prospects for informed consent and evidence-based medicine are undermined. As long as medical knowledge remains asymmetrical, with better knowledge of efficacy than of complications, medical decision making will fall short of the ideal desired by both patients and doctors. My concern should not be read as an indictment of bypass surgery or angioplasty. First, these techniques provide important benefits to many patients. Second, the deficiencies of knowledge about complications are pervasive in medicine. Instead, my concern is that all patients be provided with more realistic expectations of benefit and more nuanced explanations of risk. This is especially true for feared complications such as stroke, cognitive dysfunction, and personality change, which have such a profound impact on patients' lives and identities. Raising the bar in this way will make clinical research more complicated, more time consuming, and more expensive. Extensive discussions will be required prior to informed consent. But physicians and researchers must undertake this work. In research, as in cardiac therapeutics, trade-offs will always exist between competing priorities. If decisions about research design are made deliberately and transparently, then patients and doctors will at least know if a procedure is truly safe or if its risks have not yet been fully characterized. This knowledge alone would allow both doctors and patients to make wiser decisions.

Puzzles and Prospects

Every day, all over America, ambulances whisk people with chest pain into emergency rooms. Doctors take a history, perform a physical exam, order diagnostic tests, and, when suspicion of a heart attack is high, send the patient to coronary angiography. Once the results are available, the doctor and patient can review clinical trials, practice guidelines, and other tools of evidence-based medicine. Such knowledge should enable good decisions about aspirin, thrombolytic therapy, angioplasty, and bypass surgery. If medicine were nothing more than facts and data, the best decision would then be clear to the fully informed patient and doctor. But other factors intervene. The accessibility and quality of medical care depend on where you happen to live and what doctor you happen to see. Consider rates of coronary angioplasty. Doctors in one Ohio town in 2003 performed angioplasty at a rate ten times that in Honolulu, and even three times that in Cleveland, a mere thirty miles away. Such variation is ubiquitous in medicine. Practice varies between doctors within a hospital, between hospitals within a city, between cities within a state, and between different states or countries. A team of Dartmouth health researchers sum this up succinctly: "In health care markets, geography is destiny."[1]

If practice variation simply reflected differences in the underlying burden of disease, then it would not be a problem. Researchers have shown, however, that this is not the case. The variation is instead "unwarranted."[2] Unwarranted variation poses a dilemma for physicians. Patients hope and expect that their doctors make treatment recommendations after carefully considering the individual's disease and its possible treatments. But as I have shown in this book, knowledge of disease and therapeutics is less perfect than patients and doctors realize. In some situations, doctors have great confidence in what they know and can predict treatment outcomes, whether desired or adverse, with great accuracy. In many other situations, this is not the case. Ambiguity about the efficacy and safety of treatments remains common. Where evidence-based medicine falters, an extraordinary range of nonclinical factors influence medical decisions.

The problem of unwarranted variation ties together many of the themes of this book and demonstrates some of the challenges ahead. Geographic and racial variations illustrate not just inconsistencies in medical decision making, but also how difficult it can be to understand exactly how decisions get made. When should a treatment be used? What rate is the right rate for any medical procedure? What is the best way to answer these questions? At stake is the rationality and equity of health care, not just in the United States but worldwide.

Coronary artery surgery had exhibited marked geographic variations from the outset. Although surgeons in New York, Baltimore, Houston, Ann Arbor, and Leningrad dabbled with the procedure in the 1960s, the work of René Favaloro and Donald Effler at the Cleveland Clinic earned the procedure wide acclaim. When Favaloro presented the latest results from the clinic at the annual meeting of the Society of Thoracic Surgeons in 1969, Houston surgeon Edward Diethrich commented, "Even in Texas, where we are accustomed to large numbers, we are envious of the number of coronary cases cared for by the Cleveland Clinic." A map of bypass surgery in the early 1970s would have shown a highly localized distribution of the procedure, with a few medical centers doing hundreds of bypass surgeries a year, while most others did few or none at all. The pioneers, however, worked quickly to colonize the medical marketplace. Favaloro wrote in 1971, "We cannot sit and wait. A tremendous task is before us, and American technology must find a way to bring diagnostic and therapeutic measures to the community level." By 1975 nearly four hundred hospitals in the United States offered bypass surgery to their patients.[3]

However, marked geographic variation in practice patterns continued. While studying the rapid growth of bypass surgery, researchers at the University of Alabama noticed that utilization varied substantially, with rates ten times higher in the western United States than in the northeast. Subsequent researchers refined this analysis by minimizing confounding variables. One approach simplified the problem by looking within a single insurance system. Geographic disparities still appeared. A 1986 analysis of Medicare patients found a 3.1-fold variation between sites with the highest and lowest rates of bypass surgery. A 1993 analysis of Medicare data found that the bypass gap had narrowed, but that significant geographic variation had appeared in the burgeoning technique of coronary angioplasty. The situation was no different for private insurers. Within the sixteen hospitals of California's Kaiser Permanente system, angiography rates varied from 30 to 77 percent. Variation even occurred among patients enrolled in a national clinical trial of thrombolytic therapies for heart attacks. Angiography was used in 52 to 81 percent of patients, with the lowest rates in New England and the highest in the South; angioplasty varied from 22 to 35 percent of patients, and bypass surgery from 9 to 17 percent.[4]

Researchers had been finding similar evidence of geographic variations in medical practice for decades. Speaking before the Royal Society of Medicine in 1938, J. Alison Glover had described how tonsillectomy rates varied twenty-sevenfold among London neighborhoods. A 1952 study found twofold variation in appendectomy rates across eleven counties in upstate New York. Sustained interest in practice variation took root in the late 1960s. In 1968 an international team of researchers compared surgery rates in Liverpool, Uppsala, and New England and found peculiar patterns of variation. While American surgeons led the pack with tonsillectomy, prostate surgery, and hernia repair, Swedes removed the most appendices and gallbladders. Surgical practice varied on smaller scales as well. A study of Kansas Blue Cross found fourfold variations in appendectomies and two- to threefold variations in other surgeries across the state. John Wennberg and Alan Gittelsohn studied practice variations in Vermont and found substantial disparities across its thirteen service areas, with tonsillectomy rates varying more than tenfold, and many other procedures varying threefold. Follow-up studies found a similar pattern in Maine. Wennberg and Gittelsohn concluded, "Small area geographic variations in use of surgical procedures are a rule for which there is yet no exception."[5]

Most of the early studies focused on surgery. When a Stanford anesthesiologist concluded that "unnecessary surgery" was being done, Francis Moore, the powerful chairman of surgery at Boston's Peter Bent Brigham Hospital, wrote a fierce response: "The mere existence of these differences in patterns of practice does not indicate a qualitative judgment about which is superior." While it was likely that "unnecessary and meddlesome" surgery took place, the simple fact of geographic variation did not prove that it did. Moore also pointed out that the problem was not unique to surgery. A slew of studies quickly supported his intuition. Wennberg, for instance, continued to work on the problem for decades, eventually organizing a group of economists and health researchers at Dartmouth Medical School. They compiled Medicare data to study national patterns in practice variation. Their 1996 *Dartmouth Atlas of Health Care* produced colorful maps of per capita spending, physician distribution, and procedure rates that made geographic variation visible for all to see. The details of the Dartmouth group's methods and findings have been challenged vigorously.[6] But even though researchers disagree about the exact extent of variation, there is broad consensus that it exists and that it is a problem.

The emergence in the literature of practice variation maps, particularly the influential images of the *Dartmouth Atlas*, demonstrate once again the importance of visualization techniques. Throughout this book I have discussed how clinicians and researchers worked to visualize everything from atherosclerotic plaques to the subtleties of postoperative cognitive dysfunction. These techniques make the invisible visible, and the resulting images often have evocative power. In the same way that angiographic evidence of coronary stenoses elicited the oculostenotic reflex, the maps of practice variation captured attention and demanded explanation. Concerned by evidence of variation and the implication that "the quality of care we deliver varies," the American College of Cardiology and the Society of Thoracic Surgeons commissioned the 1999 *Dartmouth Atlas of Cardiovascular Health Care*. With 288 pages of maps and charts, the *Atlas* exposed the problem in bright colors. The prevalence of cardiac surgeons varied fivefold between Salinas, California, and Ridgewood, New Jersey. Interventional cardiologist prevalence varied a bit more, with a sixfold difference between Houma, Louisiana, and Buffalo, New York.[7]

The variations in bypass surgery and angioplasty received special scrutiny.

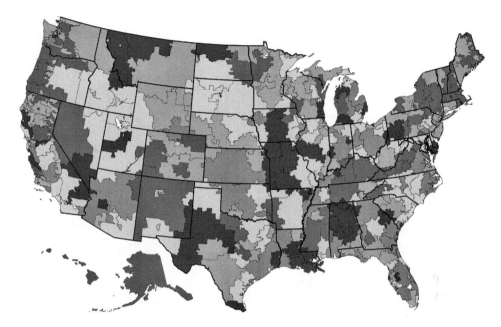

Regional variation in coronary angioplasty. The color version of this image sends an even clearer message: areas of highest use (e.g., north coast of California, west central Texas) are often adjacent to areas of lowest use (e.g., San Francisco, New Mexico). From *Dartmouth Atlas of Health Care 1999* (Chicago: AHA Press, 1999), fig. 5.4, p. 159. With permission from the Center for the Evaluative Clinical Sciences, Dartmouth Medical School.

Ratio of Rates of Percutaneous Transluminal Coronary Angioplasty Procedures to the U.S. Average
by Hospital Referral Region (1995-96)

- ■ 1.30 to 2.57 (54)
- ■ 1.10 to < 1.30 (42)
- ☐ 0.90 to < 1.10 (89)
- ▨ 0.75 to < 0.90 (60)
- ■ 0.37 to < 0.75 (61)
- ▧ Not Populated

Redding, California stood out in the original 1996 *Atlas* with a bypass rate 42 percent higher than the national average. By the 1999 second edition, Redding had the highest rate in the country. Bypass peaked there in 2001 before crashing in 2003, from first in the nation to 227th. What had happened? Long-standing suspicions about inappropriate surgery by two local doctors triggered a criminal investigation by the Federal Bureau of Investigation. Although the FBI could not prove intent to perform unnecessary surgery, Tenet Healthcare and the two physicians have had to pay millions of dollars in civil settlements. Meanwhile, the 1999 *Atlas* had identified Elyria, Ohio, as the region with the country's highest angioplasty rate. By 2003 it was four times the national average. Prompted by these data, the *New York Times* published a 2006 exposé. The president of the North Ohio Heart Center defended their aggressive approach: "With absolutely no exception, patients given aggressive treatment will come out with a better outcome." Few outside observers were convinced. The situation was clearer for Lafayette, Louisiana, which had the second highest angioplasty rate in 2003. One cardiologist there was indicted in 2006 on charges that he falsified clinical findings to justify unnecessary procedures, such as placing stents in normal coronaries. Convicted in 2009, he was sentenced to ten years in prison.[8]

As evidence of geographic variation in medical practice accumulated, a parallel body of research documented racial variations in practice. Tensions about the influence of race on therapeutic decision making surfaced early in the history of bypass surgery. The first evidence appeared in 1982, when cardiovascular epidemiologist Richard Gillum realized that there was a revascularization gap between blacks and whites. He tabulated all surgeries performed on black and white patients and found that the black-to-white ratio was 0.14, commensurate with the proportion of the populations in the United States. The ratio for bypass surgery, however, was 0.02, which "clearly represented underutilization, presumably from lack of access to cardiac revascularization, for nonwhite, primarily black, patients." Researchers at the University of Alabama confirmed this observation in 1984: although whites made up only two-thirds of the patients referred to the medical center, they received over 96 percent of the coronary angiography performed there. Disparities in bypass surgery referral rates persisted even after researchers controlled for the severity of coronary artery disease. The Alabama group found that within the subset of patients who had multivessel disease, whites were more than 2.5 times more likely to receive surgery. This discrepancy endured in the highly monitored setting of a major clinical trial. The Coronary Artery Surgery Study, funded and organized by the National Institutes of Health, found that for patients who had similar angiographic findings, surgeons recommended bypass surgery for 59.4 percent of white patients but for only 46.5 percent of black patients.[9]

Wherever researchers looked, they found evidence of race disparities in cardiac care. Whites in Massachusetts received more angiography, angioplasty, and bypass surgery than blacks, even when researchers controlled for important clinical, demographic, and socioeconomic characteristics. National Medicare data showed that whites were three times more likely than blacks to receive bypass surgery. For six southern states—Arkansas, Georgia, Louisiana,

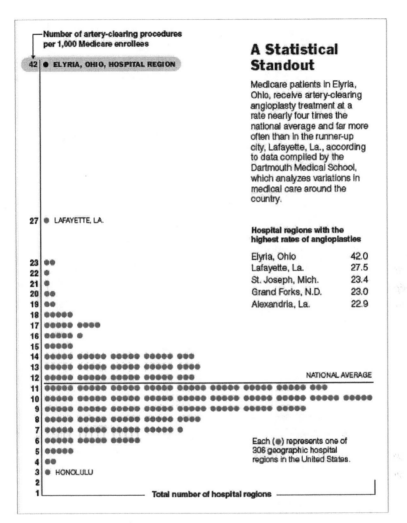

A Statistical Standout

Medicare patients in Elyria, Ohio, receive artery-clearing angioplasty treatment at a rate nearly four times the national average and far more often than in the runner-up city, Lafayette, La., according to data compiled by the Dartmouth Medical School, which analyzes variations in medical care around the country.

Number of artery-clearing procedures per 1,000 Medicare enrollees

42 ● ELYRIA, OHIO, HOSPITAL REGION

27 ● LAFAYETTE, LA.

Hospital regions with the highest rates of angioplasties

Elyria, Ohio	42.0
Lafayette, La.	27.5
St. Joseph, Mich.	23.4
Grand Forks, N.D.	23.0
Alexandria, La.	22.9

NATIONAL AVERAGE

Each (●) represents one of 306 geographic hospital regions in the United States.

3 ● HONOLULU

Total number of hospital regions

Extreme geographic variation in angioplasty rates. By arraying treatment rates in various regions along a spectrum, spindle diagrams underscore the magnitude of variation. It remains unclear what the right rate is. Reed Abelson, "Heart Procedure Is off the Charts in an Ohio City." From *The New York Times*, 18 Aug. 2006 © 2006 The New York Times. All rights reserved. Used by permission and protected by the Copyright Laws of the United States. The printing, copying, redistribution, or retransmission of this Content without express written permission is prohibited.

Mississippi, North Carolina, and South Carolina—the ratio was six to one, and for Alabama, eight to one. Growing concern about racial disparities fueled extensive research in the 1980s and 1990s. This work culminated in the landmark 2003 report from the Institute of Medicine, *Unequal Treatment*. The report's 764 pages catalogued evidence of disparities in treatment access and outcome in all areas of medicine. Cardiology and cardiac surgery figured prominently: the full report devoted more space to cardiovascular care than to any other area of medicine. The executive summary explained that racial differences in cardiac care "provide some of the most convincing evidence of healthcare disparities."[10]

As evidence of geographic and racial variation emerged between the 1960s and 1990s, physicians and other observers struggled to understand what it meant. Some reasons for variation were obvious. It was no mystery why the Cleveland Clinic had dominated the early bypass surgery landscape. A jury had no doubt about the cause of the epidemic of angioplasty in Lafayette, Louisiana. But what about the pervasive two- to threefold variation seen throughout medicine? In 1983 the prestigious Institute of Medicine convened a conference to study geographic disparities. Its president, Frederick Robbins, worried about "the potentially damning evidence that these great variations represent." He had a stark warning for his colleagues: "You can cover it up all you want, but it looks bad, and it looks bad because it is bad. It is not an appropriate way for a profession to behave." Ten years later physician and policy expert David Blumenthal was blunt. He acknowledged that variation was "natural and inevitable," and that "only in Garrison Keillor's imaginary land of Lake Wobegon, where all children are above average, is it conceivable that all patients with the same condition would be treated in precisely the same way in every place, by every provider, and at every moment." Nonetheless, variation undermined the "scientific legitimacy" of medicine: "The fact that physicians treat apparently similar patients in such widely different ways casts doubt on their knowledge base, on their competence to interpret it, or both."[11] Faced with an ever-increasing barrage of critiques, researchers set out to understand the source of unwarranted variations in health care. Their work revealed a range of factors, beyond concern with efficacy and safety, that influenced medical decision making.

Many doctors hoped that the practice variations simply reflected variations in the underlying burden of disease. But this had never seemed plausible. In

Glover's original 1938 report, he noted that tonsillectomy was more common in boys even though tonsillitis was more common in girls. As evidence of variation accumulated—between England, Sweden, and the United States, or within upstate New York or Kansas—doctors realized that "regional differences in disease cannot easily be invoked" as an explanation. Bypass surgery itself demonstrated the disconnect between disease rates and procedure rates. Gillum found that "Regional patterns of utilization of coronary artery bypass surgery and coronary angiography were opposite to patterns of mortality and morbidity from coronary heart disease." Surgeons might hope that the high rate of bypass surgery caused the low rate of heart attacks, but this was not the case: bypass surgery was not done in enough patients, and its preventive benefits were too modest, to have had such a large impact on the population's mortality rates. Other researchers found that bypass surgery was twice as common in Des Moines as in Iowa City, despite comparable rates of coronary artery disease. Similar disconnects were found from southern California to northern New England.[12]

What, then, generated the variations? Physicians worried about the specter of financial conflicts of interest. The fee-for-service insurance system in the United States attracted particular concern. Even Francis Moore, who usually rushed to the defense of his fellow surgeons, acknowledged the problem: "For the pecuniary-minded physician and surgeon alike, or for psychiatrist or pediatrician, the American population is a happy hunting ground." The dramatic spread of angiography and bypass surgery in the 1970s invigorated these concerns. In 1979 an Arizona physician cataloged stories of suspicious uses of angiography and suggested that profit motives had led to the procedure's abuse. When the Alabama group first described the geographic disparities in bypass surgery, they wondered whether reimbursement practices contributed. Canada offered a revealing comparison. Instead of generating revenue for hospitals, procedures in Canada consumed limited health care budgets. Rates of bypass surgery, not surprisingly, were much lower.[13] Financial interests, however, never offered a complete explanation. Disparities appeared within insurance systems, such as Kansas Blue Cross, Kaiser Permanente, and Medicare, where differences in financial incentives did not exist.

Analysts worried about a related problem, supplier-induced demand. Noting a correlation between surgical capacity and surgical utilization in Kansas, Boston physician Charles Lewis observed that "admissions for surgery expand to fill beds, operating suites and surgeons' time." Liverpool in the late 1960s had twelve otolaryngologists but just one urologist. Researchers wondered whether this accounted for the city's high rate of tonsillectomy and low rate of prostatectomy: "Could it be that whereas the tonsils have their predator, the prostate is left to flourish unattended?" Coronary revascularization again offered a revealing case study. Rates of bypass surgery and angioplasty both correlated closely with rates of angiography. This made sense: angiography provided the diagnosis on which both procedures relied. So what determined angiography rates? Access to the procedure proved critical. One study examined nineteen hospitals in Seattle. Some had angiography on site, others did not. If you had a heart attack, it mattered—a lot—which kind of hospital the ambulance took you to. Patients with heart attacks who arrived at hospitals with angiography were 3.21 times more likely to receive the procedure than patients who arrived at hospitals that had to refer them elsewhere for angiography. The researchers even looked specifically at the subset of doctors who worked in both types of hospitals. Any individual doctor was twice as likely to order angiography when it was readily available.[14]

But, as happened with financial conflicts of interest, the evidence for supplier-induced demand was inconsistent. With medical economics offering only a partial answer, researchers turned to medical philosophy. After studying practice variation in Vermont, John Wennberg concluded in 1973 that there must be "differences in beliefs among physicians concerning the indications for, and efficacy of, the procedure." Many researchers agreed. Canadian researchers attributed the high rates of hysterectomy in some parts of Manitoba to a "hysterectomy-prone surgical practice style." One team of researchers attributed the high rates of surgery in New England to the "activist society" in America. Comparing the United States to England, John Bunker speculated about a "genuine philosophical difference in attitudes": "In keeping with his national character, the American surgeon is more aggressive." What generated the differences in surgical philosophy? Peer pressure played a role. Consider the hospitals of the Department of Veterans Affairs. With a relatively homogeneous patient population, salaried physicians, and strong central oversight, researchers expected that the VA system would not exhibit the marked practice disparities seen elsewhere. What they found, however, surprised them: practice in any individual VA hospital mirrored practice patterns in the local community.[15]

What accounted for the variations in local enthusiasm? David Eddy, a pioneer of decision analysis and evidence-based medicine, offered his diagnosis in 1984. He attributed the problem to a delicate balance between certainty and uncertainty: "Community standards themselves exist because enough is known to enable the leaders of a community to develop opinions which, when followed by their colleagues, become community standards. The differences between community standards exist because not enough is known to establish which opinion is correct." Doctors had long known that uncertainty introduced variability into medical decisions. In 1970 one team of researchers presented case vignettes to Maryland physicians and asked them how they would treat the patients. The researchers found "marked divergence of opinion concerning the need for surgery," and concluded that "surgical decision-making is a semi-exact scientific process." Francis Moore, commenting on the study, did not defend surgeons but instead expanded the critique beyond medicine to all life sciences: "Uncertainty is a major feature of the biologic sciences today." Eddy agreed: "Uncertainty creeps into medical practice through every pore." Faced with complex and uncertain information, it was easy "for honest people to come to different conclusions." Close study of the patterns of variation was revealing. Geographic variations were largest where medical consensus was weakest. Tonsillectomy, about which few physicians agreed, topped the list of high-variation procedures. Where medical consensus was strong, as it was for appendectomy or hernia repair, variation was lower.[16]

Uncertainty emerged at another register as well. As more researchers studied practice variation and identified additional possible explanations, they became increasingly uncertain about which ones were relevant. A 1987 review of 250 articles on geographic variation cataloged the variables that researchers had studied, everything from clinical measures to race, income, marital status, living arrangements, physician availability, community tax base, migration patterns, and even climate. Amid this embarrassment of explanatory riches, the researchers concluded that it was "difficult to ascertain the degree to which any one of these major components influences a community's use rates." Some researchers turned to multivariate analysis to quantify the most important causes of variation, only to find that "the explanatory power of the regression models was low, meaning that all of the variables included in the equations together did not explain a high proportion of the variation in utilization rates." When researchers looked at angioplasty variations in Europe, they found that the "more obvious explanations, such as the incidence of coronary heart disease, income per capita, and the number of facilities and practitioners available, proved not to be valid." This left them searching for

"Other potential explanations, more difficult to quantify." [17] Decades of effort had left variation researchers without clear answers. Some things definitely mattered: practice patterns were influenced by the supply of physicians, reimbursement policies, local standards of care, and degree of therapeutic uncertainty. But the impact of each factor varied from place to place, and many other factors played a role. There was no single cause that policy makers could target to eliminate unwarranted variation.

The situation with racial disparities was just as frustrating. The initial studies had found disparities even when researchers controlled for the severity of disease. One popular theory attributed the disparities to differences in the willingness of black and white patients to accept referral for bypass. In the Coronary Artery Surgery Study 90.4 percent of white patients followed through with referral to surgery, compared with only 80.5 percent of black patients. This finding was often cited as evidence of differences in patient preferences. But when researchers interviewed black and white patients about treatment preferences, they either found no differences, or learned that blacks were skeptical about the procedures because they were not familiar with them. It was a vicious spiral: fewer black patients had had bypass surgery, so there was less familiarity with bypass surgery in African American communities, so fewer black patients were willing to consent to surgery. Questions of racial bias also emerged. As Gillum noted in 1982, "Fewer than 30 black cardiovascular surgeons and 200 black cardiologists are estimated to be practicing in the United States . . . angiography and surgery for coronary artery disease in black patients is largely in the hands of white physicians." Did this matter? Experts feared that it did and that the mostly white profession provided worse care to black patients. In 1990 the American Medical Association's Council on Ethical and Judicial Affairs warned that, despite progress since the 1960s, the health care system had not "fully eradicated this prejudice."[18]

But even when physicians suspected that prejudice played a role, their suspicions proved difficult to substantiate. One 1999 study, which reported a 40 percent disparity in referral rates for coronary angiography based on race and gender, received extensive media coverage. Reanalysis, however, showed that the researchers had presented the results in a misleading way and that the disparity—just 7 percent— existed only for black women and not black men.[19] The critics positioned their critique carefully. They did not deny that racial bias existed in medicine. Instead, they simply noted that the study overstated the evidence. Research on disparities had once again yielded an uncertain result.

Had the geographic and racial disparities simply been academic questions, then the uncertain answers would have been part of the familiar fabric of science: some questions prove difficult to answer, as this book has shown repeatedly. But as with the debates about the causes of heart attacks or the prevalence of the cerebral complications of bypass surgery, the debates about practice variation had important, pragmatic consequences. Coronary revascularization, let alone medicine more broadly, is a vast enterprise. Determining ideal utilization rates and ensuring that the right patients receive the treatments are key challenges for medicine, with important implications for rationality, policy, and health equity.

Every era has had advocates calling for the establishment of a more rational basis for medical practice. This took several forms in the decades after World War II. Some reformers, backed by the Food and Drug Administration, focused on randomized clinical trials as the most reliable guide for treatment decisions. Others, working in parallel, developed a new science of clinical epidemiology, which applied rigorous statistical thinking to clinical decision making. Enthusiasm for these techniques grew steadily in the 1970s and 1980s. By the 1990s a new field emerged: evidence-based medicine. Despite its appeal, evidence-based medicine has proven difficult to implement. Randomized trials reflect social, economic, and political interests and often propagate the controversies they are intended to resolve. Subsequent trials often reverse the findings of their predecessors, a problem that has plagued studies of diet, hormone replacement therapy, and cancer treatments. Even when a trial manages to produce a clear result, critics assail it furiously, often with credible concerns. For instance, when the first randomized trial of bypass surgery found in 1977 that the operation provided only a modest benefit only for the sickest patients, surgeons critiqued and condemned the findings. More recently, a study that demonstrated the benefit of statins was interpreted differently in different countries, leading to different treatment guidelines in Europe, Canada, and the United States—each of the guidelines supposedly evidence based.[20]

It was in the 1970s and 1980s, just as evidence-based medicine struggled to take hold, that the problem of unwarranted variation received widespread attention. On one hand, evidence of practice variation strengthened the position of therapeutic reformers. Evidence of tenfold variation showed how much need there was for improved rationality and discipline in medical decisions. On the other hand, as the variation persisted decade after decade, it became an affront to the ambitions of evidence-based medicine, a testimony to how far medicine still was from being a fully rational enterprise.

Researchers never gave up trying. One of the most persistent efforts came from the RAND Corporation. Established to design military strategy during the Cold War—most famously its doctrine of nuclear deterrence by mutually assured destruction—RAND has applied its analytic acumen to health care since the 1960s. In the 1980s it set out to define more precisely what counts as appropriate treatment. Its researchers reviewed the literature, assembled panels of experts, and sought consensus about the appropriateness of specific procedures in hundreds of clinical situations. Testing these hypothetical standards against practice, they found widespread problems. An analysis of coronary angiography showed that 17.4 percent of cases were "inappropriate" and another 8.5 percent "equivocal." For bypass surgery, 14 percent of cases were inappropriate and 30 percent equivocal. However, this approach raised as many questions as it answered. The RAND method, which relied on the ability of clinicians to reach consensus about appropriateness, inadvertently confirmed the difficulty of doing so. Their experts agreed in only half the cases and actively disagreed in nearly one-third.[21]

Physicians and policy makers had more pragmatic concerns as well. Between the 1960s and the 1990s, health care grew into a trillion-dollar industry. This growth attracted unprecedented scrutiny to medical practice. Recognition of the problem of practice variation led repeatedly to calls for specific policy interventions. For instance, the passage of Medicare and Medicaid and the advent of expensive new technologies had all increased the costs of health care in the late 1960s. The costs of Medicare alone doubled in the program's first five years. Congress enacted programs to monitor health care costs and the efficiency of health care delivery. These, in turn, produced much of the early data on practice variations. Practice variation rose to the fore again in the early 1990s as the Clinton Administration debated health care reform. Some researchers looked to Canada as a model for cost containment. Cardiac care provided a useful test case. A slew of studies found dramatically lower rates of angiography, angioplasty, and bypass surgery in Canada, yet Canadian cardiac patients fared nearly as well as their southern neighbors.[22]

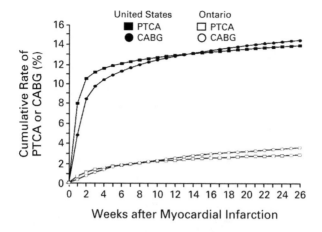

Cumulative rates of percutaneous transluminal coronary angioplasty (PTCA) and coronary artery bypass grafting (CABG) after acute myocardial infarction among elderly patients in the United States and Ontario, 1991. Even though revascularization rates are five to six times higher in the United States than in Ontario, patients in the two regions have comparable survival rates after heart attacks. From Jack V. Tu et al., "Use of Cardiac Procedures and Outcomes in Elderly Patients with Myocardial Infarction in the United States and Canada," *New England Journal of Medicine* 336 (22 May 1997): 1500–1505, fig. 2, p. 1503. With permission from the Massachusetts Medical Society.

When the Obama Administration took up health care reform, geographic variation again took center stage. The spotlight fell on McAllen, Texas, which the *Dartmouth Atlas* had identified as one of the highest spending Medicare regions in the country, with spending per enrollee twice that of residents in nearby El Paso. Residents of McAllen were more likely to see specialists, more likely to have diagnostic tests, and more likely to have surgery, all without evidence of better outcomes. Surgeon and *New Yorker* essayist Atul Gawande concluded, "The primary cause of McAllen's extreme costs was, very simply, across the board overuse of medicine." Gawande's essay "The Cost Conundrum" received enthusiastic attention at the White House.[23] Obama's team latched onto the claim, first made by the Dartmouth team in 1997, that it would be possible to eliminate overuse and save 20 to 30 percent of Medicare spending without compromising the health of Americans. The proposal received fierce critique. A *Wall Street Journal* editorial argued that "Medicare reflects the entire practice of medicine only as a funhouse mirror." John Kerry rushed to the defense of his high-spending Massachusetts hospitals, arguing that they were "concentrated centers of innovation." They might be more expensive, but the whole country benefited.[24] Amid the controversy about rationing and death panels in summer 2009, the Obama administration backed away from the goal of cutting costs by eliminating geographic variation in spending.

Concerns about equity followed closely on the heels of concerns about rationality and policy. Unwarranted variation in medical practice, particularly within Medicare, meant that resources were being distributed unfairly. As the *Dartmouth Atlas* pointed out, "Is it fair for citizens living in regions with low per capita health care costs to subsidize the greater (and more costly) use of care by people living in high resource and high utilization regions?"[25] Reformers did not just want good health care; they wanted equitable health care. But what would equitable care mean? Should a uniform practice rate be imposed nationwide, and could this even be done? Should physicians target underuse in black patients or overuse in white patients? Can knowledge of efficacy and complications provide a clear guide for decisions? As this book has shown, achieving solid knowledge of efficacy and safety is never as easy as it ought to be.

The impact of nonmedical considerations is demonstrated in an especially troubling way by the huge disparities that exist with coronary revascularization on a global scale. These disparities may even have consequences for how we understand our own commitments to these technologies in the United States.

Coronary artery bypass grafting spread quickly from its origins in Cleveland. René Favaloro returned to his home country, Argentina, to establish a cardiac surgery center in 1971. At the Cleveland Clinic's celebration of the twenty-fifth anniversary of bypass surgery in 1992, its surgeons boasted that surgeons they had trained practiced "in every corner of the globe," in Canada, Ecuador, Guyana, Suriname, Peru, Brazil, Buenos Aires, Chile, South Africa, Egypt, Israel, Syria, Greece, Italy, England, Netherlands, Poland, Japan, Taiwan, India, Pakistan, Thailand, Malaysia, and Australia.[26] No country, however, initially embraced the operation as enthusiastically as the United States. In 1982, for instance, the bypass rate was 10.7/100,000 in England, 41.0 in

Australia, and 75.0 in the United States. Angioplasty showed similar early disparities. Developed by a German angiologist working in Switzerland, it was first implemented widely in the United States. By 1991 the rate in the United States was nearly twice that of Belgium, its nearest competitor. Significant disparities also existed within Europe, with the rate in Belgium fourteen times higher than that in Portugal. Such disparities have persisted. By 2009 rates of angioplasty in Belgium and Germany exceeded those in the United States. Ireland, England, and Portugal, meanwhile, lagged far behind.[27]

An even more alarming disparity is nearly hidden from view. Each year the Organization for Economic Co-operation and Development prepares a bar graph of revascularization rates. Its 2009 estimates ranged from 582/100,000 in Germany to 377 in the United States and 82 in Ireland. Barely visible at the bottom of the chart sat Mexico, with a revascularization rate (2/100,000) barely 0.5 percent that of the United States. Mexico is emblematic of middle- and low-income countries' limited access to the procedures. Although angioplasty arrived in India in 1986 and is now available at over four hundred centers, half of these centers are located in only six cities; fewer than 10 percent of heart attack patients have access to revascularization. In China, with a population of well over 1 billion, fewer than 20,000 people received angioplasty in 2010 (versus 300 million Americans with 1 million procedures). Nigeria performed its first angioplasty in 2009.[28]

Low rates of coronary revascularization in these countries would be appropriate if rates of coronary artery disease were also low, but this is not the case. Although many people do not think of heart disease when they think of China, India, or other middle- and low-income countries, urbanization and industrialization in the late twentieth century transformed populations, diets, lifestyles, and patterns of disease throughout the world. Indian cardiologist Srinath Reddy explained that these changes "propelled the developing countries into the vortex of the global CVD epidemic." When the World Bank surveyed global health conditions in 1993, it found that cardiovascular disease caused 28.8 percent of all deaths worldwide. Coronary artery disease, responsible for 10.3 percent of all deaths, was the leading single cause. In 1999 the World Heart Federation warned of an "impending global pandemic." By 2002 India had become the country with the largest number of heart attack deaths (1,500,000), followed by Russia, China, and then the United States (540,000). Ukraine had the highest rate of heart attack mortality; nine of the top ten countries by this measure were in the former Soviet Union. The United States

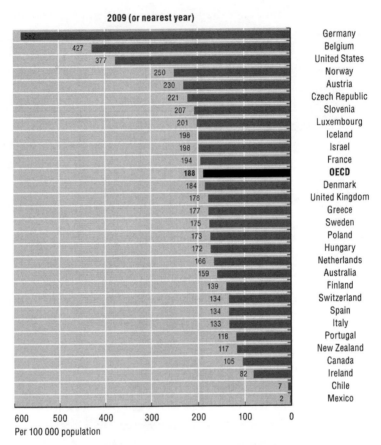

2009 (or nearest year)

582	Germany
427	Belgium
377	United States
250	Norway
230	Austria
221	Czech Republic
207	Slovenia
201	Luxembourg
198	Iceland
198	Israel
194	France
188	**OECD**
184	Denmark
178	United Kingdom
177	Greece
175	Sweden
173	Poland
172	Hungary
166	Netherlands
159	Australia
139	Finland
134	Switzerland
134	Spain
133	Italy
118	Portugal
117	New Zealand
105	Canada
82	Ireland
7	Chile
2	Mexico

600   500   400   300   200   100   0
Per 100 000 population

Global incidence of coronary angioplasty. Angioplasty rates vary sevenfold between Ireland and Germany, and nearly two-hundredfold between the United States and Mexico. From OECD (2011), *Health at a Glance 2011: OECD Indicators,* OECD Publishing, fig. 4.6.1, p. 91. http://dx/doi.org/10.1787/health_glance-2011-en.

ranked thirty-third. A critical threshold was reached in 2011 when the United Nations held a summit on noncommunicable diseases, only the second time in its history that heads of state have met specifically to discuss disease.[29]

Why has coronary revascularization not followed the rise of coronary artery disease and heart attacks in low- and middle-income countries? Although debates about the appropriateness of coronary revascularization have raged for decades, by the 1990s one stable island of consensus had emerged: patients with heart attacks have improved survival if they receive angioplasty within 90

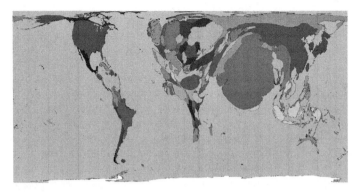

Heart attack deaths, 2002. The area (and shape) of each country is morphed so that the size is proportional to that country's incidence of heart attack deaths. India, to the surprise of many, leads the world in number of deaths from heart attack. From *Worldmapper* (2010). © Copyright SASI Group (University of Sheffield) and Mark Newman (University of Michigan).

minutes of the onset of symptoms. Once doctors agreed about this point, they focused on questions of cost and feasibility. Health officials in Europe worried in the 1990s that it would be prohibitively expensive to make angioplasty available to all patients suffering heart attacks.[30] In other countries, where health resources were scarcer, angioplasty and bypass surgery were dismissed altogether. The World Bank in 1992 described cardiac surgery, along with treatment of hypertension or elevated cholesterol, as "immensely unattractive investments for public funds." It moderated its position in 1993, acknowledging the value of "relatively cost-effective measures" such as aspirin and blood pressure medicines, but it held the line at heart surgery: "Many health services have such low cost-effectiveness that governments will need to consider excluding them from the essential clinical package. In low-income countries these might include heart surgery." In 1998 the Institute of Medicine recommended prevention and low-cost medications and advised poor countries to avoid "sophisticated, expensive technologies," including angiography, angioplasty, and bypass surgery. One Boston cardiologist ranked procedures by cost per disability-adjusted-life-year averted, with interventions ranging from $3 to $52 for increased tobacco taxes, $9 to $20 for aspirin to prevent heart attacks, $25 to $2300 for bans on trans fats, $16,000 for thrombolysis, and $24,000 to $72,000 for the full package of bypass surgery and drugs. The report from the United

Nations' 2011 summit focused, not surprisingly, on prevention, behavior, and lifestyle, especially tobacco and diet: "Prevention must be the cornerstone of the global response to noncommunicable diseases."[31]

All these groups argued, in effect, that a life-saving treatment was not an appropriate technology for resource-poor settings. The idea of appropriate technology had emerged in the aftermath of World War II. The shock of the atomic bomb had shattered long-standing faith in the unquestioned value of scientific and technological progress. Similar disillusionment appeared in medicine, especially with concern about how new life support technologies had confounded the dying process. Scholars and activists sought ways to re-align technology with human relations and wondered, in particular, what technologies might be appropriate for specific societies. For instance, after working on the Navajo Reservation in the 1950s and 1960s, a team of Cornell physicians concluded that state-of-the-art medical technology was not well suited for the impoverished reservation. There was a "technological misfit" between what medicine could offer and what was needed to improve health conditions. Advocates of "appropriate technology" hoped that decisions about technology would be informed by frank discussions of its value.[32] Since the 1990s, however, skeptics have increasingly worried that the idea was being invoked to withhold expensive technologies from resource-poor populations.

Consider HIV/AIDS. From the late 1990s until 2003, physicians and public health experts debated whether to bring antiretroviral therapy—at a cost of $20,000 per person per year—to low-income countries. Many argued that they should not. Prevention was far more cost effective. Given the prohibitive costs of the medications, they were not an appropriate technology. Treatment activists rejected this logic. They argued that the rhetoric of "appropriate technology" simply provided an excuse for denying medical care for people in poor countries. They challenged the prevention consensus by asserting that access to life-saving therapy was a human right. They also showed that it was possible to decrease the cost of AIDS drugs to less than $100 per person per year. This price drop changed the calculus of cost effectiveness and trans-formed understandings of appropriateness. In 2003 the United States and the World Health Organization rolled out large-scale treatment programs, in-cluding the President's Emergency Program for AIDS Relief and the Global Fund to Fight AIDS, Tuberculosis, and Malaria.[33]

The success of global HIV treatment opens the door to future challenges. Even though coronary artery disease has become the leading cause of death worldwide, few resources exist for its treatment. In 2007 less than 3 percent of international health aid was spent on noncommunicable diseases. Even as advocates called for help for these diseases at the 2011 UN Summit, they knew that resources would not be forthcoming. The United States, for instance, had no plans to increase its overall health assistance, and it was unwilling to divert resources from its existing priorities, including HIV/AIDS and maternal and child health. The relative neglect was not difficult to understand. The international health community had traditionally assumed that coronary artery disease, a problem for affluent and aged populations, would not be a significant problem for poor countries. Where coronary artery disease was recognized, it was often characterized as the consequence of individual choices—about diet, smoking, and exercise—that did not merit government intervention. As one group of cardiovascular advocates concluded, heart disease "has few of the features that attract international sympathy or support."[34]

Can the conditions of possibility be changed again? Motivated by the history of AIDS activism, some cardiologists and public health officials in recent years began to push for a more serious commitment to cardiac care in the developing world, one that would not stop at prevention and low-cost medications.[35] Replacing a damaged cardiac valve, at a cost of several thousand dollars, can extend a patient's life by fifteen or twenty years. Timely angioplasty can saves the lives of heart attack victims. But even amid rising calls for global health equity, no one made a serious call for global access to cardiac surgery or interventional cardiology. There is no Global Fund or President's Emergency Program for heart attacks, hypertension, or rheumatic heart disease. The magnitude of the need is great, with heart disease killing more people each year than AIDS, tuberculosis, and malaria combined. Technology exists that might save the lives of millions of people each year. And yet no one has identified access to angioplasty or cardiac surgery as a human rights crisis.

Yes, the obstacles would be daunting. It would be expensive to train the personnel and create necessary infrastructure. It would be difficult to design health care systems and emergency medical services that would make timely care available to all patients. It would be important to prevent the overuse and unsustainable costs that have become a problem in the United States. However, the high costs of cardiac surgery and interventional cardiology might be malleable. Heart attack therapy does not demand rigorous compliance with daily medications (though these certainly help to prevent subsequent attacks).

The cost is concentrated in the acute intervention of angioplasty and a brief hospitalization. Much of this cost goes to the device companies that manufacture the catheters and stents at extraordinary profits. If Indian or Chinese device companies followed the lead of Indian and Chinese pharmaceutical companies and began to produce generic stents, costs could fall substantially.

This proposal provides a valuable thought experiment. When I have pitched the idea to global health experts, no one has been enthusiastic. Some argue that AIDS deserves treatment in a way that heart attacks do not because the victims of AIDS are younger. Others argue that heart disease in developing countries will never motivate sustained international aid because it is not contagious and poses no direct threat to donor nations. Still others remind that funding is a zero-sum game and that investments could be better spent elsewhere. Skeptics of revascularization add that the procedures are overused in the United States and that it would be wrong to export such overuse elsewhere.

But these arguments dodge a fundamental challenge. Even though the United States and other wealthy countries invest heavily in cardiac care, especially in angioplasty, they have not made a commitment to fund such care elsewhere. This discrepancy means one of two things. It might mean that we are committed to coronary revascularization but do not think people in resource-poor settings deserve comparable access (or that it is not our responsibility to provide it). Or it might mean that we are committed to providing essential medical care to resource-poor settings but do not consider angioplasty and bypass surgery to be essential treatments. Neither option holds tremendous appeal. What rate is the right rate for coronary revascularization? Equity and social justice would suggest that one target be set for all people, not just in the United States but worldwide. It remains to be seen whether that rate will be high or low.

Unwarranted variations and tensions in the global response to heart disease illustrate long-standing challenges in medical decision making. Practice variation in the United States illustrates the unstable configurations of economic and social pressures that shape decisions. Decisions about revascularization in low- and middle-income countries are not based on knowledge of efficacy and safety, but on other values, specifically on how societies choose to invest scarce health care resources and on the resources that donor nations are willing to make available. Some policy experts have argued that variation is inevitable. A less accepting attitude may be warranted. Progress against the problem, however, will require a complex understanding of the origins of the problem and a serious commitment to finding creative solutions.

John Hunter felt powerless to stop the angina that tormented him for decades. He could neither alter the course of his disease nor control the behavior of the rascals who surrounded him. People with coronary artery disease today face vastly better prospects, in part because of the extraordinary diversity of treatments that physicians developed in the twentieth century. Bill Clinton and his doctors decided to make a preemptive strike and insert four grafts to bypass plaques in his coronary arteries. They hoped that this new blood supply would spare Clinton the fate of another former president, Dwight Eisenhower, who suffered one heart attack after another before finally succumbing to heart failure. Some experts, however, have second-guessed Clinton's decision. One cardiologist posted an open letter that invoked both the plaque rupture hypothesis and the risk of cerebral complications as part of his argument that Clinton should have focused his efforts on diet and other lifestyle changes.[36]

It is easy to sympathize with the challenge Clinton faced. Doctors and patients now face an embarrassment of therapeutic riches. Because of the complexities of bodies, diseases, and therapies, they often lack thorough knowledge of their options. Doctors have also come to respect that patients, even if presented with the same information, can have different preferences. Some women want aggressive prenatal testing to help with their decisions about a pregnancy, but not all do. Some cancer patients want nothing but the most aggressive care, while others are willing to tolerate the ambiguity of "watchful waiting."[37]

Recent developments send conflicting messages about the broad trajectory of medical decision making. The United States Preventive Services Task Force kicked open a hornet's nest in 2009 when it recommended less aggressive screening for breast cancer. But when it made a similar recommendation for prostate cancer in 2011, the outcry was muted. Has medical culture subtly shifted away from doing everything possible, toward a more guarded use of medical technology?[38] In other areas the opposite has happened. Following the lead of cardiac surgeons and cardiologists, neurosurgeons, neurologists, and interventional radiologists began to use bypass grafts and balloon angioplasty for stroke prevention, by treating atherosclerotic plaques in carotid and cerebral arteries. The intuitive appeal of revascularization motivated aggressive intervention in the absence of clear evidence of efficacy. Now that large trials have cast doubt on cerebral revascularization, it remains to be seen whether patients and doctors will back away from these treatments.[39]

Broken Hearts has shown why medical decisions can be so hard to make. Why do we think treatments might work? Expectations of therapeutic benefit arise from personal experience, clinical trials, or hypotheses about disease mechanisms. To make matters worse, doctors and patients must grapple with competing disease models and with inconsistent outcomes from clinical trials. Beliefs about therapeutic efficacy can cycle back and change how doctors think about disease mechanisms. This exposes medical knowledge to the vagaries of therapeutic enthusiasm. What can be done? Patients and doctors must approach medical knowledge with a dose of healthy skepticism. Proponents of evidence-based medicine must understand the power of other ways of thinking about efficacy. If a treatment, especially a visible, mechanistic one like bypass surgery or angioplasty, directly addresses a widely held disease model, then that treatment will be given more than just the benefit of the doubt. Only by understanding the complexity of our thinking about efficacy will it be possible to guide how we make treatment decisions.

Why do we think treatments are safe? Doctors can predict some complications, based on their knowledge of therapeutic mechanisms, but this is an imperfect science. And even when complications are foreseen, they sometimes receive little attention. Many factors conspire to direct researchers' attention away from treatment complications and preclude resolution of uncertainty. Solutions are easy to design but difficult to implement. Researchers could collect data about every measurable physiological parameter. They could move past their focus on expedient surrogate markers and turn to open-ended, patient-centered outcomes. They could collect long-term follow-up data on all patients involved in clinical research. Yet no studies go to such lengths. The problem is one of prioritization, which reflects the values of patients and doctors and the requirements set by hospitals, professional societies, journal editors, and government regulators.

To make good medical decisions, we need more than just an improved knowledge of efficacy and safety. We need to clarify the values that shape decisions, and we need to improve our decision-making processes. Many models have been proposed for establishing a more rational basis for medical practice. Physicians sought therapeutic reform through evidence-based medicine and comparative effectiveness research. Employers and insurers implemented techniques of managed care to regulate (and limit) medical practice. Economists and policy makers introduced the economic rationality of cost-effectiveness analysis, which considers not just the medical interests of the individual patient but also the financial health of the broader society. Advocates of patient-centered care articulated models of shared decision making based on serious efforts to provide patients with the best possible knowledge of realistic treatment outcomes. Studies have found that patients, once well informed, are less likely to pursue aggressive interventions. Unfortunately other studies have found that doctors do not give patients complete information about their treatment options, and one survey revealed that only 16 percent of angioplasty patients were even asked about their treatment preferences.[40]

All these reforms have bred resistance. Evidence-based medicine relies on the fickle results of clinical trials. Managed care earned the wrath of patients and doctors who resent the intrusion of utilization reviewers. Cost-effectiveness analysis raises fears of rationing. Shared decision making places burdens on patients, some of whom would rather have their physicians make the decision. These reforms also rely on a shared hope: that if doctors and patients know the right information (whether about safety, efficacy, or cost), they will make the right decisions. But the situation is more complex than this.

There are different sources of information about efficacy, from disease models to clinical trials. Which should be weighted more? There are different ways of assessing the relevance of treatment complications. There are different ways to make a decision. What kind of decision is the best decision: an evidence-based one, a cost-effective one, or a patient-centered one? These are questions of values. Reasonable people will disagree. As a result, reforms cannot simply target a single thread in a web. Doctors and policy makers need to develop sophisticated ways of thinking about the complexity of disease and therapeutics in order to place medical decisions on a more stable foundation. Historical analysis can play a valuable role by clarifying the interests and stakes that influence medical decisions and the values that motivate them.

What will the future bring? It is likely that rates of coronary revascularization will decrease in the future, at least in the United States and Europe. This could happen for different reasons. One important factor is the decline in coronary artery disease. Cardiac mortality rates in the United States have fallen by more than 60 percent since their peak in the 1960s. Changes in diet, exercise, smoking, and medical care, and possibly even broader changes in society and the economy, have all contributed. If these trends continue, then heart disease will lose its status as the leading cause of death in the near future. Further declines are possible if populations become better at using what is already known about coronary artery disease, especially the roles of diet, exercise, smoking, and other risk factors, to reduce the burden of the disease.[41] Pressure on revascularization could also come from another direction. Increasingly constrained resources might force us to reassess our current health care priorities and take cost effectiveness into consideration. This would shift resources away from coronary revascularization and toward other treatments and preventive strategies.

The future is more difficult to see for low- and middle-income countries. Epidemiologists once dreamed of an alternative epidemiological transition in which developing countries would shed their traditional burden of infectious diseases without acquiring our burden of chronic and degenerative diseases. This has not happened. Instead, old epidemics persist even as the diseases of civilization take root.[42] Although cardiovascular disease has finally begun to receive attention from global health policy makers, the enormous disparities that exist in access to treatments have not yet provoked a sense of injustice. As the epidemic of heart disease tightens its grip, demands for coronary revascularization will inevitably increase. This will create difficult dilemmas for health and social justice. If we really expect developing countries to set aside expensive, technologically intensive cardiac medicine in favor of prevention and low-cost pharmaceutical therapies, then we will need to think hard about the precedents our own choices set.

The decisions we make for ourselves and the decisions we make as part of global communities must take advantage of all that is known about medicine and disease. Biomedical research, from molecular biology to comparative effectiveness research, has much to offer, but it cannot be the whole story. Disease and therapeutics are social processes that reflect the structures and values of our society. The messy details of the complicated worlds in which we live leave doctors and policy makers as vexed as Hunter was by his rascals. But we cannot turn our backs on complexity. We need to confront head on the complex social dynamics that influence the diseases we suffer, the treatments we can access, the outcomes of those treatments, and our knowledge of those processes. The decisions we make about how best to respond to the global challenge of heart disease will be defining decisions for the twenty-first century. We need to ensure that we make them well.

Notes

INTRODUCTION: An Embarrassment of Riches

1. Heberden, "Some Account" (1772). See also Proudfit, "Origin of Concept" (1983); Proudfit, "John Hunter" (1986).

2. Quoted in Sanders, "Inner World" (1962): 691.

3. Proudfit, "John Hunter" (1986); Heberden, "Some Account" (1772); Alberti, "Bodies, Hearts, and Minds" (2009): 799.

4. McFadden and Altman, "Clinton Is Given Bypass" (2004) [cigar, cliff]; McFadden, "Clinton Suffers Pains" (2004) [potentially].

5. Baker and Macropoulos, "Bill Clinton" (2010).

6. Effler, comment in Favaloro and others, "Myocardial Revascularization" (1967): 370.

7. Stallones, "Rise and Fall" (1980).

8. Early history: Michaels, *Eighteenth-Century Origins* (2002). Mortality data: CDC, "Leading Causes" (2009). American disease: Leary, "Mallory Institute Dedication" (1933): 7. Skeptics: Bolduan and Bolduan, "Is the 'Appalling Increase' Real?" (1932).

9. NHI: Fye, *American Cardiology* (1996). Eisenhower: Lasby, *Eisenhower's Heart Attack* (1997). Reversal: Havlik and Feinleib, *Proceedings* (1979); Jones and Greene, "Contributions of Prevention and Treatment" (2012). Global burden: Mathers and Loncar, "Projections" (2006).

10. Evans, "Losing Touch" (1993); Greene, "Releasing the Flood Waters" (2005); Jones, Podolsky, and Greene, "Burden of Disease" (2012).

11. Rosenberg, "Pathologies of Progress" (1998).

12. Rosenberg, "Disease and Social Order" (1986); Brandt, *No Magic Bullet* (1987); Hacking, *Social Construction* (2000): 125–62; Duffin, "Disease Game" (2005).

13. Erickson, *Language of the Heart* (1997); Blair, *Victorian Poetry* (2006); Alberti, *Matters of the Heart* (2010), especially pp. 140–56. The shift in attention is paralleled by the paucity of work on the history of heart disease compared with that of other ailments. This is a blindspot in the medical social sciences. The few good works on the subject include Aronowitz, *Making Sense of Illness* (1998); Fye, *American Cardiology* (1996); Howell, *Heart Attack* (forthcoming).

14. Effler, "Surgery" (1968): 43.

15. Fye, "History of the Origin" (1994).

16. Histories: Fye, "Coronary Arteriography" (1984); King, "Development" (1998). Early attempts: Radner, "Attempt" (1945); Lemmon and others, "Suprasternal" (1959).

17. Sones and Shirey, "Cine Coronary Angiography" (1962); Sones to Roy (1963); Mearns, "Effler and Sones" (1973).

18. Shumacker, *Evolution* (1992): 129–42; Jones, *On the Origins* (in preparation).

19. Favaloro, "Saphenous Vein Autograft" (1968); Grüntzig, "Transluminal" (1978).

20. Lawrence and Hall, *1997 Summary* (1999); Hall and others, *National Hospital— 2007* (2010).

21. Ford and others, "Explaining the Decrease" (2007); Jones and Greene, "Contributions of Prevention and Treatment" (2012).

22. Therapeutics in context: Rosenberg, "Therapeutic Revolution" (1977); Pernick, *Calculus of Suffering* (1985); Warner, *Therapeutic Perspective* (1986); Pressman, *Last Resort* (1998); Lerner, *Breast Cancer Wars* (2001). Social efficacy: Rosenberg, *Our Present Complaint* (2007): 9–10; Aronowitz, "Converged Experience" (2009). Therapeutics gone astray: Scull, *Madhouse* (2005).

23. Surveys and oral histories: Baldry, *Battle Against Heart Disease* (1971); Shumacker, *Evolution* (1992); Weisse, *Heart to Heart* (2002); Stoney, *Pioneers* (2008). Transplantation and pacemakers: Fox and Swazey, *Courage to Fail* (1974); Jeffrey, *Machines in Our Hearts* (2001); Pollock, "Internal Cardiac Defibrillator" (2008). Angioplasty: Monagan and Williams, *Journey into the Heart* (2007). Critics: Millman, *Unkindest Cut* (1977); Ozner, *Heart Hoax* (2008). Patient memoirs: MacKenzie, "Risk" (1970); Stutz, "Pump Head" (2003). The literature remains much thinner than that on many other medical topics.

24. Jones, "Visions" (2000).

25. Eight-five percent: Feldman and others, "Comparison of Outcomes" (2006): 1336. Fifteen percent: Chan and others, "Appropriateness" (2011).

26. McGlynn and others, "Quality of Health Care" (2003).

27. Overuse: Brownlee, *Overtreated* (2008); Hadler, *Worried Sick* (2008); Lenzer and Brownlee, "Reckless Medicine" (2010); Freedman, "Lies, Damned Lies" (2010). Angioplasty survey: Fowler and others, "Decision-Making Process" (2012). Clinical cascade: Epstein, "Use of Diagnostic Tests" (1996): 1197; Topol and Nissen, "Our Preoccupation" (1995).

28. Nostalgia: Brandt and Gardner, "Golden Age" (2000). Evidence-based medicine: Timmermans and Berg, *Gold Standard* (2003); Daly, *Evidence-Based Medicine* (2005); Weisz and others, "Emergence" (2007); Marks, "What Does Evidence Do?" (2009). Other authorities: Isaacs and Fitzgerald, "Seven Alternatives" (1999). Shared decision making: Sepucha and Mulley, "Perspective on the Patient's Role" (2009). Insider perspectives: Thomas, *Youngest Science* (1983); Gawande, *Complications* (2002); Groopman, *How Doctors Think* (2007); Gawande, *Checklist Manifesto* (2010); Groopman and Hartzband, *Your Medical Mind* (2011).

29. Institute for Motivational Research, *Research Study on Pharmaceutical Advertising* (1955), quoted in Greene and Podolsky, "Keeping Modern" (2009): 342 [misery]; *Dartmouth Atlas 1998* (1998): 114 [dilemma].

30. Rapp, *Testing Women* (1999): 220–36.

31. Brandt, *Cigarette Century* (2007): 159–210; Proctor and Schiebinger, *Agnotology* (2008); Conway and Oreskes, *Merchants of Doubt* (2011). See also Jasanoff, "Songlines of Risk" (1999).

32. As demonstrated well by organ transplantation and reproductive technology: Fox

and Swazey, *Courage to Fail* (1974); Lock, *Twice Dead* (2002); Thompson, *Making Parents* (2005).

33. "Doctor-Owned Specialty Hospitals" (2003); Casalino and others, "Focused Factories?" (2003); Desmon and Little, "St. Joe's" (2009).

34. Bosk, *Forgive and Remember* (1979); Good, "How Medicine Constructs" (1994); Mol, *Body Multiple* (2002); Schlich, "Emergence of Modern Surgery" (2004); Saunders, *CT Suite* (2008).

35. Sheldon to Farmer (1979); Sheldon, "1979 Annual Report" (1980); Monagan and Williams, *Journey into the Heart* (2007).

36. Clough, *To Act as a Unit* (2004); Sweeney and Lautzenheiser, *Neuroscience* (2006); Sheldon, *Pathfinders of the Heart* (2008).

CHAPTER ONE: The Mysteries of Heart Attacks

1. Kee and others, "Risks and Benefits" (1997); Whittle and others, "Understanding the Benefits" (2007); Rothberg and others, "Patients' and Cardiologists' Perceptions" (2010).

2. Mount Sinai Hospital, "Ironic" (2011). For a parallel discussion of disease model and therapeutic mechanism, see Duffin and Campling, "Therapy and Disease Concepts" (2002).

3. Robbins, "Cardiac Pathology" (1974): 9, 21.

4. Fleck, *Genesis and Development* (1935); Kuhn, *Structure* (1962); Latour and Woolgar, *Laboratory Life* (1979); Shapin and Schaeffer, *Leviathan* (1985); Collins, *Changing Order* (1985); Galison, *How Experiments End* (1987); Latour, *Pasteurization* (1988); Biagioli, *Galileo, Courtier* (1994); Shapin, *Social History of Truth* (1995); Daston, "Coming into Being" (2000); Shapiro, *A Culture of Fact* (2000); Winter, *Mesmerized* (2000); Secord, *Victorian Sensation* (2000); Browne, *Charles Darwin* (2002).

5. Egyptian Mummies: Allam and others, "Atherosclerosis" (2011). Osler, *Principles and Practice* (1892): 640–41, 655–59, 669.

6. "Heart attack" appeared in passing in various medical reports: Minard, "Pathological Anteflexion" (1891); Winslow, "Outbreak of Tonsillitis" (1911); "Therapeutics: Disturbances of the Heart" (14 Dec. 1912); "Therapeutics: Disturbances of the Heart" (21 Dec. 1912). It appeared frequently in doctors' obituaries, for example Emerson, "William Palmer Bolles" (1917): 363. "Heart attack" also appeared in general periodicals. For examples from the *New York Times*, see "Early Tricks on the Comstock" (1895); "His Joy Killed Him" (1903). It is difficult to know whether any of these early instances correspond to the modern diagnosis. In 1927 the *Boston Medical and Surgical Journal* published its first article devoted to heart attacks (no earlier one appeared in *JAMA*). The author protested the "multitude of various diagnoses" all lumped under that term. See Sprague, "'Heart Attacks'" (1927): 472. For the problem of changing disease names over time, see Aronowitz, *Making Sense of Illness* (1998); Rosenberg, "Tyranny of Diagnosis" (2002).

7. Herrick, "Clinical Features" (1912); Herrick, "Thrombosis" (1919): 389 [doctor], 388 [Atlantic City]. Herrick did not use "heart attack"; instead he described coronary thromboses or obstructions. Although Americans often credit Herrick for describing the phenomenon of nonfatal myocardial infarction, priority belongs to two Russian physicians, V. P. Obraztsov and N. D. Strazhesko, who described the syndrome in 1910. Herrick

cited a German translation of this Russian article in his 1912 account. See Muller, "Diagnosis" (1977).

8. Herrick, "Thrombosis" (1919), 387 [doctor]; Herrick, "Clinical Features" (1912) [literature review].

9. Blumgart and others, "Studies" (1940): 81 [site]; Blumgart and others, "Angina Pectoris" (1941): 96 [there is no].

10. Hamman, "Symptoms" (1926).

11. Braunwald, "Heart" (1966).

12. Gofman and others, "Role of Lipids" (1950); Constantinides, "Experimental Atherosclerosis in the Rabbit" (1965); Steinberg, "Interpretive History" (2004); Konstantinov and others, "Nikolai N. Anichkov" (2006); Li, *Triumph of the Heart* (2009): 11–13.

13. By 1965 researchers had published case reports of spontaneous myocardial infarctions in many species, including dogs, a chimpanzee, a bald eagle, a white collared mangabey, and even a sperm whale. But it was not always clear if the report was credible or if atherosclerosis was the cause. See Strauss and others, "Spontaneous Myocardial Infarcts" (1965). For the rabbit case, see Taylor and others, "Fatal Myocardial Infarction" (1959); Taylor and others, "Atherosclerosis" (1962): 32 [undomesticatable]; Taylor, "Experimentally Induced Arteriosclerosis" (1965).

14. Leary, "Research in the Hospital" (1939): 6.

15. Leary, "Coronary Spasm" (1935): 342–43 [violinist, bookie]; Leary, "Pathology" (1935): 333 [other triggers]. Boston doctors: Fitzhugh and Hamilton, "Coronary Occlusion" (1933). Details of the "unusual sexual activity" were deliberately left vague: "Questioning on this subject was not done in a routine manner because we feel that the difficulties in obtaining full information would not yield enough useful instruction to justify the effort and annoyance of collecting it" (p. 480).

16. Leary, "Coronary Spasm" (1935): 342–43. Skeptics: Master and others, "Activities Associated" (1939)—this is discussed in detail in chapter 5.

17. Friedberg and Horn, "Acute Myocardial Infarction" (1939).

18. Clark and others, "Thrombosis" (1936).

19. Paterson, "Vascularization and Hemorrhage" (1936); Paterson, "Capillary Rupture" (1938); Paterson, "Capillary Rupture" (1939).

20. The lesions: Leary, "Pathology" (1935): 334 [cysts], 337 [rupture]. Rabbit atherosclerosis: Leary, "Experimental Atherosclerosis" (1934). As he explained, "If a lesion of similar degree were discovered in the left coronary artery of a human being who was found dead with evidence of indigestion there is little doubt that the cause of death would be properly referred to [as] coronary sclerosis and insufficiency" (p. 489). Disasters: Leary, "Cocoanut Grove Club" (no date); Leary, "Disasters" (no date).

21. Praise: Timmins to Leary (1934); Banting to Leary (1934). Critiques: Leary, "Vascularization" (1938).

22. Periodic wars: Forrester and others, "Perspective" (1987): 505. Naming: Master and others, "Coronary Occlusion" (1944): 814.

23. Vague consensus: American Heart Association, "Report of the Committee" (1962): 615. Vexation: Robbins, "Cardiac Pathology" (1974): 11.

CHAPTER TWO: The Case for Plaque Rupture

1. For Thorvaldsen's death, see Falk, "Why Do Plaques Rupture" (1992): III-30. As with "heart attacks," the terminology is tricky here. Over the decades, researchers have used different terms to describe what a modern medical scientist would see as the same process: plaque fissures, plaque ruptures, fragile plaques, vulnerable plaques, and others. I use "plaque rupture" as the umbrella term but pay close attention to the words in use at a particular time.

2. For the conference, see Roberts and others, *Comparative Atherosclerosis* (1965). For orcas and dolphins, Roberts and others, "Spontaneous Atherosclerosis" (1965): 153–55. For the emotional monkey, Taylor, "Experimentally Induced Atherosclerosis" (1965): 229.

3. Constantinides and others, "Production" (1960); Constantinides, *Experimental Atherosclerosis* (1965); Constantinides, "Cause of Thrombosis" (1990): 37G [detail].

4. Constantinides and Lawder, "Experimental Thrombosis" (1963); Constantinides, "Experimental Atherosclerosis in the Rabbit" (1965): 285 [violent].

5. "Coronary Thrombosis" (1964): 35 [mystery], 37 [while fissures]; Constantinides, "Plaque Fissures" (1966): 1 [speculation], 5 [since we did not].

6. Friedman, *Pathogenesis* (1969). See also Aronowitz, *Making Sense of Illness* (1998): 145–55.

7. Friedman, *Pathogenesis* (1969): 172.

8. Friedman and van den Bovenkamp, "Pathogenesis of Coronary Intramural Hemorrhages" (1966): 347–48 [opinions], 348 [luminal catastrophe]; Friedman and van den Bovenkamp, "Pathogenesis of a Coronary Thrombus" (1966); Friedman, *Pathogenesis* (1969): 75 [animal models], 164–94; Friedman, "Coronary Thrombus" (1971).

9. Constantinides: "Coronary Thrombosis" (1964): 37 [only a few]; Constantinides, "Plaque Fissures" (1966): 16 [Miller]. Friedman: Friedman, *Pathogenesis* (1969): 173 Other groups: Chapman, "Morphogenesis" (1965); Bouch and Montgomery, "Cardiac Lesions" (1970).

10. Roberts, "Coronary Arteries" (1972): 221 [even by this, interpretation]; Roberts and Buja, "Frequency and Significance" (1972): 435 [cracks].

11. Roberts and Buja, "Frequency and Significance" (1972): 427; Roberts, "Coronary Arteries" (1972): 223 [chance]; Friedman, "Pathogenesis" (1975): 39.

12. Latour and Woolgar, *Laboratory Life* (1979); Lynch, "Discipline" (1985); Lynch, "Science in the Age" (1991); Rasmussen, *Picture Control* (1997); Galison, *Image and Logic* (1997); Daston and Galison, *Objectivity* (2007).

13. Clinical gaze: Foucault, *Birth of the Clinic* (1973). Medical visualization: Saunders, *CT Suite* (2008), especially pp. 15–17 [autopsy].

14. Friedman, "Pathogenesis" (1975): 39.

CHAPTER THREE: The Case against Plaque Rupture

1. Galison, *How Experiments End* (1987); Collins and Pinch, *Golem* (1998); Jones, "Virgin Soils Revisited" (2003); Jones, *Rationalizing Epidemics* (2004); Jones, "Persistence" (2006).

2. Leary, "Coronary Spasm" (1935): 344.

3. Variant asthma: Prinzmetal and others, "Angina Pectoris" (1959). Nitrates: Sewell, "Coronary Spasm" (1966).

4. Lange and others, "Nonatheromatous Ischemic Heart Disease" (1972).

5. Reviews: Davis, "Incidence" (1970); Robbins, "Cardiac Pathology" (1974). Edinburgh: Branwood and Montgomery, "Observations" (1956): 368. Oxford: Mitchell and Schwartz, "Relation" (1963).

6. Baroldi, "Lack of Correlation" (1969): 520.

7. Initial study: Spain and Bradess, "Relationship" (1960). Follow up: Spain and Bradess, "Sudden Death" (1970): 110.

8. Erhardt and others, "Incorporation" (1973).

9. Ibid., 390.

10. Lerner, "Perils" (1992); Howell, *Technology* (1995); Dumit, *Picturing Personhood* (2004); Saunders, *CT Suite* (2008).

11. Effler, "Operative Report" (1962); Effler, "Surgical Treatment" (1969): 8 [frequently].

12. Mayo: Eusterman and others, "Atherosclerotic Disease" (1962). Minnesota: Vlodaver and others, "Correlation" (1973). Cleveland: Zimmerman and Demany, "Coronary Artery Visualization" (1966): 725.

13. Sones, "1963 Annual Report" (no date). For photographic realism, see Daston and Galison, *Objectivity* (2007).

14. Safari: Mearns, "Effler and Sones" (1973): 4. Effler's assessment: Effler, "Surgery" (1968): 38 [printing press], 39 [visual diagnosis]; Effler, "Surgical Treatment" (1969): 3 [select patients]. Lasker Award: DeBakey, "Albert Lasker" (1983). Patient education: Richards, *Heart to Heart* (1987): 99, which echoes Sones: see Mearns, "Effler and Sones" (1973): 5.

15. The 1967 study: Likoff and others, "Paradox" (1967). The 1974 study: Khan and Haywood, "Myocardial Infarction" (1974). Denver studies: Oliva and others, "Coronary Arterial Spasm" (1973); Oliva and Breckinridge, "Arteriographic Evidence" (1977).

16. Anarchy: Prioreschi, "Myocardial Infarction" (1966): 370. Discord: Friedberg, "Symposium" (1972): 181. Ignorance: Robbins, "Cardiac Pathology" (1974). Heretics: Baroldi, "Reply" (1977): 392.

CHAPTER FOUR: Learning by Doing

1. Classification in science: Foucault, *Order of Things* (1966); Hacking, *Social Construction* (2000); Bowker and Star, *Sorting Things Out* (2000). Classification of disease: Sicherman, "Uses of a Diagnosis" (1977); Hacking, *Mad Travelers* (1998); Rosenberg, "Tyranny of Diagnosis" (2002).

2. Herrick, "Clinical Features" (1912): 2016 [clinical manifestations]; Herrick, "Thrombosis" (1919): 387 [other quotations].

3. For example, Miller and others (1951); Robbins and Cotran, "Heart" (1979): 653–56.

4. Wide estimates: Davis, "Incidence" (1970); Robbins, "Cardiac Pathology" (1974). First series: Davies and others, "Pathology" (1976): 659 [hardly credible]. Sudden death series: Davies and Thomas, "Thrombosis" (1984): 1140 [virtually all].

5. Littman, "The Second" (1972); Lasby, *Eisenhower's Heart Attack* (1997): 113–54; Reiser, "Intensive Care Unit" (1992).

6. For the prehistory, see Jones, *On the Origins* (forthcoming). For surgery for heart attacks, see Cohn and others, "Aorto-coronary Bypass" (1972), especially comments by Floyd Loop (p. 510) and Favaloro (pp. 512–13).

7. Many names: Braunwald, "Unstable Angina" (1989). Preinfarction: Harrison and Shumway, "Evaluation and Surgery" (1972): 50 [presupposes, clinical state]. Los Angeles meeting: Favaloro, comment in Cohn and others, "Aorto-coronary Bypass" (1972): 513. Cleveland series: Loop, comment in Cohn and others, "Aorto-coronary Bypass" (1972): 510.

8. Effler, in discussion in Johnson and Kayser, "Expanded Indication" (1973): 6. See also Lin and others, "Why Physicians Favor" (2008).

9. Old fears of acute angiography: DeWood and others, "Prevalence" (1980): 897. "Sine qua non": Effler, "New Era" (1966): 1311. Cleveland series: Favaloro and others, "Acute Coronary Insufficiency" (1971): 598; Loop, comment in Cohn and others, "Aorto-coronary Bypass" (1972): 510. Other centers: Cohn and others, "Aorto-coronary Bypass" (1972).

10. DeWood and others, "Prevalence" (1980): 901.

11. Ibid., 899.

12. Fletcher and others, "Maintenance" (1959): 1112; Sikri and Bardia, "History of Streptokinase" (2007).

13. One study: Rentrop and others, "Selective Intracoronary Thrombolysis" (1981): 315 [in agreement]. Italian trial: Gruppo Italiano, "Effectiveness" (1986). Success: Ganz and others, "Intracoronary Thrombolysis" (1981): 10.

14. Rentrop and others, "Selective Intracoronary Thrombolysis" (1981).

15. Heart attack angioplasty: Rentrop and others, "Initial Experience" (1979). Comparison: O'Neill and others, "Prospective Randomized Trial" (1986): 812.

16. Leinbach and Gold, "Coronary Angiography" (1982): 771.

17. Glagov, Preface (1990): xii; Harker, in "Roundtable Discussion" (1995): 97B.

18. For a discussion of clinical practice as experiment, see Fox and Swazey, *Courage to Fail* (1974), especially pp. 60–83.

19. Gregory, "Cell Antagonism" (1887): 647.

20. Nutton, "From Medical Certainty" (1991).

21. Many atheroscleroses: Mol, *Body Multiple* (2002). Galison, "Trading Zone" (1999).

22. O'Connor and others, "Azithromycin" (2003).

23. Collins, *Changing Order* (1992).

24. Healy, *Antidepressant Era* (1997); Healy, *Creation of Psychopharmacology* (2002); Carlat, *Unhinged* (2010).

CHAPTER FIVE: The Plaque Rupture Consensus

1. Falk, "Plaque Rupture" (1983); Gorlin and others, "Anatomic-Physiologic Links" (1986); Fuster and others, "Atherosclerotic Plaque Rupture" (1990); Fuster and others, "Pathogenesis [Part 1]" (1992); Fuster and others, "Pathogenesis [Part 2]" (1992).

2. Falk, "Unstable Angina" (1985): 706.

3. Success story: Libby, "Molecular Bases" (1995). Thermal probes: Casscells and others, "Thermal Detection" (1996).

4. Forrester and others, "Perspective" (1987): 505.

5. Kuhn, *Structure* (1962). For discussions, see Fuller, *Thomas Kuhn* (2000); Baake, *Metaphor and Knowledge* (2003).

6. Obrzatov and Strazhesko: Muller, "Diagnosis" (1977); Muller and others, "Circadian Variation and Triggers" (1989). Against triggering: Master and others, "Activities Associated" (1939); Master, "Role of Effort" (1960).

7. Muller and others, "Circadian Variations in Frequency" (1985): 1315.

8. Muller and others, "Circadian Variation and Triggers" (1989): 741.

9. Falk, "Plaque Rupture" (1983): 133. Muller and Tofler, "Triggering" (1992): 400–401. Macrophages: Lendon and others, "Atherosclerotic Plaque Caps" (1991). Molecular mechanisms: Libby and others, "Inflammation and Atherosclerosis" (2002).

10. Muller and others, "Circadian Variation and Triggers" (1989): 739; Muller and others, "Triggers" (1994).

11. Naghavi and others, "From Vulnerable Plaque [Part 1]" (2003): 1665.

12. Identifying patients: Naghavi and others, "From Vulnerable Plaque [Part 1]" (2003), especially p. 1665 [future genomic]; Naghavi and others, "From Vulnerable Plaque [Part 2]" (2003): 1777 [bioinformatics]; Muller and others, "New Opportunities" (2006).

13. Fortun, "Mediated Speculations" (2001); Fortun, *Promising Genomics* (2008); Hedgecoe, "Terminology and the Construction" (2003).

14. Prominent reviews: Fuster and others, "Atherosclerotic Plaque Rupture" (1990); Fuster and others, "Pathogenesis [Part 1]" (1992). Authorial authority: Naghavi and others, "From Vulnerable Plaque [Part 1]" (2003).

CHAPTER SIX: Rupture Therapeutics

1. Facts become most interesting when they become relevant. Daston examines salience, productivity, and embeddedness: Daston, "Coming into Being" (2000). Latour suggests leaving aside debates about the reality of facts to focus on what makes them "matters of concern": Latour, "Why Has Critique" (2004); Latour, *What Is the Style* (2005); Latour, *Reassembling* (2007), 114–15. This echoes Kleinman's call for anthropologists to focus on what is "at stake" for people: Kleinman, *Writing at the Margin* (1995).

2. Aspirin's potential: Mustard, "Platelets and Thrombosis" (1972). Disappointing results: Aspirin Myocardial Infarction Study Research Group, "Randomized Controlled Trial" (1980).

3. Unstable angina: Lewis and others, "Protective Effects" (1983). Heart attack treatment: "Randomised Trial of Intravenous Streptokinase" (1988). Heart attack prevention: Hennekens and others, "Aspirin" (1989); Ridker and others, "Low-dose Aspirin" (1991). Broad efficacy: Fuster and others, "Aspirin" (1993).

4. Alternatives: FitzGerald, "Ticlopidine" (1990). Clopidogrel: CAPRIE Steering Committee, "Randomized, Blinded Trial" (1996). Marketing Plavix: Iskowitz, "Cardiovascular" (2011): 39.

5. Early interest: Boyd, "Inflammatory Basis" (1928). Cholesterol's shadow: Crawford,

"Morphological Aspects" (1961). Renewed interest: Moreno and others, "Macrophage Infiltration" (1994); Falk and others, "Coronary Plaque Disruption" (1995).

6. Inconclusive trials: Marks, *Progress* (1997); Greene, *Prescribing by Numbers* (2007): 164–67. Modest results: Rafflenbeul and others, "Quantitative" (1979).

7. Statin history: Steinberg, "Interpretive History" (2006); Greene, *Prescribing by Numbers* (2007): 177–219; Li, *Triumph of the Heart* (2009). Statin results: Brown and others, "Regression" (1990). Medicating life: Greene, "Abnormal and the Pathological" (2007); Dumit, "Pharmaceutical Witnessing" (2010); Dumit, "Prescription Maximization" (2012).

8. Brown and others, "Regression" (1990).

9. Statin mechanisms: Loscalzo, "Regression" (1990); Brown and others, "Lipid Lowering" (1993); Rabbani and Topol, "Strategies" (1999). Statin sales: "1988 Midyear" (1988); "1992 Midyear" (1992); Madden and others, "Business Watch 2000" (2001); Liebman, "DTC's Role" (2001); Iskowitz, "Cardiovascular" (2011): 39.

10. Rosenberg, "Therapeutic Revolution" (1977); Pellegrino, "Sociocultural Impact" (1979); Warner, *Therapeutic Perspective* (1986); Paul, "Relentless Therapeutic Imperative" (2004).

11. Early textbooks: Ross, "Ischemic Heart Disease" (1974); Julian, "Myocardial Infarction" (1979); Braunwald and Cohn, "Ischemic Heart Disease" (1983). Interested reviews: Fuster and others, "Pathogenesis [Part 1]" (1992); Sobel, "Acute Myocardial Infarction" (1992); Libby, "Molecular Bases" (1995); Libby, "Atherosclerosis" (1998); Fuster, "Atherosclerosis-Thrombosis" (2000). "recent": Libby and others, "Current Concepts" (1998): 14S. "new and emerging": Kolata, "New Heart Studies" (2004).

12. Molecular therapies: Rabbani and Topol, "Strategies" (1999). Five strategies: Forrester, "Prevention" (2002), especially p. 830 [Cell biology].

13. Daston, "Coming Into Being" (2000).

14. Germ theory: Warner, *Therapeutic Perspective* (1986); Tomes, *Gospel of Germs* (1998); Barnes, *Great Stink* (2006). Hypertension: Evans, "Losing Touch" (1993); Porter, "Life Insurance" (2000); Greene, "Releasing the Flood Waters" (2005).

15. Physicians have long used "therapeutic enthusiasm" to label unsubstantiated faith in the effectiveness of treatments: Gregory, "Cell Antagonism" (1887); Woods, "Use of Diaphoresis" (1904); D., "Therapeutic Enthusiasm" (1911). Concerns became more pronounced in the 1970s: Cornfield, "Approaches to Assessment" (1972): 1126; Haggerty, "Effectiveness" (1973): 372; Kaplan, "Support of Continuing" (1979); McKinlay, "From 'Promising Report'" (1981). For historical analyses, see Pellegrino, "Sociocultural Impact" (1979); Kawachi and Conrad, "Medicalization" (1996); Greene, *Prescribing by Numbers* (2007); Greene and Podolsky, "Keeping Modern" (2009). Case studies: Pressman, *Last Resort* (1998); Lerner, *Breast Cancer Wars* (2001); Watkins, *Estrogen Elixir* (2007).

CHAPTER SEVEN: Therapeutic Ruptures

1. Plumbing job: Sones or Effler [unclear from context], quoted in Mearns, "Effler and Sones" (1973): 6. Mearns also writes about "repairing the plumbing in hearts" (p. 4). Critique: Sanghavi, "Plumber's Butt?" (2007). Advertisement: Mount Sinai Hospital, "Ironic" (2011). The real irony is that the *New York Times* had described persistent belief

in progressive obstruction as a "popular misconception." See Kolata, "How It Happens" (2007).

2. Surgeon's confidence: Effler, "Role of Surgery" (1969): 377. The debates: Murphy and others, "Treatment" (1977); Jones, "Visions" (2000).

3. Cohn and others, "Aorto-coronary Bypass" (1972): 506.

4. Prophylaxis: Favaloro, in discussion of Cohn and others, "Aorto-coronary Bypass" (1972): 513. Zapped: Mearns, "Effler and Sones" (1973). For the spread, see the range of authors in the discussion of the Cohn article, pp. 509–13.

5. Cosgrove and others, "Should Coronary Arteries" (1981): 525.

6. Early hopes: Grüntzig, "Transluminal" (1978): 263 [the atheroma]; Grüntzig and others, "Nonoperative Dilatation" (1979): 65 [comparatively]. Broadening indications: Ischinger and others, "Should Coronary Arteries" (1983): 148 [preventive], 152 [timely].

7. Bypass peak: Lawrence and Hall, *1997 Summary* (1999). Most recent data: Hall and others, *National Hospital—2007* (2010).

8. Cleveland: Kramer and others, "Segmental Analysis" (1983). Dutch: Bruschke and others, "Dynamics" (1989): 296.

9. Shub and others, "Unpredictable Progression" (1981): 158 [almost], 155 [our findings]. For unpredictability and cancer, see Aronowitz, *Unnatural History* (2007).

10. Autopsies: Davies and Thomas, "Thrombosis" (1984): 1139. Prior angiograms: Ambrose and others, "Angiographic Progression" (1988): 56. Small plaque: Falk and others, "Coronary Plaque Disruption" (1995).

11. Comparisons: Vlodaver and others, "Correlation" (1973). Ultrasound: McPherson and others, "Delineation" (1987): 304. Luminograms: Topol and Nissen, "Our Preoccupation" (1995).

12. White and others, "Does Visual Interpretation" (1984): 823.

13. Maseri and Fuster, "Is There" (2003): 2068.

14. Little and others, "Can Coronary Angiography" (1988): 1165–66.

15. Bypass trials: Murphy and others, "Treatment" (1977); CASS Principal Investigators, "Coronary Artery Surgery Study: Survival Data" (1983); Jones, "Visions" (2000). A recent angioplasty trial: Boden and others, "Optimal Medical Therapy" (2007). Responsible plaques: Little and others, "Cause of Acute" (1990); Libby, "Molecular Bases" (1995).

16. Disturbing: White and others, "Does Visual Interpretation" (1984): 822. Dogma: Libby, "Molecular Bases" (1995). Fixation: King, "Development" (1998): 80B. Illusion: Forrester and Shah, "Lipid Lowering" (1997). Koestler wrote about communism; I do not know whether Forrester and Shaw meant to invoke such connotations.

17. Reflex: Topol, "Coronary Angioplasty" (1988): 975. Other comments: Topol and Nissen, "Our Preoccupation" (1995).

18. Proliferation: Casscells and others, "Thermal Detection" (1996): 1449. Image analysis: Brown and others, "Quantitative Coronary" (1977). Eccentric lesions: Ambrose and others, "Angiographic Morphology" (1985). Intravascular ultrasound: Nissen and others, "Application" (1990): 665. Angioscopy: Forrester and others, "Perspective" (1987). Spectroscopy: Waxman and others, "Detection and Treatment" (2006). Thin-capped plaques: Stone and others, "Prospective" (2011); this study was funded by Volcano Corporation to demonstrate the value of its imaging catheters.

19. Lerner, *Breast Cancer Wars* (2001); Aronowitz, *Unnatural History* (2007).

20. Focal enough for surgery: Bailey and others, "Survival" (1957); Blumgart and others, "Discussion" (1959). Diffuse enough to be problematic: Szilagyi and others, "Applicability" (1958); Crawford and others, "Coronary-Artery Pathology" (1961). Effler's operative notes could be evocative: "By palpation, the coronary arteries, both right and left, feel like uncooked noodles. There is obvious extensive atherosclerosis in all of the vessels." See Effler, "Operative Report" (1962). See also Jones, *On the Origins* (forthcoming).

21. Michigan: Goldstein and others, "Multiple Complex" (2000): 920. Macrophages: Buffon and others, "Widespread" (2002): 5. AHA: Smith, "Risk-Reduction" (1996).

22. Hot spots: Cheruvu and others, "Frequency" (2007); Serruys and others, "From Postmortem" (2007). Buy time: Casscells and others, "Vulnerable" (2003): 2073. Entire tree: Little and others, "Can Coronary Angiography" (1988): 1157. Whack-a-mole: Feder, "In Quest to Improve" (2006).

23. The exchange: Boden and others, "Optimal Medical Therapy" (2007); Wharton and others, "PCI" (2007); Boden and others, "Authors Reply" (2007). High estimate: Feldman and others, "Comparison of Outcomes" (2006). Low estimate: Chan and others, "Appropriateness" (2011).

24. International Study of Comparative Health Effectiveness with Medical and Invasive Approaches (2012).

25. Kee and others, "Risks and Benefits" (1997); Whittle and others, "Understanding the Benefits" (2007); Rothberg and others, "Patients' and Cardiologists' Perceptions" (2010). "Therapeutic autonomy" has many meanings, from patient autonomy in a world of informed consent to physicians' efforts to preserve autonomy in a world of managed care. See Starr, *Social Transformation* (1982); Weisz, "Origins of Medical Ethics" (1990). I focus on how treatments achieve inertia that buffers them as medical theories change.

26. Effler's income: Mearns, "Effler and Sones" (1973): 10. Surgeon's fees: McIntosh and Garcia, "First Decade" (1978): 406. Cooley's income: "In re: Denton A. Cooley" (1988): 48. Cooley's estates: Conniff and Dominis, "Profiles" (1987).

27. Angioplasty fees: Preston, "Marketing" (1984): 36. Rising salaries: Monagan, "American Cardiac Surgeons" (2002). One-third of revenue: "Doctor-Owned Specialty Hospitals" (2003); Casalino and others, "Focused Factories?" (2003), 62; Desmon and Little, "St. Joe's" (2009). Stone's consulting: Herper and Langreth, "Dangerous Devices" (2006). By 2011, Stone could list paid consultancies with twenty-three pharmaceutical and medical device companies. See the ICMJE Form for Disclosure of Potential Conflicts of Interest included in the Supplementary Material to Stone and others, "Prospective" (2011).

28. California: "Tenet Physicians" (2005); Klaidman, *Coronary* (2007). Louisiana: Abelson, "Heart Procedure" (2006); "Louisiana Cardiologist" (2009).

29. Harris, "Doctor Faces Suits" (2010).

30. Lin and others, "Cardiologists' Use" (2007): 1606–7. See also Lin and others, "Why Physicians Favor" (2008). Powerful emotions also motivate cancer surgeons to make sure they leave no tumor cells behind. See Aronowitz, *Unnatural History* (2007).

31. Swiss: Goy and Eeckhout, "Intracoronary Stenting" (1998): 1948. Adrenaline: Pande, quoted in Kolata, "How It Happens" (2007). Financial rush: Sanghavi, "Plumber's Butt?" (2007).

32. Heroic bleeding: Sullivan, "Sanguine Practices" (1994). Heroic oncology: Lerner, *Breast Cancer Wars* (2001): 76–77. Daring surgeons: Fox and Swazey, *Courage to Fail* (1974); Ratcliff, "'Stopped Heart'" (1956). American psyches: David Hillis, quoted in Kolata, "New Heart Studies" (2004).

33. Morality: Brandt, *No Magic Bullet* (1987). Responsibility: Knowles, "Responsibility" (1977); Brandt, *Cigarette Century* (2007); Brownell and others, "Personal Responsibility" (2010); Ingelfinger, "Bariatric Surgery" (2011). Difficulty of change: Friedman, *Pathogenesis* (1969): 223. What was Friedman's tongue-in-cheek advice? "If Western man (1) ate the very low cholesterol-fat diet of the Japanese villager, (2) smoked no cigarettes, (3) indulged in as much physical activity as the Masai Negroes, and (4) lived in a milieu similar to that of the Navajo and responded to this milieu as the latter does (i.e., with a minimum of competitive zeal, aggressive drive, and sense of time urgency), his chances of suffering from clinical coronary artery disease would be almost nil until his eighth or ninth decade."

34. Stanford surgeons: Harrison and Shumway, "Evaluation and Surgery" (1972): 57 [three quotations]. Clinton: Martin, "From Omnivore to Vegan" (2011). Statin adherence: Kulik and others, "Adherence to Statin Therapy" (2011).

CHAPTER EIGHT: Fear and Unpredictability

1. Rosenberg, "Pathologies of Progress" (1998); Rothstein, *Public Health* (2003); Greene, *Prescribing by Numbers* (2007); Aronowitz, "Converged Experience" (2009); Dumit, "Pharmaceutical Witnessing" (2010). A broader anxiety in response to the many new risks introduced by modern, technological society has fostered the emergence of a "risk society." See Beck, *Risk Society* (1992). By focusing on fear, I hope to capture the visceral experience of patients and doctors rather than just the calculated, rationalized notions of risk.

2. Sudden death: Eggleston, "Diseases" (1928): 1021. Unexpectedly: Harrison and Resnik, "Etiologic Aspects" (1950): 1286. Distressing: Robbins, "Heart" (1967): 518. Dramatic: Libby, "Atherosclerosis" (1998): 1346.

3. For example, Millman, "Top 10 Reasons" (2000).

4. Morbus medicorum: Osler, "Angina Pectoris" (1910): 698. Herrick's physician patient: Herrick, "Thrombosis" (1919): 389. Met Life: Dublin and Spiegelman, "Longevity" (1947): 1215. Hazard: Thompson and Plachta, "Experiences" (1953): 678.

5. Braunwald, "Direct Coronary Revascularization" (1971): 9.

6. Harbinger: Kattus, "Physiologic Management" (1973): 47. Risk factor advice: Council on Foods and Nutrition (AMA), "Diet" (1965). For context, see Rothstein, *Public Health* (2003). Researchers continue to produce ever more precise forecasts based on risk factors, for example Berry and others, "Lifetime Risks" (2012).

7. Prima facie: Effler, "Role of Surgery" (1969): 377. Anxiety management: Lin and others, "Cardiologists' Use" (2007); Lin and others, "Why Physicians Favor" (2008).

8. Schoen and Contran, "Blood Vessels" (1999): 507 [may develop]; Schoen, "Heart" (1999): 551 [unpredictable], 553 [asymptomatic]; Schoen, "Heart" (2005): 573 [regrettably].

9. Fissures: Falk, "Plaque Rupture" (1983): 133. Smoldering: de Boer and others,

"Leucocyte Recruitment" (1999): 448. Diagrams, for example, in Friedman and van den Bovenkamp, "Pathogenesis of a Coronary Thrombus" (1966): 25; Libby, "Current Concepts" (2001): 366; Libby and others, "Inflammation and Atherosclerosis" (2002): 1136; Libby and Theroux, "Pathophysiology" (2005): 3484. Sometimes the analogy is explicit. One pathologist, describing how plaques could erode, noted that "the term 'erosion' is appropriate in both geologic and pathologic senses." See Flory, "Arterial Occlusions" (1945): 549.

10. The company: Casscells and others, "Vulnerable" (2003): 2072. One of the founders, David Brown, told me that the name was chosen "to evoke the comparison of a rupturing plaque to an erupting volcano" (e-mail message to author, 2 June 2011). Santorini: Schaar and others, "Terminology for High Risk" (2004): 1077. Deliberate choice: Ton van der Steen, e-mail message to author (26 July 2011).

11. Paris Constantinides described how thrombi "can mushroom to occlude the whole lumen" ("Atherosclerosis" [1984]: 486), though this usage is admittedly ambiguous. For "vulnerable plaque," see Muller and others, "Circadian Variation and Triggers" (1989); Muller and others, "Triggers" (1994). For the linkage, see: Miller, "Anti-Arms Groups" (1982); "Warfare, Medicine Converge" (1999); Muller, conversation with the author (7 July 2011).

12. Little bombs: Sanghavi, "Plumber's Butt?" (2007). Walking bombs: Roger S. Newton, quoted in Rozhon, "Pfizer to Buy" (2003). Scare tactic: David Hillis, quoted in Kolata, "New Heart Studies" (2004).

13. Greene, *Prescribing by Numbers* (2007); Dumit, "Pharmaceutical Witnessing" (2010). Aronowitz has described this process as "a problematic, self-reinforcing cycle of fear promotion followed by the marketing of tests and products that promise some means to reassert control over fear." See Aronowitz, "Converged Experience" (2009): 437.

14. Marks, "What Does Evidence Do?" (2009).

CHAPTER NINE: Surgical Ambition and Fear

1. Effler to Vineberg (Dec. 1960); Vineberg to Effler (1960).

2. Billroth, quoted in Monagan and Williams, *Journey into the Heart* (2007): 13. Some authorities doubt this quotation. See Blatchford, "Ludwig Rehn" (1985); Paget quoted on 492. For an overview, see Schlich, "Emergence of Modern Surgery" (2004), especially pp. 75–87.

3. Blatchford, "Ludwig Rehn" (1985); Wi, "History" (2003); Shumacker, *Evolution* (1992): 11–17.

4. Moore, "Rheumatic Fever" (1909).

5. Amenable, mechanical: Souttar, "Surgical Treatment" (1925): 603. Unjustifiable: Souttar to Harken (1961). No special risk: Cutler and Levine, "Cardiotomy" (1923): 1027. Nine deaths: Glover and others, "Commissurotomy" (1950): 330. Heroic, disrepute: Levine, *Clinical Heart Disease* (1937): 313, 314. See also Shumacker, *Evolution* (1992): 31–40.

6. Shumacker, *Evolution* (1992): 66–71; Wi, "History" (2003); Stoney, *Pioneers* (2008): 12–18.

7. War: Harken and Zoll, "Foreign Bodies" (1946). Valvuloplasty: Harken and others, "Surgical Treatment" (1948); Shumacker, *Evolution* (1992): 107–14.

8. Bailey's mortality: Miller, *King of Hearts* (2000): 57–60; Stoney, *Pioneers* (2008): 20–21. Harken's experience: Harken and Curtis, "Heart Surgery" (1967). Harken's son: Alden Harken, "Oral History [2001]" (2008): 341, 348.

9. Overviews: Romaine-Davis, *John Gibbon* (1991): 19–61; Shumacker, *Evolution* (1992): 242–53; Stammers, "Historical Aspects" (1997); Stoney, *Pioneers* (2008): 24–27. Unfitted: O'Shaughnessy, "Future" (1939): 971.

10. 200%: Conrad Lim, quoted in Gott, "Oral History [1997]" (2008): 333. First attempt: Lillehei and others, "Direct Vision Closure" (1955). Significant stroke: Gott, "Oral History [1997]" (2008): 333. See also Shumacker, *Evolution* (1992): 262–65; DeWall, "Oral History [2003]" (2008): 102–5; Stoney, *Pioneers* (2008): 31–33.

11. Miller, *King of Hearts* (2000): 147–48.

12. DeBakey, "Oral History [1998]" (2008): 153; Lillehei, "Oral History [1998]" (2008): 91–92; Miller, *King of Hearts* (2000): 92; DeWall, "Oral History [2003]" (2008): 104–7; Stoney, *Pioneers* (2008): 33–34.

13. Oxygenator options: Shumacker, *Evolution* (1992): 266–79. Plasma foam: Roe Wells, "Cardiovascular Lab," in Thorn, "Report" (1967): 28.

14. Gibbon and Kirklin: Romaine-Davis, *John Gibbon* (1991): 66–131, 138–42; Shumacker, *Evolution* (1992): 253–55; Shumacker, *Dream of the Heart* (1999): 175–90; Stoney, *Pioneers* (2008): 24–30. Detroit: Dodrill and others, "Temporary Mechanical Substitute" (1952); Shumacker, *Evolution* (1992): 258; Wi, "History" (2003).

15. Lillehei, "Oral History [1998]" (2008): 92; DeWall, "Oral History [2003]" (2008): 106–7; Stoney, *Pioneers* (2008): 33–34.

16. Monster: Favaloro, "Oral History [1997]" (2008): 362. Blood: Fraser, "Retrospective" (2003).

17. Surgery-industry collaborations: Romaine-Davis, *John Gibbon* (1991): 147–54; see also Jeffrey, *Machines in Our Hearts* (2001). Can opener: Cooley (1986), quoted in Stoney, *Pioneers* (2008): 34. Exporting cardiac surgery: Stoney, *Pioneers* (2008): 36, 54.

18. *Reader's Digest*: Ratcliff, "'Stopped Heart'" (1956): 29 [required, Effler quotation], 31 [fluttery]. *Time*: Cooley, quoted in "Surgery in the Heart" (1956).

19. Dogs: Magovern, "Oral History [1997]" (2008): 297. Houston mortality: Ochsner, "Oral History [2000]" (2008): 393–94.

20. Derrida, "Plato's Pharmacy" (1981).

21. Anesthesia: Pernick, *Calculus of Suffering* (1985). Lobotomy: Pressman, *Last Resort* (1998).

22. Cardiac cripples: Bailey and others, "Exploratory Surgery" (1952): 640; Effler and others, "Surgical Repair" (1965): 42; Effler and others, "Heart Valve" (1965): 21. Cardiac derelicts: Mason, "Myocardial Ischemia" (1951): 366. Wait-list mortality: Wooler, "Oral History [1997]" (2008): 143. Effler's experience: Ratcliff, "'Stopped Heart'" (1956): 33 [poor score, desperate cases, Effler quotation].

23. Courage to fail: "Cooley Criticizes" (1970); Fox and Swazey, *Courage to Fail* (1974): xl. Mortality of 15–35%: Dexter, "Surgery" (1956).

24. "Letters to the Editors" (1970).

25. Mortality under 5%: McGoon and others, "Decreased Mortality" (1964). Effler's

success: Effler to LeFevre and others (1966). Effler's complaint: Effler to Hewitt (1966). Stanford: Pillsbury and others, "Four Hundred and Fifty" (1965): 181.

CHAPTER TEN: Suffering Cerebrums

1. Never justifiable: Souttar, "Surgical Treatment" (1925): 603. No incentive: O'Shaughnessy, "Future" (1939): 969.

2. Bailey's first 235: Bailey and others, "Commissurotomy" (1952): 1086. Chicago: Alfano and others, "Visual Loss" (1957). Harken's patients: Fox and others, "Psychological" (1954). Math teacher: Priest and others, "Neurologic" (1957): 159.

3. Olsen and others, "Cerebral Embolization" (1952): 171.

4. Gentle motions: Harken and others, "Surgery" (1951): 734. Dire consequences: Olsen and others, "Cerebral Embolization" (1952): 173; see also Bailey and others, "Commissurotomy" (1952): 1090. Harken's failure: Ellis and Harken, "Clinical Results" (1955): 638.

5. Harken and Black, "Improved" (1955): 671.

6. Real key: O'Shaughnessy, "Future" (1939): 969. Terrible danger: Lillehei, "Oral History [1998]" (2008): 91.

7. Dog and cat: O'Shaughnessy, "Future" (1939): 971. Case Western: Clowes and others, "Factors Contributing" (1954): 559. Pennsylvania: Swank and Hain, "Effect" (1952): 294. Lillehei's exam: DeWall, "Oral History [2003]" (2008): 107.

8. Importance: O'Shaughnessy, "Future" (1939): 969. John Gibbon's "Postoperative Note": reprinted in Romaine-Davis, *John Gibbon* (1991): 121–22. Kirklin's team: Patrick and others, "Effects" (1958): 272.

9. Cleveland outcomes: Effler and others, "Heart Valve" (1965): 17. Lost to us: Effler to Vineberg (April 1960).

10. Iowa City: Ehrenhaft and Claman, "Cerebral Complications" (1961): 505–6. Boston: Gilman, "Cerebral Disorders" (1965): 490. New York: Fred Plum, in discussion of Silverstein and Krieger, "Neurologic Complications" (1960): 153.

11. New illnesses: Miller, "Some Neurological" (1964): 143. Hazard: Ellis and Harken, "Arterial Embolization" (1961): 611. Long-term danger: Ellis and Harken, "Clinical Results" (1955).

12. Cohen, "Neurological" (1964): 578.

13. "Cerebral Injury" (1964): 90.

14. Operative risk: Harken and others, "Surgery" (1951); Bailey and Likoff, "Surgical Treatment" (1955). At risk of disease: Rothstein, *Public Health* (2003); Greene, *Prescribing by Numbers* (2007), especially pp. 7–14. Risk factors: Oppenheimer, "Becoming" (2005). Aortic vents: McGoon and others, "Decreased Mortality" (1964).

15. Van Schaik, " 'Taking' " (2010).

16. Physical exams: Atwater, "Touching" (1985). Clinical gaze: Foucault, *Birth of the Clinic* (1973); Taussig, "Reification" (1980).

17. Tracings and inscriptions: Latour and Woolgar, *Laboratory Life* (1986); Lynch, "Externalized Retina" (1988); Latour, "Drawing Things Together" (1990). Image and objectivity: Daston and Galison, *Objectivity* (2007); see also: Brain, "Representation" (2002);

Kaiser, "Stick-Figure Realism" (2000). Clinical imaging: Evans, "Losing Touch" (1993); Howell, *Technology* (1995); Dumit, *Picturing Personhood* (2004); Greene, "Releasing the Flood Waters" (2005); Saunders, *CT Suite* (2008).

18. EEG history and brainscripts: Brock, "Recording" (2008). Mayo: Theye and others, "Electro-encephalogram" (1957): 714. Effler: Effler and others, "Elective Cardiac Arrest" (1957): 506. Montreal: Davenport and others, "Electroencephalogram" (1959): 679, 683.

19. Air bubbles, micro-emboli: Spencer and others, "Use of Ultrasonics" (1969): 491, 495. Complacency: G. Hugh Lawrence, in discussion of Javid and others, "Neurological Abnormalities" (1969): 508.

20. Williams, "Intravascular Changes" (1971): 690.

21. Lee and others, "Effects on Personality and Cerebration" (1969).

22. Surgeon's appreciation: Lee and others, "Effects on Personality and Cerebration" (1969); Harken's discussion on p. 567. Draw-a-Person: Zaks and others, "Neuropsychiatric" (1960). For a discussion of the ambitions and limits of psychological testing, see Galison, "Image of the Self" (2004).

23. Sponge: Theye and others, "Electro-encephalogram" (1957). 714. Antifoam: Penry and others, "Cerebral Embolism" (1960). Fat: Hill and others, "Neuropathological Manifestations" (1969): 414.

24. Sloan and others, "Open Heart Surgery" (1962): 135.

25. Zero to 100%: Sotaniemi, "Cerebral Outcome" (1983). Prospective studies: Kornfeld and others, "Psychiatric Complications" (1965); Javid and others, "Neurological Abnormalities" (1969). Relevance: Gilberstadt and Sako, "Intellectual" (1967): 213.

26. Aberg and Kihlgren, "Cerebral Protection" (1977): 525.

27. Cerebral suffering, disturbance: Davenport and others, "Electroencephalogram" (1959): 679, 683. Injury, damage: "Cerebral Injury" (1964): 89. Deficit, dysfunction: Lee and others, "Effects on Personality and Cerebration" (1969): 563. Brain damage: Gilberstadt and Sako, "Intellectual" (1967): 213; Brierly, "Brain Damage" (1967).

28. British team: Shaw and others, "Early Neurological" (1985): 1386. It is difficult to determine when this use of "pump head" began. The phrase has been used since the seventeenth century to describe the top piece of a water pump: see the *Oxford English Dictionary*. As surgeons and engineers developed heart-lung machines, they described the pump heads in their devices: Dennis and others, "Development" (1951): 711; Liddicoat and others, "Membrane vs. Bubble" (1975): 748; Stammers, "Trends" (1992). In one case, a pump head malfunction nearly caused a catastrophic air emboli: Kurusz and others, "Runaway Pump Head" (1979). In others, pump heads did contribute to harmful emboli: Orenstein and others, "Microemboli" (1982); Neema and others, "Systemic Air Embolization" (2007). The phrase is not used in the medical literature to describe the cerebral complications of heart-lung machines (at least not in the full-text searchable runs of *Circulation*, the *American Heart Journal*, the *Journal of Cardiothoracic and Vascular Anesthesia*, the *New England Journal of Medicine*, or *JAMA*). The earliest occurrence I have found is a 1999 *New Yorker* article— Groopman, "Heart Surgery, Unplugged". Such familiar usage between two doctors (the author and the informant) suggests that the phrase had become common by that time. It subsequently appeared regularly in articles by science journalists: Jauhar, "Saving the Heart" (2000); Stutz, "Pump Head" (2003); Foreman, "Heart Patients' Mental Decline" (2004).

CHAPTER ELEVEN: Deliriogenic Personalities

1. Freyhan and others, "Psychiatric Complications" (1971): 194 [no other], 187 [kalei-doscopic, psychedelic].

2. Lab technician: Kornfeld and others, "Psychiatric Complications" (1965): 288. MacKenzie's memoir: MacKenzie, "Risk" (1970); MacKenzie, *Risk* (1971).

3. Reiser, "Intensive Care Unit" (1992).

4. Lazarus and Hagens, "Prevention" (1968): 1191.

5. Lunbeck, *Psychiatric Persuasion* (1994); Hale, *Rise and Crisis* (1995).

6. First patient: Kaplan, "Psychological Aspects" (1956): 222 [neurotic], 226 [sexual role]. Second patient: Galdston, "Psychotic Reaction" (1970): 371.

7. Cultural demands: Danilowicz and Gabriel, "Postcardiotomy Psychosis" (1971): 318. Dominance: Kornfeld and others, "Personality" (1974): 251. Ordinarily assets: Heller and others, "Delirium" (1980): 159.

8. Kennedy and Bakst, "Influence" (1966): 816; Kornfeld, "Influence" (1966): 846.

9. Farmer, "Social Scientists" (1997).

10. Kornfeld, "Influence" (1966): 847.

11. Brandt, "Cigarette, Risk" (1990); Brandt, *Cigarette Century* (2007).

12. Sheps and Shapiro, "Physician's Responsibility" (1962): 406.

13. Membrane oxygenators: Carlson and others, "Total Cardiopulmonary" (1972); Carlson and others, "Landé-Edwards" (1973). Filters: Dutton and others, "Platelet Aggregate" (1974); Patterson and others, "Effect" (1974). Reduced incidence: Branthwaite, "Prevention" (1975). No complacency: "Brain Damage" (1975): 400.

CHAPTER TWELVE: The Case of the Missing Complications

1. McIntosh and Garcia, "First Decade" (1978).

2. Cleveland's single mention: Favaloro and others, "Combined Simultaneous" (1969): 23. Other complications: Favaloro and others, "Direct Myocardial Revascularization" (1969). Milwaukee: Johnson and others, "Direct Coronary Surgery" (1970). New York: Green and others, "Anastomosis" (1970). Texas: Reul and others, "Long-Term Survival" (1975).

3. The full list of two hundred is available from the author. The four: Urschel and others, "Combined Gas" (1970); Hallin and others, "Revascularization" (1971); Vontz and others, "One Hundred Consecutive" (1972); Wilson and others, "Increased Safety" (1972). Textbooks—No mention: Killip, "Angina Pectoris" (1975): 1005–6. No cerebral complications: Sabiston, "Coronary Circulation" (1972): 2034–35.

4. May and others, "1971 Reflection" (1972): 475.

5. Main report: Murphy and others, "Treatment" (1977). Quality of life: Hultgren and others, "Medical versus Surgical" (1976). Non-compliance: Parisi and others, "Characteristics" (1984). Angiogaphy: Takaro and others, "Analysis of Deaths" (1973). One mention: Takaro and others, "The VA Cooperative" (1976): 111–13.

6. Primary reports: CASS Principal Investigators, "Coronary Artery Surgery Study: Survival Data" (1983); CASS Principal Investigators, "Coronary Artery Surgery Study: Quality of Life" (1983). Stroke: Frye and others, "Stroke" (1992): 218.

7. Bypass surgery debates: Jones, "Visions" (2000). Editorials, for example: Spodick, "Revascularization" (1971). Critics: Braunwald, "Direct Coronary Revascularization" (1971); Preston, "Hazards" (1978); McIntosh and Garcia, "First Decade" (1978).

8. Breuer and others, "Neurologic Complications" (1980): 109. See also Breuer and others, "Brain and Nerve" (1980): 42; Breuer and others, "Neurologic Complications" (1981): 205.

9. Pre-WWII consent: Wilde, "Truth, Trust" (2009); Lederer, *Subjected to Science* (1995). Postwar changes: Faden and Beauchamp, *History and Theory* (1986): 53–113; Weindling, "Origins" (2001).

10. Did not alter: Surgical Committee, "Minutes" (1969). Not protect: Hoerr to Members (1969).

11. Methodist: Committee for Clinical Investigation, Memo (1968). St. Luke's: Konnagan to Blattner and Seybold, "Patient Consent" (1970).

12. Ochsner's form: Ochsner and Mills, *Coronary Artery Surgery* (1978): 103. Cooley's patient: Gene Austin, quoted in "Gene Austin Undergoes" (1987): 12. Brown's patient: Derrick, "How Open Heart" (1979): 278. Preston's critique: Preston, "Marketing" (1984): 36.

13. Alfidi, "Informed Consent" (1971).

14. Robinson and Merav, "Informed Consent" (1976), quotation on p. 212.

15. DeBakey, "Oral History [1998]" (2008): 153.

CHAPTER THIRTEEN: Selective Inattention

1. Sabiston, "Coronary Circulation" (1974): 320–21.

2. Bounds and others, "Fatal Cerebral" (1976).

3. Breuer and others, "Central Nervous System" (1983): 682.

4. Anomalies: Kuhn, *Structure* (1962). Relevance: Daston, "Coming into Being" (2000); Latour, *What Is the Style* (2004); Latour, "Why Has Critique" (2004). Stakes: Kleinman, *Writing at the Margin* (1995); Kleinman, *What Really Matters* (2006). Sickle cell anemia offers an interesting case study: only when the disease served the purpose of both the new field of medical genetics and the burgeoning civil rights movements of the 1960s did it receive substantial attention. See Wailoo, *Dying in the City of Blues* (2000).

5. Selective inattention: Schön, *Reflective Practitioner* (1983): 44. Taxonomy of ignorance: Proctor, "Agnotology" (2008): 3. Creating controversy: Brandt, *Cigarette Century* (2007); Proctor and Schiebinger, *Agnotology* (2008); Conway and Oreskes, *Merchants of Doubt* (2011). Another example of such "nonknowledge": Petryna, *Life Exposed* (2002).

6. The study: Layne and Yudofsky, "Postoperative Psychosis" (1971). Notifying DeBakey: Bruch to DeBakey (1969). Firing Cooley: DeBakey to Cooley (1969). National attention: "Surgery: The Texas Tornado" (1965); Thompson, "Texas Tornado vs. Doctor Wonderful" (1970). See also Fox and Swazey, *Courage to Fail* (1974): 135–98.

7. DeBakey's response: DeBakey to Bruch (1970). Yudofsky's perspective: Yudofsky, "Hilde" (1987); Yudofsky, conversation with author (8 Nov. 2011). Cover letter: Bruch to Ingelfinger (1970).

8. Damper, misgivings, "Mighty Arm": Bruch to Yudofsky (1970). Telling the students: Yudofsky, "Hilde" (1987). Telling the journal: Bruch to Malt (1970).

9. DeBakey's approval: [Note, no identifiers] (1970). Yudofsky's recollection: Yudofsky, conversation with author (8 Nov. 2011). Even this is ambiguous: in his 1987 account, Yudofsky noted that DeBakey "recommended that the paper not be submitted in its present form to a juried journal" (Yudofsky, "Hilde" [1987]). DeBakey evidently held no grudge: in 1991 he actively recruited Yudofsky to return to Houston to be chair of the Department of Psychiatry at the Baylor College of Medicine. For a more recent case of an author being forced to remove her name from a paper, see the story of Carolyn Cannuscio and rofecoxib: Avorn, "Introduction" (2005): xv; Solomon and others, "Relationship" (2004); Burton, "Merck Takes Author's Name" (2004).

10. Surgeon's incomes: Mearns, "Effler and Sones" (1973); *In re:* Denton A. Cooley (1988). Recent scandals: Avorn, "Dangerous Deception" (2006).

11. Rheumatic brains: Dencker and Sandahl, "Mental Disease" (1961); Rabiner and others, "Psychiatric Complications" (1975); Meyendorf, "Causes" (1993).

12. Favaloro's move to heart-lung machines: Favaloro and others, "Direct Myocardial Revascularization" (1970). Survey: Miller and others, "Current Practice" (1977). Comparable risks: Bounds and others, "Fatal Cerebral" (1976): 611.

13. Mayo: Bounds and others, "Fatal Cerebral" (1976): 611. Reminiscences: DeVries, "Oral History [1999]" (2008): 475.

14. Nursing: Effler to Hoerr (1969). Scheduling: Effler to Hoffman (1970). Cardiologists: Proudfit to the Board of Governors (1971). Waitlists: Department of Cardiology, "1976 Annual Report" (1977). Slowdown: Sheldon, "1977 Annual Report" (1978). Recovery: Shirey, "1978 Annual Report" (1979).

15. Four neurologists: Sweeney and Lautzenheiser, *Neuroscience* (2006): 51. Their research: Williams to the Board of Governors (1971).

16. Unsupported research: Gilman, "Cerebral Disorders" (1965); Breuer and others, "Central Nervous System" (1983). Brigham-MIT project: Thorn, "Report" (1969): 205–6; Thorn, "Report" (1970): 82.

17. For a discussion of how physicists found common ground, see Galison, "Trading Zone" (1999).

18. Mearns, "Effler and Sones" (1973).

19. Two hours: Javid and others, "Neurological Abnormalities" (1969): 506. Effler's advice: quoted in Mearns, "Effler and Sones" (1973): 8. Fifteen to twenty minutes: Cooley, "Oral History [1997]" (2008): 249. It is noteworthy that major studies often did not report pump times. See: Reul and others, "Long-Term survival" (1975); Murphy and others, "Treatment" (1977); Loop and others, "11-Year Evolution" (1979); Breuer and others, "Central Nervous System" (1983); CASS Principal Investigators, "Coronary Artery Surgery Study: Survival Data" (1983). Breuer and others did note that some of their patients had pump times in excess of two hours, far beyond Cooley's fifteen-to-twenty-minute ideal.

20. Pernick, *Calculus of Suffering* (1985).

21. Fifty percent incidence: Carlson and others, "Landé-Edwards" (1973): 894. For another comparative study, see Liddicoat and others, "Membrane vs. Bubble" (1975). Cleveland's choice: Breuer and others, "Central Nervous System" (1983). Persistence of bubble oxygenators: Pokar, "Study of Equipment" (1990); McKhann and others, "Cognitive Outcome" (1997).

22. Green, "Microvascular Technique" (1970).
23. Yudofsky, conversation with the author (8 Nov. 2011).
24. Eddy, "Variations" (1984): 80–81.

CHAPTER FOURTEEN: The Cerebral Complications of Coronary Artery
Bypass Surgery

1. 1962 stroke: Sabiston, "Coronary Circulation" (1974). Mayo: Bounds and others, "Fatal Cerebral" (1976). Psychiatric: Rabiner and others, "Psychiatric Complications" (1975). Swedish series: Aberg, "Effect" (1974). Cleveland series: Loop and others, "11-Year Evolution" (1979).

2. Distinct pathologies: Rabiner and others, "Psychiatric Complications" (1975): 342. Mayo Clinic: Bounds and others, "Fatal Cerebral" (1976): 611.

3. 1980 abstract: Breuer and others, "Brain and Nerve" (1980): 42. 1983 report: Breuer and others, "Central Nervous System" (1983): 682. Concern about overuse: Braunwald, "Coronary-Artery" (1977).

4. New York: Freyhan and others, "Psychiatric Complications" (1971): 181. Cleveland: Furlan and Breuer, "Central Nervous System" (1984): 915. CABG-stroke model: Mangano, "Cardiovascular Morbidity" (1995): 367.

5. First study: Rabiner and others, "Psychiatric Complications" (1975): 347. Other study: Merwin and Abram, "Psychologic Responses" (1977).

6. 13 neurologists: Sweeney and Lautzenheiser, *Neuroscience* (2006): 59–63. Advantage: Breuer and others, "Neurologic Complications" (1980): 109. Templates: Breuer and others, "Neurology of Open Heart Surgery" (1979), Computer Record Codes A-L. See also Breuer and others, "Central Nervous System" (1983): 682.

7. Results: Breuer and others, "Central Nervous System" (1983); audible crunch on p. 686. Neuropsychologist: Department of Neurology, "1980 Annual Report" (1980): 5.

8. Cleveland Clinic Foundation, *Challenge* (1986): cover, 2 [exploring].

9. Newcastle study: Shaw and others, "Early Neurological" (1985). Six-month follow-up: Shaw and others, "Neurological Complications" (1986): 165. Cognitive dysfunction: Shaw and others, "Early Intellectual" (1986).

10. Hazards: Shaw and others, "Early Neurological" (1985): 1386. Comparison: Shaw and others, "Neurologic and Neuropsychological" (1987): 700.

11. Roach and others, "Adverse Cerebral Outcomes" (1996): 1857.

12. Special issue: Murkin, "Introduction" (1996). Book: Newman and others, *Brain and Cardiac Surgery* (2000). Funded research: McKhann and others, "Stroke and Encephalopathy" (2006).

13. Doppler flurries: Barbut and Gold, "Aortic Atheromatosis" (1996): 26. Snowstorm: Barbut, quoted in Jauhar, "Saving the Heart" (2000). Retina as window: Blauth and others, "Retinal Microembolism" (1986): 839. Retina findings: Blauth and others, "Cerebral Microembolism" (1988). MRI findings: Restrepo and others, "Diffusion- and Perfusion-Weighted" (2002): 2909. Blood markers: Ramlawi and others, "C-Reactive Protein" (2006). SCADs: Moody and others, "Brain Microemboli" (1990).

CHAPTER FIFTEEN: A Taxonomy of Inattention

1. Kirklin, chair, "ACC/AHA Guidelines" (1991): 1126 [become], 1132 [damaging, gross]. But many bypass surgery patients—31.9% in the largest study—were older than 70: Roach and others, "Adverse Cerebral Outcomes" (1996): 1859.

2. No mention: Davis and Sabiston, "Coronary Circulation" (1997): 2091–92; Selwyn and Braunwald, "Ischemic Heart Disease" (1998): 1372–73. Passing: Galloway and others, "Acquired Heart Disease" (1994): 862; Cohn, "Surgical Treatment" (1996): 318.

3. Falling mortality: Taylor, "Cardiac Surgery" (1993): 1. Testimony: Stump, "Selection" (1995): 1344. Roach and others, "Adverse Cerebral Outcomes" (1996): 1857.

4. Stable estimates: Aberg and Kihlgren, "Cerebral Protection" (1977); Loop and others, "11-Year Evolution" (1979); Ivert, "Coronary Bypass" (1981); Furlan and Breuer, "Central Nervous System" (1984); Gardner and others, "Stroke Following" (1985); Roach and others, "Adverse Cerebral Outcomes" (1996). Balanced changes: Selnes and Mc-Khann, "Neurocognitive Complications" (2005). The stroke rate, however, may have decreased in the past decade. See Selnes and others, "Cognitive and Neurologic" (2012).

5. Complaint: Kuroda and others, "Central Nervous System" (1993): 222. 30 years: van der Mast and Roest, "Delirium" (1996): 27. 40 years: Elkins and Johnston, "Twinning" (2004): 2211. Unreliable patients: Newman and others, "Subjective Reports" (1989): 227. Imperfect tests: Smith, "Cerebral Dysfunction" (1995): 1360.

6. Perhaps neurologists had an interest in the success of bypass surgery because revenues from cardiac surgery sustained so many medical centers.

7. Tolerance: Spencer and others, "Use of Ultrasonics" (1969): 495. Bombardment: Loop and others, "Events" (1976): 415. London anesthesiologist: Gilston, "Brain Damage" (1986): 1323. Eloquent: Bendszus and Stoll, "Silent Cerebral" (2006): 369.

8. Daily activities: Shaw and others, "Early Neurological" (1985). Six-month importance: Shaw and others, "Neurological Complications" (1986): 165. Overtly disabled: Shaw and others, "Early Intellectual" (1986): 59. Oxford surgeon: Taggart and Westaby, "Neurological and Cognitive" (2001): 274.

9. 1993 psychiatrist: Willner, "Use of Neuropsychological Tests" (1993): 195. *Time:* Gorman, "Hearts and Minds" (2001).

10. Cleveland: Breuer and others, "Central Nervous System" (1983). Hopkins: McKhann and others, "Predictors" (1997). Pennsylvania: Lynn and others, "Risk Factors" (1992): 1522.

11. Genetic interests: Nelkin and Lindee, *DNA Mystique* (2004). Duke study: Tardiff and others, "Preliminary Report" (1997): 715. Gene chip: Ramlawi and others, "Genomic Expression Pathways" (2007): 996.

12. McKhann and others, "Stroke and Encephalopathy" (2006): 568. See also Selnes and others, "Cognitive and Neurologic" (2012).

CHAPTER SIXTEEN: Competition's Complications

1. Lawrence and Hall, *1997 Summary* (1999).

2. Modest ambitions: Grüntzig and others, "Nonoperative Dilatation" (1979). 1996 data: Graves and Owings, *1996 Summary* (1998).

3. Comparative studies: King and others, "Randomized Trial" (1994); Hamm and others, "Randomized Study" (1994); Bypass Angioplasty Revascularization Investigation, "Comparison" (1996); Bravata and others, "Systematic Review" (2007). 1979 outcomes: Grüntzig and others, "Nonoperative Dilatation" (1979): 64.

4. Skeptics: Spodick, "Percutaneous" (1979): 1659. Registry report: Dorros and others, "Percutaneous" (1983). Cleveland experience: Galbreath and others, "Central Nervous System" (1986): 616.

5. Three trials: King and others (EAST), "Randomized Trial" (1994) [1.5% vs. 0.5%]; Hamm and others (GABI), "Randomized Study" (1994) [1.2% vs. 0%]; Bypass Angioplasty Revascularization Investigation, "Comparison" (1996) [0.8% vs. 0.2%]. Meta-analysis: Bravata and others, "Systematic Review" (2007). The age distributions of patients in the trials are unclear. The 1996 multicenter study of bypass surgery noted only that 31.9% of the patients were older than 70 (Roach and others, "Adverse Cerebral Outcomes" [1996]: 1859). In the EAST comparison of angioplasty and surgery, the mean age of each group was 61.4 and 61.8; in GABI, all patients were under 75; in BARI, 39% were older than 65. With such age distributions, it is possible that cerebral complication rates were low, making a difference between the two treatments difficult to detect.

6. Rubens and Nathan, "Lessons Learned" (2007): 153.

7. Cooley, "Oral History [1997]" (2008): 254.

8. Favaloro's early surgeries: Favaloro, "Saphenous Vein Graft" (1969): 182–84. Texas: Archer and others, "Coronary Artery Revascularization" (1984). Sao Paolo motivations: Buffolo and others, "Direct Myocardial Revascularization" (1985). Their results: Buffolo and others, "Myocardial Revascularization" (1990); see p. 507 for Thomas Treasure [amazement].

9. Surgeon's hopes: Murkin and others, "Beating Heart Surgery" (1999): 1500. Early results: Diegeler and others, "Neuromonitoring" (2000): 1166.

10. Groopman, "Heart Surgery, Unplugged" (1999): 43 [inconceivable], 51 [others]. As noted earlier (chapter 10, note 28), this *New Yorker* article is the first occurrence of "pump head" that I have found in the medical or popular literature.

11. Karagoz, Sönmez, and others, "Coronary Artery" (2000): 95 [strategy]. Patient toleration: Karagoz, Kurtoglu, and others, "Coronary Artery" (2003): 1404. Stanford rebuttal: Mora Mangano, "Risky Business" (2003): 1206.

12. Complications and distinctions: Aronowitz, "Converged Experience" (2009): 437. Rofecoxib: Avorn, "Dangerous Deception" (2006). Antipsychotics: Wilson, "Side Effects" (2010).

13. BARI: Hlatky and others, "Cognitive Function" (1997): 13. On- and off-pump: Taggart and others, "Is Cardiopulmonary Bypass" (1999): 414.

14. Bypass vs. angioplasty: Serruys and others (SYNTAX), "Percutaneous Coronary Intervention" (2009). On- vs. off-pump: Shroyer and others (ROOBY), "On-Pump" (2009), quotation on p. 1836. Patient factors: Selnes and others, "Cognitive and Neurologic" (2012): 255.

15. Duke study: Newman and others, "Longitudinal Assessment" (2001): 400. Critics argue that this study, which lacked a control group, "reinforced misconceptions." See Selnes and others, "Cognitive and Neurologic" (2012): 254. Emboli and outcomes: Stygall

and others, "Cognitive Change" (2003). MRI: Bendszus and Stoll, "Silent Cerebral" (2006): 370. Hopkins rates: McKhann and others, "Stroke and Encephalopathy" (2006): 563. Lower survival: Gottesman and others, "Delirium" (2010). Consensus guidelines: Eagle and Guyton, co-chairs, "ACC/AHA" (2004). One textbook: Lytle, "Surgical Treatment" (2004): 429.

16. Former president: Hanlon, "Oral History [2000]" (2008): 171. Testimonial: Stutz, "Pump Head" (2003): 78.

17. Fifty-two patients: Müllges and others, "Cognitive Performance" (2002). Twin study: Potter and others, "Age Effects" (2004). Cognitive improvement: Jeffrey Gold, quoted in Jauhar, "Saving the Heart" (2000): F4.

18. No differences: Selnes and others, "Cognition 6 Years" (2008): 581. Comparisons: Selnes and others, "Do Management Strategies" (2009). Rebranding: Selnes and others, "Determinants of Cognitive Change" (1999).

19. Media debates: Foreman, "Heart Patients' Mental Decline" (2004); Vedantam, "Clinton's Heart" (2004). Speculations: McDougall, "Bill Clinton's Madness" (2008); Purdum, "Bubba Trouble" (2008). Rebuttal: Smith, "Clinton Attacks" (2008).

20. STS monitoring: Shahian and others, "Quality Measurement" (2007). The STS does not, however, specify how information about surgeons' outcomes should be ascertained. Moreover, the STS guidelines do not mention cognitive dysfunction or other cerebral complications. New England collaborative: Likosky and others, "Determination" (2003): 2834. Lack of precautions: Caplan, "Translating" (2009): 1064. Not standard of care: Selnes and others, "Cognitive and Neurologic" (2012): 250.

21. Mateen and others, "Neurological Injury" (2011).

22. Rief and others, "Medication-Attributed" (2006); Rief and others, "Differences" (2009).

23. Multi-center study: Roach and others, "Adverse Cerebral Outcomes" (1996). The review: Selnes and others, "Cognitive and Neurologic" (2012). No mention: Khan and others, "Randomized Comparison" (2004); Desai and others, "Randomized Comparison" (2004); Hannan and others, "Long-Term Outcomes" (2005); Hannan and others, "Drug-Eluting Stents" (2008); Velazquez and others, "Coronary-Artery" (2011); Bonow and others, "Myocardial Viability" (2011). In passing: Diegeler and others, "Comparison" (2002); Cohen and others, "Quality of Life" (2011). Secondary or composite endpoints: Serruys and others, "Comparison" (2001); Nathoe and others, "Comparison" (2003); Seung and others, "Stents versus Coronary-Artery" (2008); Serruys and others (SYNTAX), "Percutaneous Coronary Intervention" (2009); Jones and others, "Coronary Bypass" (2009); Shroyer and others (ROOBY), "On-Pump" (2009); Park and others, "Randomized Trial" (2011). Focused report: Newman and others, "Longitudinal Assessment" (2001).

CONCLUSION: Puzzles and Prospects

1. Ohio: Abelson, "Heart Procedure" (2006). Destiny: *Dartmouth Atlas* (1996): 2. See also Wennberg, *Tracking Medicine* (2010); Brownlee, *Overtreated* (2008): 13–42.

2. Torch and others, "Timing of Neonatal Discharge" (2001); Wennberg, "Unwarranted Variations" (2002); Mulley, "Need to Confront" (2009).

3. Early attempts: Mueller and others, "History of Surgery" (1997). Cleveland's

prowess: Edward Diethrich, in discussion of Favaloro and others, "Combined Simultaneous" (1969): 27. Community hospitals: Favaloro, "Surgical Treatment" (1971): 495. Four hundred hospitals: Hiatt, "Protecting the Medical Commons" (1975).

4. Ten times: Cutter and others, "Epidemiologic Study" (1982). 1986 Medicare: Chassin and others, "Variations in the Use" (1986): 297. 1993 Medicare: Boutwell and Mitchell, "Diffusion of New Technologies" (1993). Kaiser: Selby and others, "Variation among Hospitals" (1996). Thrombolytic trial: Pilote and others, "Regional Variation" (1995).

5. London: Glover, "Incidence of Tonsillectomy" (1938). New York: Lembcke, "Measuring the Quality" (1952). International: Pearson and others, "Hospital Caseloads" (1968). Kansas: Lewis, "Variations" (1969). Vermont: Wennberg and Gittelsohn, "Small Area Variations" (1973). Maine: Wennberg and Gittelsohn, "Health Care Delivery" (1975): 127.

6. Moore, "What Puts the Surge" (1970): 163. *Dartmouth Atlas* (1996). Critiques: Bach, "Map to Bad Policy" (2010); Chernew and others, "Geographic Correlation" (2010); Epstein, "Geographic Variation" (2010).

7. *Dartmouth Atlas of Cardiovascular Care* (1999): preface [the quality], 18–19, 22–23.

8. Redding: "Dartmouth Atlas of Health Care: Studies of Surgical Variation—Cardiac Surgery" (2005): 13; "Tenet Physicians Settle Case" (2005); Klaidman, *Coronary* (2007). Elyria: *Dartmouth Atlas of Cardiovascular Care* (1999), on p. 76; "Dartmouth Atlas of Health Care: Studies" (2005): 16; Abelson, "Heart Procedure" (2006) [quotation from John W. Schaeffer]. Lafayette: Abelson, "Heart Procedure" (2006); "Louisiana Cardiologist" (2009).

9. Gillum, "Coronary Heart Disease" (1982): 849. Alabama group: Oberman and Cutter, "Issues in the Natural History" (1984): 688. CASS Trial: Maynard and others, "Blacks in the Coronary Artery Surgery Study" (1986): 1447.

10. Massachusetts: Wenneker and Epstein, "Racial Inequalities" (1989). National data: Goldberg and others, "Racial and Community Factors" (1992). Institute of Medicine: Smedley and others, *Unequal Treatment* (2003): 1–3, 5 [convincing evidence], 39–74.

11. Robbins, quoted in Iglehart, "From the Editor" (1984); Blumenthal, "Variation Phenomenon" (1994): 1017.

12. Glover, "Incidence of Tonsillectomy" (1938): 1223. "Regional differences": Bunker, "Surgical Manpower" (1970): 141. "Regional patterns": Gillum, "Coronary Artery Bypass Surgery" (1987): 1258. Iowa and California: Wennberg, "Which Rate is Right" (1986). Northern New England: Wennberg and others, "Relationship between Supply" (1997). Continuing debate on this point: Zuckerman and others, "Clarifying Sources" (2010).

13. Fee-for-service: Bunker, "Surgical Manpower" (1970). Defense: Moore, "What Puts the Surge" (1970): 163. Bypass critics: Braunwald, "Coronary-Artery" (1977); Preston, "Hazards" (1978); McIntosh and Garcia, "First Decade" (1978). Arizona physician: Phibbs, "Abuse" (1979): 1394. Alabama group: Cutter and others, "Epidemiologic Study" (1982). Canada: Rouleau and others, "Comparison of Management" (1993).

14. Kansas: Lewis, "Variations" (1969): 884. Liverpool: Pearson and others, "Hospital Caseloads" (1968): 564. Relation of angiography and revascularization: Wennberg and others, "Association" (1996); Kuhn and others, "Correlation of Rates" (1995). Seattle: Every and others, "Association" (1993).

15. Inconsistent evidence: McPherson and others, "Regional Variations" (1981). Ver-

mont: Wennberg and Gittelsohn, "Small Area Variations" (1973): 1105. Manitoba: Roos, "Hysterectomy" (1984): 332. New England: Pearson and others, "Hospital Caseloads" (1968): 565. Aggression: Bunker, "Surgical Manpower" (1970): 140. VA System: Ashton and others, "Geographic Variations" (1999); Ashton and others, "Hospital Use and Survival" (2003).

16. Community standards and uncertainty: Eddy, "Variations" (1984): 86 [community standards], 75 [uncertainty creeps, honest people]. Maryland physicians: Rutkow and others, "Surgical Decision Making" (1970): 409 [marked divergence, semi-exact], 419 [Moore comment]. Variation and uncertainty: Wennberg and others, "Professional Uncertainty" (1982); Barnes and others, "Report on Variation" (1985).

17. Review: Paul-Schaheen and others, "Small Area Analysis" (1987): 765 [ascertain]. Multivariate analysis: Holahan and others, "Area Variations" (1990): 170 [explanatory power]. Europe: van den Brand and the European Angioplasty Survey Group, "Utilization" (1993): 395.

18. Patient preference: Maynard and others, "Blacks in the Coronary Artery Surgery Study" (1986). Citations of this: Wenneker and Epstein, "Racial Inequalities" (1989); Goldberg and others, "Racial and Community Factors" (1992). Patient interviews: Crawford and others, "Do Blacks and Whites Differ" (1994); Whittle and others, "Do Patient Preferences Contribute" (1997). Black physicians: Gillum, "Coronary Heart Disease" (1982): 849. Prejudice: Council on Ethical and Judicial Affairs, "Black-White Disparities" (1990): 2346.

19. Elusive prejudice: Hannan and others, "Interracial Access" (1991); Ayanian and others, "Racial Differences" (1993). Referral disparity: Schulman and others, "Effect of Race and Sex" (1999). Reanalysis: Schwartz and others, "Misunderstandings" (1999).

20. Bodewitz and others, "Regulatory Science" (1987); Richards, *Vitamin C and Cancer* (1991); Epstein, *Impure Science* (1996); Marks, *Progress* (1997); Berg, *Rationalizing Medical Work* (1997); Timmermans and Berg, *Gold Standard* (2003); Daly, *Evidence-Based Medicine* (2005); Carpenter, *Reputation and Power* (2010). Reversed findings: Lerner, *Breast Cancer Wars* (2001); Watkins, *Estrogen Elixir* (2007); Taubes, *Good Calories, Bad Calories* (2007). Bypass surgery: Jones, "Visions" (2000). Statin standards: Ridker, "Tale of Three Labels" (2011).

21. The RAND Project: Chassin and others, "Does Inappropriate Use" (1987). Angiography: Chassin and others, "How Coronary Angiography" (1987). Bypass surgery: Winslow and others, "The Appropriateness" (1988). Expert disagreement: Park and others, "Physician Ratings" (1986); Mulley and Eagle, "What Is Inappropriate Care?" (1988).

22. Medicare costs: Gornick, "Medical Patients" (1975). Costs and scrutiny: Wennberg, *Tracking Medicine* (2010); Weisz and others, "Emergence" (2007): 706–9. Canada studies: Rouleau and others, "Comparison of Management" (1993); Anderson and others, "Use of Coronary Artery Bypass" (1993); McGlynn and others, "Comparison of Appropriateness" (1994); Mark and others, "Use of Medical Resources" (1994); Tu and others, "Use of Cardiac Procedures" (1997).

23. Gawande, "Cost Conundrum" (2009): 39. White House interest: Pear, "Health Care Spending" (2009). The McAllen story may be more complicated than Gawande sug-

gest: it is not a high-spending region for private insurance: Chernew and others, "Geographic Correlation" (2010).

24. Cutting Medicare: Skinner and Fisher, "Regional Disparities" (1997); Fisher and others, "Implications of Regional Variations [Part 1]" (2003); Fisher and others, "Implications of Regional Variations [Part 2]" (2003). Critiques: Pear, "Health Care Spending" (2009) [Kerry quotation]; "Obama's Health Cost Illusion" (2009) [funhouse mirror]; Bach, "Map to Bad Policy" (2010); Zuckerman and others, "Clarifying Sources" (2010).

25. *Dartmouth Atlas* (1996): 4.

26. Department of Thoracic and Cardiovascular Surgery, "1995 Calendar."

27. England and Australia: English and others, "UK Cardiac Surgical Register" (1984). Angioplasty disparities: van den Brand and the European Angioplasty Survey Group, "Utilization" (1993). 2009 data: Organization for Economic Co-operation and Development, *Health at a Glance 2011* (2011): 90–91. See also: Widimsky and others, "Reperfusion Therapy" (2010).

28. Bar graph: Organization for Economic Co-operation and Development, *Health at a Glance 2011* (2011): 91. Mexico: Ban-Hayashi and others, "Interventional Cardiology" (1995). India: Padmavati, "Interventional Cardiology" (1995); Association of Physicians of India, "Expert Consensus" (2006); Hiremath, "Future of Thrombolytic Therapy" (2011). China: Zhang and Huo, "Early Reperfusion" (2011). Nigeria: Muanya, "Nigerian Scores First" (2009).

29. Vortex: Reddy, Foreword (2004): vi. 1993 report: World Bank, *Investing in Health* (1993): 225. 1999 report: Chockalingam and others, "World Heart Federation's White Book" (2000): 227. 2002 data: "Heart Attack Deaths (2002)," *Worldmapper*. UN summit: Fink and Rabinowitz, "UN's Battle" (2011).

30. Consensus: Bradley and others, "Strategies" (2006). European concerns: de Jaegere and Simoons, "Immediate Angioplasty" (1995).

31. World Bank 1992: Phillips and others, "Emerging Agenda" (1992): 273. World Bank 1993: World Bank, *Investing in Health* (1993): 10. IOM 1998: Howso and others, *Control of Cardiovascular Diseases* (1998): 32. Cost effectiveness: Gaziano, "Reducing the Growing Burden" (2007): 18. UN summit: General Assembly, "Political Declaration" (2011).

32. Disillusionment: Smith, "Technological Determinism" (1994); Marx, "Technology" (2010). Navajo health care: McDermott and others, "Health Care Experiment" (1972); Jones, "Health Care Experiments" (2002). Advocates: Long and Oleson, *Appropriate Technology* (1980).

33. Cost-effective prevention: Marseille and others, "HIV Prevention before HAART" (2002). Critique of cost-effectiveness analysis: Farmer, "Listening for Prophetic Voices" (1997). HIV treatment: Cohen, "New World of Global Health" (2006); Kim and Farmer, "AIDS in 2006" (2006).

34. Limited funding: Fink and Rabinowitz, "UN's Battle" (2011). Neglect of cardiovascular disease: Yach, Foreword (2004): iii; Leeder and others, *Race Against Time* (2004): 14 [sympathy].

35. Bukhman and Kidder, "Cardiovascular Disease " (2008); Narayan and others, "Global Noncommunicable Diseases—Lessons" (2011).

36. McDougall to Clinton (2004).

37. Pregnancy: Rapp, *Testing Women* (1999). Cancer: Barry and others, "Watchful Waiting" (1988); Sepucha and Mulley, "Perspective on the Patient's Role" (2009).

38. Kolata, "Considering" (2011).

39. Feder, "Stent vs. Scalpel" (2005); Davis and Donnan, "Carotid-Artery Stenting" (2010); Broderick, "Challenges" (2011); Broderick and Meyers, "Acute Stroke Therapy" (2011).

40. Barry and others, "Patient Reactions" (1995); Morgan and others, "Randomized, Controlled Trial" (2000); Fowler and others, "Medical-Decision Process" (2012).

41. Explaining the decline: Jones and Greene, "Contributions of Prevention and Treatment" (2012). Potential for more decline: Bjorck and others, "Increasing Evidence-Based" (2011); Ford and Capewell, "Proportion of the Decline" (2011). However, early signs warn that the decline may soon stop or even reverse: Jones and Greene, "Fall and Rise" (in preparation).

42. Frenk and others, "Health Transition" (1989).

Bibliography

Archival Collections

Brigham and Women's Hospital Archives. Countway Library, Center for the History of Medicine, Boston, Massachusetts. [BWH Archives]

Bruch, Hilde, Papers. John P. McGovern Historical Collections and Research Center, Houston, Texas. Manuscript Collection 7. [HB Papers]

Cleveland Clinic Foundation Archives. Cleveland Clinic, Cleveland, Ohio. [CCF Archives]

Cooley, Denton A., Papers. John P. McGovern Historical Collections and Research Center, Houston, Texas. Manuscript Collection 43. [DC Papers]

Fonds, Arthur Vineberg. Osler Library Archive Collections, McGill University, Montreal, Quebec. Collection P126. [AV Fonds]

Harken, Dwight E., Papers. Countway Library, Center for the History of Medicine, Boston, Massachusetts. HC 48 BR. [DH Papers]

Leary, Timothy, Papers. Countway Library, Center for the History of Medicine, Boston, Massachusetts. B MS c40. [TL Papers]

Seybold, William, Papers. John P. McGovern Historical Collection and Research Center, Houston, Texas. Manuscript Collection 4. [WS Papers]

Texas Heart Institute Papers. John P. McGovern Historical Collections and Research Center, Houston, Texas. Institutional Collection 43. [THI Papers]

Primary and Secondary Sources

"1988 Midyear Health Care Advertising Review: Pharmacy." *Medical Marketing and Media* 23 (20 Sept. 1988): 40–46.

"1992 Midyear Health Care Advertising Review: Pharmacy." *Medical Marketing and Media* 27 (Oct. 1992): 80–83.

"2000 Calendar, Department of Thoracic and Cardiovascular Surgery." CCF Archives, box 1152, 21-PT, "Thoracic & CV Surgery Publications."

Abelson, Reed. "Heart Procedure Is off the Charts in an Ohio City." *New York Times*, 18 Aug. 2006.

Aberg, Torkel. "Effect of Open Heart Surgery on Intellectual Function." *Scandinavian Journal of Thoracic and Cardiovascular Surgery* suppl. 15 (1974): 1–63.

Aberg, Torkel, and Margareta Kihlgren. "Cerebral Protection during Open-Heart Surgery." *Thorax* 32 (1977): 525–33.

Alberti, Fay Bound. "Bodies, Hearts, and Minds: Why Emotions Matter to Historians of Science and Medicine." *Isis* 100 (2009): 798–810.

———. *Matters of the Heart: History, Medicine, and Emotion.* New York: Oxford University Press, 2010.

Alfano, Joseph E., Ruth E. Fabritius, and Marvin A. Garland. "Visual Loss Following Mitral Commissurotomy for Mitral Stenosis." *American Journal of Ophthalmology* 44 (1957): 213–16.

Alfidi, Ralph J. "Informed Consent: A Study of Patient Reaction." *JAMA* 216 (1971): 1325–29.

Allam, Adel H., Randall C. Thompson, L. Samuel Wann, Michael I. Miyamoto, Abd el-Halim Nur el-Din, Gomaa Abd el-Maksoud, Muhammad Al-Tohamy Soliman, Ibrahem Badr, Hany Abd el-Rahman Amer, M. Linda Sutherland, James D. Sutherland, Gregory S. Thomas. "Atherosclerosis in Ancient Egyptian Mummies." *JACC: Cardiovascular Imaging* 4 (2011): 315–27.

Ambrose, John A., Mark A. Tannenbaum, Dimitrios Alexopoulos, Craig E. Hjemdahl-Monsen, Jeffrey Leavy, Melvin Weiss, Susan Borrico, Richard Gorlin, and Valentin Fuster. "Angiographic Progression of Coronary Artery Disease and the Development of Myocardial Infarction." *Journal of the American College of Cardiology* 12 (July 1988): 56–62.

Ambrose, John A., Stephen L. Winters, Audrey Stern, Angie Eng, Louis E. Teichholz, Richard Gorlin, and Valentin Fuster. "Angiographic Morphology and the Pathogenesis of Unstable Angina Pectoris." *Journal of the American College of Cardiology* 5 (March 1985): 609–16.

American Heart Association. "Report of the Committee on the Effect of Strain and Trauma on the Heart and Great Vessels." *Circulation* 26 (Oct. 1962): 612–22.

Anderson, Geoffrey M., Kevin Grumbach, Harold S. Luft, Leslie L. Roos, Cameron Mustard, and Robert Brook. "Use of Coronary Artery Bypass Surgery in the United States and Canada." *JAMA* 269 (7 April 1993): 1661–66.

Archer, Robert, David A. Ott, Robert Parravicini, Denton A. Cooley, George J. Reul, Howard Frazier, J. Michael Duncan, James J. Livesay, and William E. Walker. "Coronary Artery Revascularization without Cardiopulmonary Bypass." *Texas Heart Institute Journal* 11 (March 1984): 52–57.

Aronowitz, Robert A. "The Converged Experience of Risk and Disease." *Milbank Quarterly* 87 (2009): 417–42.

———. *Making Sense of Illness: Science, Society, and Disease.* Cambridge: Cambridge University Press, 1998.

———. *Unnatural History: Breast Cancer and American Society.* Cambridge: Cambridge University Press, 2007.

Ashton, Carol M., Nancy J. Petersen, Julianne Souchek, Terri J. Menke, Hong-Jen Yu, Kenneth Pietz, Marsha L. Eigenbrodt, Galen Barbour, Kenneth W. Kizer, and Nelda P. Wray. "Geographic Variations in Utilization Rates in Veterans Affairs Hospitals and Clinics." *New England Journal of Medicine* 340 (7 Jan. 1999): 32–39.

Ashton, Carol M., Julianne Souchek, Nancy J. Petersen, Terri J. Menke, Tracie C. Collins, Kenneth W. Kizer, Steven M. Wright, and Nelda P. Wray. "Hospital Use and Survival among Veterans Affairs Beneficiaries." *New England Journal of Medicine* 349 (23 Oct. 2003): 1637–46.

Aspirin Myocardial Infarction Study Research Group. "A Randomized Controlled Trial of Aspirin in Persons Recovered from Myocardial Infarction." *JAMA* 243 (15 Feb. 1980): 661–69.

Association of Physicians of India. "API Expert Consensus Document on Management of Ischemic Heart Disease." *Journal of the Association of Physicians of India* 54 (2006): 469–80.

Atwater, Edward C. "Touching the Patient: The Teaching of Internal Medicine in America." In *Sickness and Health in America: Readings in the History of Medicine and Public Health*, 2nd ed., ed. Judith Walzer Leavitt and Ronald L. Numbers, 129–47. Madison: University of Wisconsin Press, 1985.

Avorn, Jerry. "Dangerous Deception—Hiding the Evidence of Adverse Drug Effects." *New England Journal of Medicine* 355 (23 Nov. 2006): 2169–71.

———. Introduction to the Updated Vintage Books Edition. *Powerful Medicine: The Benefits, Risks, and Costs of Prescription Drugs.* New York: Vintage Books, 2005.

Ayanian, John Z., I. Steven Udvarhelyi, Constantine A. Gatsonis, Chris L. Pashos, and Arnold M. Epstein. "Racial Differences in the Use of Revsacularization Procedures after Coronary Angiography." *JAMA* 269 (26 May 1993): 2642–46.

Baake, Ken. *Metaphor and Knowledge: The Challenges of Science Writing.* Albany: State University of New York Press, 2003.

Bach, Peter B. "A Map to Bad Policy—Hospital Efficiency Measures in the Dartmouth Atlas." *New England Journal of Medicine* 362 (18 Feb. 2010): 569–73.

Bailey, Charles P., R. P. Glover, and T. J. O'Neill. "Exploratory Surgery of the Heart." *Diseases of the Chest* 22 (Dec. 1952): 640–70.

Bailey, Charles P., and William Likoff. "The Surgical Treatment of Coronary Insufficiency." *Chest* 27 (1955): 477–514.

Bailey, Charles P., Angelo May, and William M. Lemmon. "Survival after Coronary Endarterectomy in Man." *JAMA* 164 (8 June 1957): 641–46.

Bailey, Charles P., Axel K. Olsen, Kenneth K. Keown, and Henry T. Nichols. "Commissurotomy for Mitral Stenosis: Technique for Prevention of Cerebral Complications." *JAMA* 149 (1952): 1085–91.

Baker, Peter, and Angela Macropoulos. "Bill Clinton Released from Hospital." *New York Times*, 11 Feb. 2010.

Baldry, P. E. *The Battle Against Heart Disease.* Cambridge: Cambridge University Press, 1971.

Ban-Hayashi, Ernesto, Jorge Gaspar, Ramón Villavicencio, Manuel Gil, and Marco A. Martinez-Rios. "Interventional Cardiology in Mexico: A Perspective from the Instituto Nacional de Cardiologia Ignacio Chavez." *Journal of Interventional Cardiology* 8 (1995): 23–27.

Banting, F. G., to Timothy Leary, 4 April 1934. TL Papers, box 2.

Barbut, Denise, and Jeffrey P. Gold. "Aortic Atheromatosis and Risks of Cerebral Embolization." *Journal of Cardiothoracic and Vascular Anesthesia* 10 (Jan. 1996): 24–30.

Barnes, Benjamin A., Elizabeth O'Brien, Catherine Comstock, Deborah G. D'Arpa,

Charles L. Donahue. "Report on Variation in Rates of Utilization of Surgical Services in the Commonwealth of Massachusetts." *JAMA* 254 (19 July 1985): 371–75.

Barnes, David S. *The Great Stink of Paris and the Nineteenth-Century Struggle against Filth and Germs.* Baltimore, MD: Johns Hopkins University Press, 2006.

Baroldi, Giorgio. "Lack of Correlation between Coronary Thrombosis and Myocardial Infarction or Sudden 'Coronary' Heart Death." *Annals of the New York Academy of Science* 156 (Jan. 1969): 504–25.

———. "Reply." *American Heart Journal* 94 (Sept. 1977): 392–93.

Barry, Michael J., Floyd J. Fowler, Albert G. Mulley, Joseph V. Henderson, and John E. Wennberg. "Patient Reactions to a Program Designed to Facilitate Patient Participation in Treatment Decisions for Benign Prostatic Hyperplasia." *Medical Care* 33 (Aug. 1995): 771–82.

Barry, Michael J., Albert G. Mulley, Floyd J. Fowler, and John W. Wennberg. "Watchful Waiting vs. Immediate Transurethral Resection for Symptomatic Prostatism." *JAMA* 259 (27 May 1988): 3010–17.

Beck, Ulrich. *Risk Society: Towards a New Modernity.* Trans. M. Ritter. London: Sage, 1992.

Bendszus, Martin, and Guido Stoll. "Silent Cerebral Ischaemia: Hidden Fingerprints of Invasive Medical Procedures." *Lancet Neurology* 5 (April 2006): 364–72.

Berg, Marc. *Rationalizing Medical Work: Decision-Support Techniques and Medical Practices.* Cambridge, MA: MIT Press, 1997.

Berry, Jarett D., Alan Dyer, Xuan Cai, Daniel B. Garside, Hongyan Ning, Avis Thomas, Philip Greenland, Linda Van Horn, Russell P. Tracy, and Donald M. Lloyd-Jones. "Lifetime Risks of Cardiovascular Disease." *New England Journal of Medicine* 366 (26 Jan. 2012): 321–29.

Biagioli, Mario. *Galileo, Courtier: The Practice of Science in the Culture of Absolutism.* Chicago: University of Chicago Press, 1994.

Bjorck, L., S. Capewell, K. Bennett, G. Lappas, and A. Rosengren. "Increasing Evidence-Based Treatments to Reduce Coronary Heart Disease Mortality in Sweden: Quantifying the Potential Gains." *Journal of Internal Medicine* 269 (2011): 452–67.

Blair, Kirstie. *Victorian Poetry and the Culture of the Heart.* New York: Oxford University Press, 2006.

Blatchford, James W. "Ludwig Rehn: The First Successful Cardiorrhaphy." *Annals of Thoracic Surgery* 39 (1985): 492–95.

Blauth, C., J. Arnold, E. M. Kohner, and K. M. Taylor. "Retinal Microembolism during Cardiopulmonary Bypass Demonstrated by Fluorescein Angiography." *Lancet* 328 (11 Oct. 1986): 837–39.

Blauth, Christopher I., John V. Arnold, W. Edmund Schulenberg, Alison C. McCartney, and Kenneth M. Taylor, sponsored by Floyd D. Loop. "Cerebral Microembolism during Cardiopulmonary Bypass: Retinal Microvascular Studies in vivo with Fluorsceine Angiography." *Journal of Thoracic and Cardiovascular Surgery* 95 (April 1988): 668–76.

Blumenthal, David. "The Variation Phenomenon in 1994." *New England Journal of Medicine* 331 (13 Oct. 1994): 1017–18.

Blumgart, Herrman L., Monroe J. Schlesinger, and David Davis. "Studies on the Relation of the Clinical Manifestations of Angina Pectoris, Coronary Thrombosis, and Myocardial Infarction to the Pathologic Findings, with Particular Reference to the Signifi-cance of the Collateral Circulation." *American Heart Journal* 19 (Jan. 1940): 1–91.

Blumgart, Herrman L., Monroe J. Schlesinger, and Paul M. Zoll. "Angina Pectoris, Coro-nary Failure and Acute Myocardial Infarction: The Role of Coronary Occlusions and Collateral Circulation." *JAMA* 116 (11 Jan. 1941): 91–96.

Blumgart, Herrman L., Paul M. Zoll, and George S. Kurland. "Discussion of Direct Relief of Coronary Occlusion." *Archives of Internal Medicine* 104 (1959): 862–69.

Boden, William E., Robert A. O'Rourke, Koon K. Teo, Pamela M. Hartigan, David J. Maron, William J. Kostuk, Merril Knudtson, Marcin Dada, Paul Casperson, Crystal L. Harris, Bernard R. Chaitman, Leslee Shaw, Gilbert Gosselin, Shah Nawaz, Lawrence M. Title, Gerald Gau, Alvin S. Blaustein, David C. Booth, Eric R. Bates, John A. Spertus, Daniel S. Berman, John Mancini, and William S. Weintraub, for the COURAGE Trial Research Group. "Optimal Medical Therapy with or without PCI for Stable Coronary Disease." *New England Journal of Medicine* 356 (12 April 2007): 1503–16.

Boden, William E., Koon K. Teo, and William S. Weintraub, for the COURAGE Trial Research Group. "The Authors Reply." *New England Journal of Medicine* 357 (26 July 2007): 417–18.

Bodewitz, Henk J. H. W., Henk Buurma, and Gerard H. de Vries. "Regulatory Science and the Social Management of Trust in Medicine." In *The Social Construction of Technologi-cal Systems,* ed. Wiebe E. Bijker, Thomas P. Hughes, and Trevor J. Pinch, 243–59. Cam-bridge, MA: MIT Press, 1987.

Bolduan, Charles F., and Nils W. Bolduan. "Is the 'Appalling Increase' in Heart Disease Real?" *Journal of Preventive Medicine* 6 (1932): 321–33.

Bonow, Robert O., Gerald Maurer, Kerry L. Lee, Thomas A. Holly, Philip F. Binkley, Patrice Desvigne-Nickens, Jaroslaw Drozdz, Pedro S. Farsky, Arthur M. Feldman, Tor-sten Doenst, Robert E. Michler, Daniel S. Berman, Jose C. Nicolau, Patricia A. Pel-likka, Krzysztof Wrobel, Nasri Alotti, Federico M. Asch, Liliana E. Favaloro, Lilin She, Eric J. Velazquez, Robert H. Jones, and Julio A. Panza, for the STICH Trial Investiga-tors. "Myocardial Viability and Survival in Ischemic Left Ventricular Dysfunction." *New England Journal of Medicine* 364 (2011): 1617–25.

Bosk, Charles L. *Forgive and Remember: Managing Medical Failure.* Chicago: University of Chicago Press, 1979.

Bouch, D. C., and G. L. Montgomery. "Cardiac Lesions in Fatal Cases if Recent Myocardial Ischemia from a Coronary Care Unit." *British Heart Journal* 32 (1970): 795–803.

Bounds, James V., Burton A. Sandok, and Donald A. Barnhorst. "Fatal Cerebral Embolism Following Aorto-Coronary Bypass Graft Surgery." *Stroke* 7 (Nov.–Dec. 1976): 611–14.

Boutwell, Robert C., and Janet B. Mitchell. "Diffusion of New Technologies in the Treat-ment of the Medicare Population." *International Journal of Technology Assessment in Health Care* 9 (1993): 62–75.

Bowker, Geoffrey C., and Susan Leigh Star. *Sorting Things Out: Classification and Its Con-sequences.* Cambridge, MA: MIT Press, 2000.

Boyd, Adam. "An Inflammatory Basis for Coronary Thrombosis." *American Journal of Pa-thology* 4 (1928): 159–66.

Bradley, Elizabeth H., Jeph Herrin, Yongfei Wang, Barbara A. Barton, Tashonna R. Webster,

Jennifer A. Mattera, Sarah A. Roumanis, Jeptha P. Curtis, Brahmajee K. Nallamothu, David J. Magid, Robert L. McNamara, Janet Parkosewich, Jerod M. Loeb, and Harlan M. Krumholz. "Strategies for Reducing the Door-to-Balloon Time in Acute Myocardial Infarction." *New England Journal of Medicine* 355 (30 Nov. 2006): 2308–20.

"Brain Damage after Open-Heart Surgery." *Lancet* 306 (30 Aug. 1975): 399–400.

Brain, Robert M. "Representation on the Line: The Graphic Method and the Instruments of Scientific Modernism." In *From Energy to Information: Representation in Science, Art, and Literature,* ed. Bruce Clark and Linda D. Henderson, 155–78. Stanford, CA: Stanford University Press, 2002.

Brandt, Allan M. *The Cigarette Century: The Rise, Fall, and Deadly Persistence of the Product that Defined America.* New York: Basic Books, 2007.

———. "The Cigarette, Risk, and American Culture." *Daedalus* 119 (Fall 1990): 155–76.

———. *No Magic Bullet: A Social History of Venereal Disease in the United States since 1880.* Expanded ed. New York: Oxford University Press, 1987.

Brandt, Allan M., and Martha Gardner. "The Golden Age of Medicine?" In *Medicine in the Twentieth Century,* ed. Roger Cooter and John Pickstone, 21–37. Amsterdam: Harwood Academic Publishers, 2000.

Branthwaite, M. A. "Detection of Neurological Damage during Open-Heart Surgery." *Thorax* 28 (1973): 464–72.

———. "Prevention of Neurological Damage during Open-Heart Surgery." *Thorax* 30 (1975): 258–61.

Branwood, A. Whitley, and George L. Montgomery. "Observations on the Morbid Anatomy of Coronary Artery Disease." *Scottish Medical Journal* 1 (Dec. 1956): 367–75.

Braunwald, Eugene. "Coronary-Artery Surgery at the Crossroads." *New England Journal of Medicine* 297 (22 Sept. 1977): 661–63.

———. "Direct Coronary Revascularization . . . A Plea Not to Let the Genie Escape from the Bottle." *Hospital Practice* 6 (May 1971): 9–10.

———. "Heart." *Annual Review of Physiology* 28 (1966): 227–66.

———. "Unstable Angina: A Classification." *Circulation* 80 (Aug. 1989): 410–14.

Braunwald, Eugene, and Peter F. Cohn. "Ischemic Heart Disease." In *Harrison's Principle's of Internal Medicine,* 10th ed,. ed. Robert G. Petersdorf, Raymond D. Adams, Eugene Braunwald, Kurt J. Isselbacher, Joseph B. Martin, and Jean D. Wilson, 1423–32. New York: McGraw-Hill, 1983.

Bravata, Dean M., Allison L. Gienger, Kathryn M. McDonald, Vandana Sundaram, Marco V. Perez, Robin Varghese, John R. Kapoor, Reza Ardehali, Douglas K. Owens, and Mark A. Hlatky. "Systematic Review: The Comparative Effectiveness of Percutaneous Coronary Interventions and Coronary Artery Bypass Graft Surgery." *Annals of Internal Medicine* 147 (20 Nov. 2007): 703–16.

Breuer, Anthony C., Anthony J. Furlan, Maurice R. Hanson, Richard J. Lederman, Floyd D. Loop, Delos M. Cosgrove, Edwin G. Beven, Norman R. Hertzer, and F. George Estafanous. "Neurology of Open Heart Surgery," Cleveland Clinic Foundation, Feb. 1979. CCF Archives, 3-PR10, "Breuer, Anthony."

Breuer, A. C., A. J. Furlan, M. R. Hanson, R. J. Lederman, F. D. Loop, D. M. Cosgrove, and F. G. Estafanous. "Brain and Nerve Injury Associated with Coronary Artery Bypass

Graft Surgery: A Prospective Analysis of 421 Cases." *Circulation* 62 suppl. 3 (Oct. 1980): 42.

Breuer, Anthony C., Anthony J. Furlan, Maurice R. Hanson, Richard J. Lederman, Floyd D. Loop, Delos M. Cosgrove, Michel A. M. Ghattas, F. George Estafanous. "Neurologic Complications of Open-Heart Surgery: Computer Assisted Analysis of 531 Patients." Paper presented at Anesthesia and the Heart Patient: An International Symposium, 30 May–1 June 1980. CCF Archives, "Continuing Medical Education Office (May 1980– Oct. 1979)," folder "Syllabus" (#4884).

Breuer, Anthony C., Anthony J. Furlan, Maurice R. Hanson, Richard J. Lederman, Floyd D. Loop, Delos M. Cosgrove, Michel A. Ghattas, and Fawzy G. Estafanous. "Neurologic Complications of Open Heart Surgery: Computer-Assisted Analysis of 531 Patients." *Cleveland Clinic Quarterly* 48 (Spring 1981): 205–6.

Breuer, A. C., A. J. Furlan, M. R. Hanson, R. J. Lederman, F. D. Loop, D. M. Cosgrove, R. L. Greenstreet, and F. G. Estafanous. "Central Nervous System Complications of Coronary Artery Bypass Graft Surgery: Prospective Analysis of 421 Patients." *Stroke* 14 (Sept.–Oct. 1983): 682–87.

Brierly, James B. "Brain Damage Complicating Open-Heart Surgery: A Neuropathological Study of 46 Patients." *Proceedings of the Royal Society of Medicine* 60 (Sept. 1967): 858–59.

Brock, Cornelius. "Recording the Brain at Work: The Visible, the Readable, and the Invisible in Electroencephalography." *Journal of the History of the Neurosciences* 17 (2008): 367–79.

Broderick, Joseph P. "The Challenges of Intracranial Revascularization for Stroke Prevention." *New England Journal of Medicine* 365 (15 Sept. 2011): 1054–55.

Broderick, Joseph P., and Philip M. Meyers. "Acute Stroke Therapy at the Crossroads." *JAMA* 306 (9 Nov. 2011): 2026–28.

Brown, B. Greg, Edward Bolson, Morris Frimer, and Harold T. Dodge. "Quantitative Coronary Arteriography: Estimation of Dimensions, Hemodynamic Resistance, and Atheroma Mass of Coronary Artery Lesions Using the Arteriogram and Digital Computation." *Circulation* 55 (Feb. 1977): 329–37.

Brown, B. Greg, Xue-Qiao Zhao, Dianne E. Sacco, and John J. Albers. "Lipid Lowering and Plaque Regression: New Insights Into Prevention of Plaque Disruption and Clinical Events in Coronary Disease." *Circulation* 87 (June 1993): 1781–91.

Brown, Greg, John J. Albers, Lloyd D. Fisher, Susan M. Schaefer, Jiin-Tarng Lin, Cheryl Kaplan, Xue-Qiao Zhao, Brad D. Bisson, Virginia F. Fitzpatrick, and Harold T. Dodge. "Regression of Coronary Artery Disease as a Result of Intensive Lipid-Lowering Therapy in Men with High Levels of Apolipoprotein B." *New England Journal of Medicine* 323 (8 Nov. 1990): 1289–98.

Browne, Janet. *Charles Darwin: The Power of Place.* Princeton, NJ: Princeton University Press, 2002.

Brownell, Kelly D., Roger Kersh, David S. Ludwig, Robert C. Post, Rebecca M. Puhl, Marlene B. Schwartz, and Walter C. Willett. "Personal Responsibility and Obesity: A Constructive Approach to a Controversial Issue." *Health Affairs* 29 (March 2010): 379–87.

Brownlee, Shannon. *Overtreated: Why Too Much Medicine Is Making Us Sicker and Poorer.* New York: Bloomsbury, 2008.

Bruch, Hilde, to Michael DeBakey, 11 Nov. 1969. HB Papers, series V, box 6.

Bruch, Hilde, to Franz J. Ingelfinger, 25 Aug. 1970. HB Papers, series V, box 6.

Bruch, Hilde, to Ronald Malt, 22 Dec. 1970. HB Papers, series V, box 6.

Bruch, Hilde, to Stuart Yudofsky, 15 Oct. 1970. HB Papers, series V, box 6.

Bruschke, Albert V. G., John R. Kramer, Egbert T. Bal, Ihsan Ul Haque, Robert C. Detrano, and Marlene Goormastic. "The Dynamics of Progression of Coronary Atherosclerosis Studied in 168 Medically Treated Patients Who Underwent Coronary Arteriography Three Times." *American Heart Journal* 117 (Feb. 1989): 296–305.

Buffolo, E., J. C. S. Andrade, J. N. R. Branco, L. F. Aguiar, E. E. Ribeiro, and A. D. Jatene. "Myocardial Revascularization without Extracorporeal Circulation: Seven-Year Experience in 593 Cases." *European Journal of Cardio-thoracic Surgery* 4 (1990): 504–8.

Buffolo, Enio, José Carlos S. Andrade, José Ernesto Succi, Luiz Eduardo V. Leão, Clotário Cueva, Joaõ Nelson Branco, Antonio Carlos C. Carvalho, and Costabile Galluci. "Direct Myocardial Revascularization without Extracorporeal Circulation: Technique and Initial Results." *Texas Heart Institute Journal* 12 (March 1985): 33–41.

Buffon, Antonino, Luigi M. Biasucci, Giovanna Liuzzo, Giuseppe D'Onofrio, Filippo Crea, and Attilio Maseri. "Widespread Coronary Inflammation in Unstable Angina." *New England Journal of Medicine* 347 (4 July 2002): 5–12.

Bukhman, Gene, and Alice Kidder. "Cardiovascular Disease and Global Health Equity: Lessons from Tuberculosis Control Then and Now." *American Journal of Public Health* 98 (Jan. 2008): 44–54.

Bunker, John P. "Surgical Manpower: A Comparison of Operations and Surgeons in the United States and in England and Wales." *New England Journal of Medicine* 282 (15 Jan. 1970): 135–44.

Burton, Thomas. "Merck Takes Author's Name Off Vioxx Study." *Wall Street Journal,* 18 May 2004, B1.

Bypass Angioplasty Revascularization Investigation (BARI) Investigators. "Comparison of Coronary Bypass Surgery with Angioplasty in Patients with Multivessel Disease." *New England Journal of Medicine* 335 (25 July 1996): 217–25.

Caplan, Louis R. "Translating What Is Known about Neurological Complications of Coronary Artery Bypass Graft Surgery into Action." *Archives of Neurology* 66 (Sept. 2009): 1062–64.

CAPRIE Steering Committee. "A Randomized, Blinded Trial of Clopidogrel versus Aspirin in Patients at Risk of Ischaemic Events (CAPRIE)." *Lancet* 348 (16 Nov. 1996): 1329–39.

Carlat, Daniel. *Unhinged: The Trouble with Psychiatry: A Doctor's Revelations about a Profession in Crisis.* New York: Free Press, 2010.

Carlson, Robert G., Arnold J. Landé, Louis W. Ivey, Peter J. Starek, J. Richard Rees, V. A. Subramanian, Joseph Twichell, James Baxter, Jack H. Bloch, and C. Walton Lillehei. "Total Cardiopulmonary Support with Disposable Membrane Oxygenator during Aortocoronary Artery-Vein Graft Operations." *Chest* 62 (Oct. 1972): 424–32.

Carlson, Robert G., Arnold J. Landé, Bernard Landis, Barbara Rogoz, James Baxter, Russell

H. Patterson, Kurt Stenzel, and C. W. Lillehei. "The Landé-Edwards Membrane Oxygenator during Heart Surgery: Oxygen Transfer, Microemboli Counts, and Bender-Gestalt Visual Motor Test Scores." *Journal of Thoracic and Cardiovascular Surgery* 66 (Dec. 1973): 894–905.

Carpenter, Daniel. *Reputation and Power: Organization Image and Pharmaceutical Regulation at the FDA.* Princeton, NJ: Princeton University Press, 2010.

Casalino, Lawrence P., Kelly J. Devers, and Linda R. Brewster. "Focused Factories? Physician-Owned Specialty Facilities." *Health Affairs* 22 (Nov./Dec. 2003): 56–67.

Casscells, Ward, Bryan Hathorn, Tim Krabach, William K. Vaughn, Hugh A. McAllister, Greg Bearman, and James T. Willerson. "Thermal Detection of Cellular Infiltrates in Living Atherosclerotic Plaques: Possible Implications for Plaque Rupture and Thrombosis." *Lancet* 347 (25 May 1996): 1447–49.

Casscells, Ward, Morteza Naghavi, and James T. Willerson. "Vulnerable Atherosclerotic Plaque: A Multifocal Disease." *Circulation* 107 (29 April 2003): 2072–75.

CASS Principal Investigators and Their Associates. "Coronary Artery Surgery Study (CASS): A Randomized Trial of Coronary Artery Bypass Surgery: Quality of Life in Patients Randomly Assigned to Treatment Groups." *Circulation* 68 (Nov. 1983): 951–60.

CASS Principal Investigators and Their Associates. "Coronary Artery Surgery Study (CASS): A Randomized Trial of Coronary Artery Bypass Surgery: Survival Data." *Circulation* 68 (Nov. 1983): 939–50.

Centers for Disease Control and Prevention. "Leading Causes of Death, 1900–1998." Last updated 12 Nov. 2009, www.cdc.gov/nchs/nvss/mortality_historical_data.htm.

"Cerebral Injury Following Cardiac Operations." *Lancet* 283 (11 Jan. 1964): 89–90.

Chan, Paul S., Manesh R. Patel, Lloyd W. Klein, Ronald J. Krone, Gregory J. Dehmer, Kevin Kennedy, Brahmajee K. Nallamothu, W. Douglas Weaver, Frederick A. Masoudi, John S. Rumsfeld, Ralph G. Brindis, and John A. Spertus. "Appropriateness of Percutaneous Coronary Intervention." *JAMA* 306 (2011): 53–61.

Chapman, Irving. "Morphogenesis of Occluding Coronary Artery Thrombosis." *Archives of Pathology* 80 (Sept. 1965): 256–61.

Chassin, Mark R., Robert H. Brook, R. E. Park, Joan Kessey, Arlene Fink, Jacqueline Kosecoff, Katherine Kahn, Nancy Merrick, and David H. Solomon. "Variations in the Use of Medical and Surgical Services by the Medicare Population." *New England Journal of Medicine* 314 (30 Jan. 1986): 285–90.

Chassin, Mark R., Jacqueline Kosecoff, R. E. Park, Constance M. Winslow, Katherine L. Kahn, Nancy J. Merrick, Joan Keesey, Arlene Fink, David H. Solomon, and Robert H. Brook. "Does Inappropriate Use Explain Geographic Variations in the Use of Health Care Services? A Study of Three Procedures." *JAMA* 258 (13 Nov. 1987): 2533–37.

Chassin, Mark R., Jacqueline Kosecoff, David H. Solomon, and Robert H. Brook. "How Coronary Angiography Is Used: Clinical Determinants of Appropriateness." *JAMA* 258 (13 Nov. 1987): 2543–47.

Chernew, Michael E., Lindsay M. Sabik, Amitabh Chandra, Teresa B. Gibson, and Joseph P. Newhouse. "Geographic Correlation between Large-Firm Commercial Spending and Medicare Spending." *American Journal of Managed Care* 16 (2010): 131–38.

Cheruvu, Pavan K., Aloke V. Finn, Craig Gardner, Jay Caplan, James Goldstein, Gregg W.

Stone, Renu Virmani, and James E. Muller. "Frequency and Distribution of Thin-Cap Fibroatheroma and Ruptured Plaques in Human Coronary Arteries." *Journal of the American College of Cardiology* 50 (2007): 940–49.

Chockalingam, Arum, Ignasi Balaguer-Vintro, Aloysio Achutti, Antoni Bayes de Luna, John Chalmers, Eduardo Farinaro, Richard Lauzon, Ingrid Martin, Julius Gy Papp, Aldredo Postiglione, K. Srinath Reddy, and Tak-Fu Tse. "The World Heart Federation's White Book: Impending Global Pandemic of Cardiovascular Diseases: Challenges and Opportunities for the Prevention and Control of Cardiovascular Diseases in Developing Countries and Economies in Transition." *Canadian Journal of Cardiolology* 16 (Feb. 2000): 227–29.

Clark, Eugene, Irving Graef, and Herbert Chasis. "Thrombosis of the Aorta and Coronary Arteries: With Special Reference to 'Fibrinoid' Lesions." *Archives of Pathology* 22 (Aug. 1936): 183–212.

Cleveland Clinic Foundation. *Challenge* (Winter 1986), CCF Archives.

Clough, John D., ed. *To Act as a Unit: The Story of the Cleveland Clinic.* 4th ed. Cleveland, OH: Cleveland Clinic Foundation, 2004.

Clowes, George H. A., William E. Neville, Amos Hopkins, Jorge Anzola, and F. A. Simeone. "Factors Contributing to Success or Failure in the Use of a Pump Oxygenator for Complete By-pass of the Heart and Lung, Experimental and Clinical." *Surgery* 36 (Sept. 1954): 557–79.

Cohen, David J., Ben Van Hout, Patrick W. Serruys, Friedrich W. Mohr, Carlos Macaya, Peter den Heijer, M. M. Vrakking, Kaijun Wang, Elizabeth M. Mahoney, Salma Audi, Katrin Leadley, Keith D. Dawkins, and A. Pieter Kappetein, for the Synergy between PCI with Taxus and Cardiac Surgery (SYNTAX) Investigators. "Quality of Life after PCI with Drug-Eluting Stents or Coronary-Artery Bypass Surgery." *New England Journal of Medicine* 364 (2011): 1016–26.

Cohen, Jon. "The New World of Global Health." *Science* 311 (13 Jan. 2006): 162–67.

Cohen, Samuel I. "Neurological and Psychiatric Aspects of Open-Heart Surgery." *Thorax* 19 (1964): 575–78.

Cohn, Lawrence H. "Surgical Treatment of Coronary Artery Disease." In *Cecil Textbook of Medicine,* 20th ed., ed. J. Claude Bennett and Fred Plum, 316–19. Philadelphia, PA: W. B. Saunders, 1996.

Cohn, Lawrence H., Richard Gorlin, Michael V. Herman, and John J. Collins. "Aorto-coronary Bypass for Acute Coronary Occlusion." *Journal of Thoracic and Cardiovascular Surgery* 64 (Oct. 1972): 503–13.

Collins, Harry. *Changing Order: Replication and Induction in Scientific Practice.* London and Beverly Hills, CA: Sage Publications, 1985; Chicago: The University of Chicago Press, 1992.

Collins, Harry, and Trevor Pinch. *The Golem: What You Should Know about Science.* 2nd ed. Cambridge: Cambridge University Press, 1998.

Committee for Clinical Investigation Involving Human Beings, The Methodist Hospital, "Memo," 18 Nov. 1968. WS Papers, series V, box 7, folder 116, "Research Involving Human Beings."

Conniff, Richard, and John Dominis. "Profiles: Denton A. Cooley." *Architectural Digest* (May 1987): 194–201, 246.

Constantinides, Paris. "Atherosclerosis—A General Survey and Synthesis." *Survey and Synthesis of Pathology Research* 3 (1984): 477–98.

———. "Cause of Thrombosis in Human Atherosclerotic Arteries." *American Journal of Cardiology* 66 (6 Nov. 1990): 37G–40G.

———. *Experimental Atherosclerosis.* New York: Elsevier, 1965.

———. "Experimental Atherosclerosis in the Rabbit." In *Comparative Atherosclerosis: The Morphology of Spontaneous and Induced Atherosclerotic Lesions in Animals and Humans in Relation to Disease,* ed. James C. Roberts, Reuben Strauss, and Miriam S. Cooper, 276–90. New York: Harper and Row, 1965.

———. "Plaque Fissures in Coronary Thrombosis." *Journal of Atherosclerosis Research* 6 (1966): 1–17.

———. "Plaque Fissures in Human Coronary Thrombosis." *Federation Proceedings* 23 (1964): 443.

Constantinides, Paris, J. Booth, and G. Carlson. "Production of Advanced Cholesterol Atherosclerosis in the Rabbit." *Archives of Pathology* 70 (1960): 712–24.

Constantinides, Paris, and John Lawder. "Experimental Thrombosis and Hemorrhage in Atherosclerotic Arteries." *Federation Proceedings* 22 (1963): 251.

Conway, Erik M., and Naomi Oreskes. *Merchants of Doubt: How a Handful of Scientists Obscured the Truth on Issues from Tobacco Smoke to Global Warming.* New York: Bloomsbury, 2011.

"Cooley Criticizes Medical 'Rigidity, Caution.'" *Houston Chronicle,* 19 May 1970, sec. 1, p. 2.

Cooley, Denton A. "Oral History" (4 May 1997). In *Pioneers of Cardiac Surgery,* ed. William S. Stoney, 244–55. Nashville, TN: Vanderbilt University Press, 2008.

Cornfield, Jerome. "Approaches to Assessment of the Efficacy of Surgical Revascularization." *Bulletin of the New York Academy of Medicine* 48 (Oct. 1972): 1126–34.

"Coronary Thrombosis Linked to Fissure in Atherosclerotic Vessel Wall." *JAMA* 188 (11 May 1964): 35–37.

Cosgrove, Delos M., Floyd D. Loop, Craig L. Saunders, Bruce W. Lytle, and John R. Kramer. "Should Coronary Arteries with Less than Fifty Percent Stenosis Be Bypassed?" *Journal of Thoracic and Cardiovascular Surgery* 82 (Oct. 1981): 520–30.

Council on Ethical and Judicial Affairs, American Medical Association. "Black-White Disparities in Health Care." *JAMA* 263 (2 May 1990): 2344–46.

Council on Foods and Nutrition, American Medical Association. "Diet and the Possible Prevention of Coronary Atheroma." *JAMA* 194 (6 Dec. 1965): 1149–50.

Crawford, Sybil L., Sarah A. McGraw, Kevin W. Smith, John B. McKinlay, and Judith E. Pierson. "Do Blacks and Whites Differ in Their Use of Health Care for Symptoms of Coronary Heart Disease?" *American Journal of Public Health* 84 (June 1994): 957–64.

Crawford, T. "Morphological Aspects in the Pathogenesis of Atherosclerosis." *Journal of Atherosclerosis Research* 1 (1961): 3–25.

Crawford, T., D. Dexter, and R. D. Teare. "Coronary-Artery Pathology in Sudden Death from Myocardial Ischaemia." *Lancet* 277, no. 1 (28 Jan. 1961): 181–85.

Cutler, Elliott C., and Samuel A. Levine. "Cardiotomy and Valvulotomy for Mitral Steno-sis: Experimental Observations and Clinical Notes Concerning and Operated Case with Recovery." *Boston Medical and Surgical Journal* 188 (28 June 1923): 1023–27.

Cutter, Gary R., Albert Oberman, Nicholas Kouchoukos, and William Rogers. "Epidemio-logic Study of Candidates for Coronary Artery Bypass Surgery." *Circulation* 66, suppl. 3 (Nov. 1982): 6–15.

D., G. H. R. "Therapeutic Enthusiasm." *BMJ* 1 (18 Feb. 1911): 402.

Daly, Jeanne. *Evidence-Based Medicine and the Search for a Science of Clinical Care.* Berke-ley: University of California Press, 2005.

Danilowicz, Delores A., and H. Paul Gabriel. "Postcardiotomy Psychosis in Non-English-Speaking Patients." *Psychiatry in Medicine* 2 (1971): 314–20.

Dartmouth Atlas of Cardiovascular Care. Center for the Evaluative Clinical Sciences, Dart-mouth Medical School, and the Center for Outcomes Research and Evaluation, Maine Medical Center. Chicago: AHA Press, 1999.

Dartmouth Atlas of Health Care. Center for the Evaluative Clinical Sciences, Dartmouth Medical School. Chicago: American Hospital Publishing, 1996.

Dartmouth Atlas of Health Care 1998. Center for the Evaluative Clinical Sciences, Dart-mouth Medical School. Chicago: AHA Press, 1998.

"Dartmouth Atlas of Health Care: Studies of Surgical Variation—Cardiac Surgery." Center for the Evaluative Clinical Sciences, Dartmouth Medical School, 2005. www.dartmouth atlas.org/downloads/reports/Cardiac_report_2005.pdf

Daston, Lorraine. "The Coming into Being of Scientific Objects." In *Biographies of Scien-tific Objects,* ed. Lorraine Daston, 1–14. Chicago: University of Chicago Press, 2000.

Daston, Lorraine, and Peter Galison. *Objectivity.* Cambridge, MA: Zone Books, 2007.

Davenport, Harold T., G. Arfel, and F. R. Sanchez. "The Electroencephalogram in Patients Undergoing Open Heart Surgery with Heart-Lung Bypass." *Anesthesiology* 20 (Sept.–Oct. 1959): 674–84.

Davies, Michael J., and Anthony Thomas. "Thrombosis and Acute Coronary-Artery Le-sions in Sudden Cardiac Ischemic Death." *New England Journal of Medicine* 310 (3 May 1984): 1137–40.

Davies, Michael J., N. Woolf, and W. B. Robertson. "Pathology of Acute Myocardial Infarc-tion with Particular Reference to Occlusive Coronary Thrombi." *British Heart Journal* 38 (1976): 659–64.

Davis, N. A. "Incidence of Thrombosis in Myocardial Infarction." *Australian Annals of Medicine* 19 suppl. 1 (1970): 60–62.

Davis, R. Duane, and David C. Sabiston. "The Coronary Circulation." In *Textbook of Sur-gery: The Biological Basis of Modern Surgical Practice,* 15th ed., ed. David S. Sabiston and H. Kim Lyerly, 2082–3018. Philadelphia, PA: W. B. Saunders, 1997.

Davis, Stephen M., and Geoffrey A. Donnan. "Carotid-Artery Stenting in Stroke Preven-tion." *New England Journal of Medicine* 363 (1 July 2010): 80–82.

DeBakey, Michael. "Albert Lasker Clinical Medical Research Award," 11 Nov. 1983. CCF Archives, box 121, folder 3-PR10, "Sones, Mason F."

———. "Oral History" (1998). In *Pioneers of Cardiac Surgery,* ed. William S. Stoney, 148–58. Nashville, TN: Vanderbilt University Press, 2008.

DeBakey, Michael, to Hilde Bruch, 26 March 1970. HB Papers, series V, box 6.

DeBakey, Michael E., to Denton A. Cooley, 11 Sept. 1969. WS Papers, series V, box 7, folder 116, "Research Involving Human Beings."

de Boer, Onno J., Allard C. van der Wal, Peter Teeling, and Anton E. Becker. "Leucocyte Recruitment in Rupture-Prone Regions of Lipid-Rich Plaques: A Prominent Role for Neovascularization." *Cardiovascular Research* 41 (1 Feb. 1999): 443–49.

de Jaegere, Peter P., and Maartin L. Simoons. "Immediate Angioplasty: A Conservative View from Europe—Cost Effectiveness Needs to be Considered." *British Heart Journal* 73 (May 1995): 407–8.

Dencker, Sven J., and Arne Sandahl. "Mental Disease after Operations for Mitral Stenosis." *Lancet* 277 (3 June 1961): 1230–31.

Dennis, Clarence, Dwight S. Spreng, George E. Nelson, Karl E. Karlson, Russell M. Nelson, John V. Thomas, Walter Phillip Eder, and Richard L. Varco. "Development of a Pump-Oxygenator to Replace the Heart and Lungs; An Apparatus Applicable to Human Patients, and Application in One Case." *Annals of Surgery* 134 (Oct. 1951): 709–21.

Department of Cardiology. "1976 Annual Report," 1 April 1977. CCF Archives, A08-27/1, "Cardiology Annual Reports, 1975–1985" (#1554).

Department of Neurology. "1980 Annual Report." CCF Archives, 20-PN Neurology, folder "Annual Reports, 1980–1985" (#1559).

Department of Thoracic and Cardiovascular Surgery, Cleveland Clinic Foundation. "1995 Calendar." CCF Archives, 21-PT, "Thoracic and CV Surgery Publications."

Derrick, H. Fred. "How Open Heart Surgery Feels." *American Journal of Nursing* 79 (Feb. 1979): 276–85.

Derrida, Jacques. "Plato's Pharmacy." In *Dissemination*, trans. B. Johnson, 61–171. Chicago: University of Chicago Press, 1981.

Desai, Nimesh D., Eric A. Cohen, C. David Naylor, and Stephen E. Fremes, for the Radial Artery Patency Study Investigators. "A Randomized Comparison of Radial-Artery and Saphenous-Vein Coronary Bypass Grafts." *New England Journal of Medicine* 351 (25 Nov. 2004): 2302–9.

Deslauriers, Roxanne, John K. Saunders, and Michael C. McIntyre. "Magnetic Resonance Studies of the Effects of Cardiovascular Surgery on Brain Metabolism and Function." *Journal of Cardiothoracic and Vascular Anesthesia* 10 (Jan. 1996): 127–38.

Desmon, Stephanie, and Robert Little. "St. Joe's, Cardiology Group Probed: Hospital, Midatlantic Involved in Business, Legal Disputes." *Baltimore Sun*, 24 March 2009.

DeVries, William C. "Oral History" (3 June 1999). In *Pioneers of Cardiac Surgery*, ed. William S. Stoney, 469–76. Nashville, TN: Vanderbilt University Press, 2008.

DeWall, Richard A. "Oral History" (3 May 2003). In *Pioneers of Cardiac Surgery*, ed. William S. Stoney, 100–108. Nashville, TN: Vanderbilt University Press, 2008.

DeWood, Marcus A., Julie Spores, Robert Notske, Lowell T. Mouser, Robert Burroughs, Michael S. Golden, and Henry T. Lang. "Prevalence of Total Coronary Occlusion during the Early Hours of Transmural Myocardial Infarction." *New England Journal of Medicine* 303 (16 Oct. 1980): 897–902.

Dexter, Lewis. "Surgery in Heart Disease." *Disease-a-Month* 9 (1956): 1–58.

Diegeler, Anno, Robert Hirsch, Felix Schneider, Lars-Oliver Schilling, Volkmar Falk, Thomas Rauch, and Friedrich M. Mohr. "Neuromonitoring and Neurocognitive Outcome in Off-Pump Versus Conventional Coronary Bypass Operation." *Annals of Thoracic Surgery* 69 (April 2000): 1162–66.

Diegeler, Anno, Holger Thiele, Volkmar Falk, Rainer Hambrecht, Niki Spyrantis, Peter Sick, Klaus W. Diederich, Friedrich W. Mohr, and Gerhard Schuler. "Comparison of Stenting with Minimally Invasive Bypass Surgery for Stenosis of the Left Anterior Descending Coronary Artery." *New England Journal of Medicine* 347 (22 Aug. 2002): 561–66.

"Doctor-Owned Specialty Hospitals Spur Investor Interest, Capital Hill Worries." *BNA's Health Care Policy Report* 11 (2003): 532–33.

Dodrill, Forest D., Edward Hill, and Robert A. Gerisch. "Temporary Mechanical Substitute for the Left Ventricle in Man." *JAMA* 150 (18 Oct. 1952): 642–44.

Dorros, G., M. J. Cowley, J. Simpson, L. G. Bentivoglio, P. C. Block, M. Bourassa, K. Detre, A. J. Gosselin, A. R. Gruntzig, S. F. Kelsey, K. M. Kent, M. B. Mock, S. M. Mullin, R. K. Myler, E. R. Passamani, S. H. Stertzer, and D. O. Williams. "Percutaneous Transluminal Coronary Angioplasty: Report of Complications from the National Heart, Lung, and Blood Institute PTCA Registry." *Circulation* 67 (April 1983): 723–30.

Dublin, Louis I., and Mortimer Spiegelman. "The Longevity and Mortality of American Physicians, 1938–1942." *JAMA* 134 (9 Aug. 1947): 1211–15.

Duffin, Jacalyn. "The Disease Game: An Introduction to the Concepts and Construction of Disease." In *Lovers and Lives: Disease Concepts in History*, 1–36. Toronto: University of Toronto Press, 2005.

Duffin, Jacalyn, and Barbara G. Campling. "Therapy and Disease Concepts: The History (and Future?) of Antimony in Cancer." *Journal of the History of Medicine and Allied Sciences* 57 (2002): 61–78.

Dumit, Joseph. "Pharmaceutical Witnessing: Drugs for Life in an Era of Direct-to-Consumer Advertising." In *Technologized Images, Technologized Bodies*, ed. Jeanette Edwards, Penny Harvey, and Peter Wade, 37–63. New York: Berghahn Books, 2010.

——. *Picturing Personhood: Brain Scans and Biomedical Identity.* Princeton, NJ: Princeton University Press, 2004.

——. "Prescription Maximization and the Accumulation of Surplus Health in the Pharmaceutical Industry: The_Biomarx_Experiment." In *Lively Capital*, ed. Kaushik Sunder Rajan. Durham, NC: Duke University Press, 2012.

Dutton, Robert C., L. Henry Edmunds, John C. Hutchinson, and Benson B. Roe. "Platelet Aggregate Emboli Produced in Patients during Cardiopulmonary Bypass with Membrane and Bubble Oxygenators and Blood Filters." *Journal of Thoracic and Cardiovascular Surgery* 67 (Feb. 1974): 258–67.

Eagle, Kim A., and Robert A. Guyton, co-chairs. "ACC/AHA 2004 Guideline Update for Coronary Artery Bypass Graft Surgery." *Journal of the American College of Cardiology* 44 (2004): e213–310.

"Early Tricks on the Comstock: Tales of the Days When Miners Played Rough Games." *New York Times*, 4 Nov. 1895, 7.

Eddy, David M. "Variations in Physician Practice: The Role of Uncertainty." *Health Affairs* 6 (1984): 74–89.

Effler, Donald B. "A New Era of Coronary Artery Surgery." *Surgery, Gynecology and Obstetrics* 123 (1966): 1310–11.

———. "Operative Report," 19 Oct. 1962. AV Fonds, P126/C8, folder 78, "Effler."

———. "The Role of Surgery in the Treatment of Coronary Artery Disease." *Annals of Thoracic Surgery* 8 (Oct. 1969): 376–79.

———. "Surgery for Coronary Disease." *Scientific American* 210 (Oct. 1968): 36–43.

———. "Surgical Treatment of Myocardial Ischemia." *Clinical Symposia* 21 (1969): 3–17.

Effler, Donald B., René Favaloro, and Laurence K. Groves. "Heart Valve Replacement: Clinical Experience." *Annals of Thoracic Surgery* 1 (Jan. 1965): 4–24.

Effler, Donald B., Laurence K. Groves, and René Favaloro. "Surgical Repair of Ventricular Aneurysm." *Chest* 48 (July 1965): 37–43.

Effler, Donald B., Laurence K. Groves, F. Mason Sones, Harold F. Knight, and Willem J. Kolff. "Elective Cardiac Arrest—An Adjunct to Open-Heart Surgery." *Journal of Thoracic Surgery* 34 (Oct. 1957): 500–508.

Effler, Donald, to Clarence B. Hewitt, 24 Feb. 1966. CCF Archives, 21-TA, "Surgical Committee Minutes, 1966–1969," folder 1/19-6/14/66 (#2203).

Effler, Donald B., to S. O. Hoerr, 7 July 1969. CCF Archives, 21-TA, "Surgical Committee Minutes, 1966–1969," folder 8/5-12/16/69 (#2303).

Effler, Donald B., to Martha Hoffman, 17 April 1970. CCF Archives, 21-TA, "Surgical Committee Minutes, 1970–1971," folder 1/13-6/16/1970 (#2304).

Effler, Donald, to Fay LeFevre, R. A. Gottron, and J. G. Harding, 28 Jan. 1966. CCF Archives, 3-PR20, "Cardiology / Thoracic Surgery Administrative Items."

Effler, Donald, to Arthur Vineberg, 25 April 1960. AV Fonds, P126/C8, folder 78, "Effler."

Effler, Donald, to Arthur Vineberg, 13 Dec. 1960. AV Fonds, P126/C8, folder 78, "Effler."

Eggleston, Cary. "Diseases of the Myocardium." In *A Text-Book of Medicine by American Authors*, ed. Russell L. Cecil and Foster Kennedy, 1006–24. Philadelphia, PA: W. B. Saunders, 1928.

Ehrenhaft, J. L., and M. A. Claman. "Cerebral Complications of Open-Heart Surgery." *Journal of Thoracic and Cardiovascular Surgery* 41 (April 1961): 503–8.

Elkins, Jacob S., and S. Claiborne Johnston. "Twinning Hearts and Minds." *Neurology* 63 (Dec. 2004): 2211–12.

Ellis, Laurence B., and Dwight E. Harken. "Arterial Embolization in Relation to Mitral Valvuloplasty." *American Heart Journal* 62 (Nov. 1961): 611–20.

———. "The Clinical Results in the First Five Hundred Patients with Mitral Stenosis Undergoing Valvuloplasty." *Circulation* 11 (April 1955): 637–46.

Emerson, Edward Waldo. "William Palmer Bolles." *Boston Medical and Surgical Journal* 176 (8 March 1917): 362–63.

English, T. A. H., A. R. Bailey, J. F. Dark, and W. G. Williams. "The UK Cardiac Surgical Register, 1977–82." *BMJ* 289 (3 Nov. 1984): 1205–8.

Epstein, Arnold M. "Geographic Variation in Medicare Spending." *New England Journal of Medicine* 363 (1 July 2010): 85–86.

———. "Use of Diagnostic Tests and Therapeutic Procedures in a Changing Health Care Environment." *JAMA* 275 (17 April 1996): 1197–98.

Epstein, Steven. *Impure Science: AIDS, Activism, and the Politics of Knowledge*. Berkeley: University of California Press, 1996.

Erhardt, L. R., T. Lundman, and H. Mellstedt. "Incorporation of 125I-Labelled Fibrinogen into Coronary Arterial Thrombi in Acute Myocardial Infarction in Man." *Lancet* 301 (24 Feb. 1973): 387–90.

Erickson, Robert A. *The Language of the Heart, 1600–1750*. Philadelphia: University of Pennsylvania Press, 1997.

Eusterman, Joseph H., Richard W. P. Achor, Owings W. Kincaid, and Arnold L. Brown. "Atherosclerotic Disease of the Coronary Arteries: A Pathologic-Radiologic Correlative Study." *Circulation* 26 (Dec. 1962): 1288–95.

Evans, Hughes. "Losing Touch: The Controversy over the Introduction of Blood Pressure Instruments into Medicine." *Technology and Culture* 34 (Oct. 1993): 784–807.

Every, Nathan R., Eric B. Larson, Paul E. Litwin, Charles Maynard, Stephan D. Fihn, Mickey S. Eisenberg, Alfred P. Hallstrom, Jenny S. Martin, and W. Douglas Weaver, for the Myocardial Infarction Triage and Intervention Project Investigators. "The Association between On-Site Cardiac Catheterization Facilities and the Use of Coronary Angiography after Acute Myocardial Infarction." *New England Journal of Medicine* 329 (19 Aug. 1993): 546–51.

Faden, Ruth R., and Tom L. Beauchamp. *A History and Theory of Informed Consent*. New York: Oxford University Press, 1986.

Falk, Erling. "Plaque Rupture with Severe Pre-existing Stenosis Precipitating Coronary Thrombosis: Characteristics of Coronary Atherosclerotic Plaques Underlying Fatal Occlusive Thrombi." *British Heart Journal* 50 (1983): 127–34.

———. "Unstable Angina with Fatal Outcome: Dynamic Coronary Thrombosis Leading to Infarction and/or Sudden Death." *Circulation* 71 (April 1985): 699–708.

———. "Why Do Plaques Rupture?" *Circulation* 86 suppl. (1992): III-30–III-42.

Falk, Erling, Prediman K. Shah, and Valentin Fuster. "Coronary Plaque Disruption." *Circulation* 92 (1995): 657–71.

Farmer, Paul. "Listening for Prophetic Voices in Medicine." *America* 177 (5–12 July 1997): 8–13.

———. "Social Scientists and the New Tuberculosis." *Social Science and Medicine* 44 (1997): 347–58.

Favaloro, René G. "Oral History" (3 March 1997). In *Pioneers of Cardiac Surgery*, ed. William S. Stoney, 357–68. Nashville, TN: Vanderbilt University Press, 2008.

———. "Saphenous Vein Autograft Replacement of Severe Segmental Coronary Artery Occlusion." *Annals of Thoracic Surgery* 5 (1968): 334–39.

———. "Saphenous Vein Graft in the Surgical Treatment of Coronary Artery Disease." *Journal of Thoracic and Cardiovascular Surgery* 58 (Aug. 1969): 178–85.

———. "Surgical Treatment of Coronary Arteriosclerosis by the Saphenous Vein Graft Technique." *American Journal of Cardiology* 28 (Oct. 1971): 493–95.

Favaloro, René G., Donald B. Effler, Chalit Cheanvechai, Robert A. Quint, and F. Mason Sones. "Acute Coronary Insufficiency (Impending Myocardial Infarction and Myocardial Infarction): Surgical Treatment by the Saphenous Vein Graft Technique." *American Journal of Cardiology* 28 (1971): 598–607.

Favaloro, René G., Donald B. Effler, Laurence K. Groves, Mehdi Razavi, and Yair Lieberman. "Combined Simultaneous Procedures in the Surgical Treatment of Coronary Artery Disease." *Annals of Thoracic Surgery* 8 (July 1969): 20–29.

Favaloro, René G., Donald B. Effler, Laurence K. Groves, William C. Sheldon, and Mohammed Riahi. "Direct Myocardial Revascularization with Saphenous Vein Autograft: Clinical Experience in 100 Cases." *Diseases of the Chest* 56 (1969): 279–83.

Favaloro, René G., Donald B. Effler, Laurence K. Groves, William C. Sheldon, and F. Mason Sones. "Direct Myocardial Revascularization by Saphenous Vein Graft: Present Operative Technique and Indications." *Annals of Thoracic Surgery* 10 (Aug. 1970): 97–111.

Favaloro, René G., Donald B. Effler, Laurence K. Groves, F. Mason Sones, David J. G. Ferguson. "Myocardial Revascularization by Internal Mammary Artery Implant Procedures: Clinical Experience." *Journal of Thoracic and Cardiovascular Surgery* 54 (1967): 359–70.

Feder, Barnaby J. "In Quest to Improve Heart Therapies, Plaque Gets a Fresh Look." *New York Times*, 27 Nov. 2006.

———. "Stent vs. Scalpel." *New York Times*, 29 Nov. 2005.

Feldman, Dmitriy N., Christopher L. Gade, Alexander J. Slotwiner, Manish Parikh, Geoffrey Bergman, S. Chiu Wong, and Robert M. Minutello. "Comparison of Outcomes of Percutaneous Coronary Interventions in Patients of Three Age Groups (60, 60 to 80, and >80 Years) from the New York State Angioplasty Registry." *American Journal of Cardiology* 98 (2006): 1334–39.

Fink, Sheri, and Rebecca Rabinowitz. "The UN's Battle with NCDs: How Politics, Commerce, and Science Complicated the Fight Against an 'Invisible Epidemic.'" *Foreign Affairs*, 20 Sept. 2011.

Fisher, Elliott S., David E. Wennberg, Thérèse A. Stukel, Daniel J. Gottlieb, F. L. Lucas, and Étoile L. Pinder. "The Implications of Regional Variations in Medicare Spending. Part 1: The Content, Quality, and Accessibility of Care." *Annals of Internal Medicine* 138 (18 Feb. 2003): 273–87.

———. "The Implications of Regional Variations in Medicare Spending. Part 2: Health Outcomes and Satisfaction with Care." *Annals of Internal Medicine* 138 (18 Feb. 2003): 288–98.

FitzGerald, Garret A. "Ticlopidine in Unstable Angina: A More Expensive Aspirin?" *Circulation* 82 (July 1990): 296–98.

Fitzhugh, Greene, and Burton E. Hamilton. "Coronary Occlusion and Fatal Angina Pectoris: Study of the Immediate Causes and their Prevention." *JAMA* 100 (18 Feb. 1933): 475–80.

Fleck, Ludwig. *Genesis and Development of a Scientific Fact* (1935). Chicago: University of Chicago Press, 1981.

Fletcher, Anthony P., Sol Sherry, Norma Alkjaersig, Fotios E. Smyrniotis, and Sidney Jick. "The Maintenance of a Sustained Thrombolytic State in Man. II. Clinical Observations on Patients with Myocardial Infarction and Other Thromboembolic Disorders." *Journal of Clinical Investigation* 38 (1959): 1111–19.

Flory, Curtis M. "Arterial Occlusions Produced by Emboli from Eroded Aortic Atheromatous Plaques." *American Journal of Pathology* 21 (1945): 549–65.

Ford, Earl S., Umed A. Ajani, Janet B. Croft, Julia A. Critchley, Darwin L. Labarthe, Thomas E. Kottke, Wayne H. Giles, and Simon Capewell. "Explaining the Decrease in U.S. Deaths from Coronary Disease, 1980–2000." *New England Journal of Medicine* 356 (7 June 2007): 2388–98.

Ford, Earl S., and S. Capewell. "Proportion of the Decline in Cardiovascular Mortality Disease Due to Prevention versus Treatment: Public Health versus Clinical Care." *Annual Review of Public Health* 32 (2011): 5–22.

Foreman, Judy. "Heart Patients' Mental Decline Baffles Doctors." *Boston Globe*, 21 Sept. 2004.

Forrester, James S. "Prevention of Plaque Rupture: A New Paradigm of Therapy." *Annals of Internal Medicine* 137 (19 Nov. 2002): 823–33.

Forrester, James S., Frank Litvack, Warren Grundfest, and Ann Hickey. "A Perspective of Coronary Disease Seen through the Arteries of Living Man." *Circulation* 75 (March 1987): 505–13.

Forrester, James S., and Pradiman K. Shah. "Lipid Lowering versus Revascularization: An Idea Whose Time (For Testing) Has Come." *Circulation* 96 (19 Aug. 1997): 1360–62.

Fortun, Michael. "Mediated Speculations in the Genomics Futures Markets." *New Genetics and Society* 20 (2001): 139–56.

———. *Promising Genomics: Iceland and deCode Genetics in a World of Speculation.* Berkeley: University of California Press, 2008.

Foucault, Michel. *The Birth of the Clinic: An Archaeology of Medical Perception* (1963). New York: Pantheon Books, 1973.

———. *The Order of Things: An Archaeology of the Human Sciences* (1966). New York: Random House, 1970.

Fowler, Floyd J., Patricia M. Gallagher, Julie P. W. Bynum, Michael J. Barry, F. Leslie Lucas, and Jonathan S. Skinner. "Decision-Making Process Reported by Medicare Patients Who Had Coronary Artery Stenting or Surgery for Prostate Cancer." *Journal of General Internal Medicine* 27 (2012) 911–16.

Fox, Henry M., Nicholas D. Rizzo, and Sanford Gifford. "Psychological Observations of Patients Undergoing Mitral Surgery." *Psychosomatic Medicine* 16 (1954): 186–208.

Fox, Renée C., and Judith P. Swazey. *The Courage to Fail: A Social View of Organ Transplants and Dialysis* (1974). New Brunswick, NJ: Transaction Publishers, 2002.

Fraser, Joeann Guthrey Taylor. "Retrospective on Dr. Gibbon and His Heart-Lung Machine." *Annals of Thoracic Surgery* 76 (2003): S2197–S2198.

Freedman, David H. "Lies, Damned Lies, and Medical Science." *Atlantic Monthly* (Nov. 2010), www.theatlantic.com/magazine/archive/2010/11/lies-damned-lies-and-medical -science/8269.

Frenk, Julio, José L. Bobadilla, Jaime Sepúlveda, and Malaquias López Cervantes. "Health Transition in Middle-Income Countries: New Challenges for Health Care." *Health Policy and Planning* 4 (1989): 29–39.

Freyhan, F. A., S. Giannelli, R. A. O'Connell, and J. A. Mayo. "Psychiatric Complications Following Open Heart Surgery." *Comprehensive Psychiatry* 12 (May 1971): 181–95.

Friedberg, Charles K. "Symposium: Myocardial Infarction 1972 (Part 1): Introduction." *Circulation* 45 (Jan. 1972): 179–88.

Friedberg, Charles K., and Henry Horn. "Acute Myocardial Infarction Not Due to Coronary Artery Occlusion." *JAMA* 112 (29 April 1939): 1675–79.

Friedman, Meyer. "The Coronary Thrombus: Its Origin and Fate." *Human Pathology* 2 (March 1971): 81–128.

———. *Pathogenesis of Coronary Artery Disease.* New York: McGraw-Hill, 1969.

———. "The Pathogenesis of Coronary Plaques, Thromboses, and Hemorrhages: An Evaluative Review." *Circulation* 51–52 suppl. III (Dec. 1975): 34–40.

Friedman, Meyer, and G. van den Bovenkamp. "The Pathogenesis of Coronary Intramural Hemorrhages." *British Journal of Experimental Pathology* 47 (1966): 347–55.

Friedman, Meyer, and G. van den Bovenkamp. "The Pathogenesis of a Coronary Thrombus." *American Journal of Pathology* 48 (Jan. 1966): 19–44.

Frye, Robert L., Richard Kronmal, Hartzell V. Schaff, William O. Myers, Bernard J. Gersh, and the Participants in the Coronary Artery Surgery Study. "Stroke in Coronary Artery Bypass Graft Surgery: An Analysis of the CASS Experience." *International Journal of Cardiology* 36 (Aug. 1992): 213–21.

Fuller, Steve. *Thomas Kuhn: A Philosophical History for Our Times.* Chicago: University of Chicago Press, 2000.

Fulton, William F. M. *The Coronary Arteries: Arteriography, Microanatomy, and Pathogenesis of Obliterative Coronary Artery Disease.* Springfield, IL: Charles C Thomas, 1965.

Furlan, Anthony J., and Anthony C. Breuer. "Central Nervous System Complications of Open Heart Surgery." *Stroke* 15 (Sept.–Oct. 1984): 912–15.

Fuster, Valentin. "Atherosclerosis-Thrombosis and Vascular Biology." In *Cecil: Textbook of Medicine,* 21st ed., ed. Lee Goldman and J. Claude Bennett, 291–96. Philadelphia, PA: W. B. Saunders, 2000.

Fuster, Valentin, Lina Badimon, Juan J. Badimon, and James H. Chesebro. "The Pathogenesis of Coronary Artery Disease and the Acute Coronary Syndromes (First of Two Parts)." *New England Journal of Medicine* 326 (23 Jan. 1992): 242–50.

———. "The Pathogenesis of Coronary Artery Disease and the Acute Coronary Syndromes (Second of Two Parts)." *New England Journal of Medicine* 326 (30 Jan. 1992): 310–18.

Fuster, Valentin, Mark L. Dyken, Pantel S. Vokonas, and Charles Hennekens. "Aspirin as a Therapeutic Agent in Cardiovascular Disease." *Circulation* 87 (Feb. 1993): 659–75.

Fuster, Valentin, Bernardo Stein, John A. Ambrose, Lina Badimon, Juan Jose Badimon, and James H. Chesebro. "Atherosclerotic Plaque Rupture and Thrombosis: Evolving Concepts." *Circulation* 82 suppl. 2 (Sept. 1990): 47–59.

Fye, W. Bruce. *American Cardiology: The History of a Specialty and Its College.* Baltimore, MD: John Hopkins University Press, 1996.

———. "Coronary Arteriography—It Took a Long Time!" *Circulation* 70 (Nov. 1984): 781–87.

———. "A History of the Origin, Evolution, and Impact of Electrocardiography." *American Journal of Cardiology* 73 (15 May 1994): 937–49.

Galbreath, Christopher, Efrain D. Salgado, Anthony J. Furlan, and Jay Hollman. "Central Nervous System Complications of Percutaneous Transluminal Coronary Angioplasty." *Stroke* 17 (July–Aug. 1986): 616–19.

Galdston, Richard. "Psychotic Reaction to the Success of Cardiac Valvotomy: A Case Report." *Psychiatry in Medicine* 1 (1970): 367–73.

Galison, Peter. *How Experiments End.* Chicago: University of Chicago Press, 1987.

———. *Image and Logic: A Material Culture of Microphysics.* Chicago: University of Chicago Press, 1997.

———. "Image of the Self." In *Things That Talk: Object Lessons from Art and Science,* ed. Lorraine Daston, 257–94. New York: Zone Books, 2004.

———. "Trading Zone: Coordinating Action and Belief." In *The Science Studies Reader,* ed. M. Biagioli, 137–60. London: Routledge, 1999.

Galloway, Aubrey C., Stephen B. Colvin, Eugene A. Grossi, and Frank C. Spencer. "Acquired Heart Disease." In *Principles of Surgery,* 6th ed., ed. Seymour I. Schwartz, G. Tom Shires, Frank C. Spencer, and Wendy Cowles Husser, 845–902. New York: McGraw-Hill, 1994.

Ganz, William, Neil Buchbinder, Harold Marcus, Avinash Mondkar, Jamshid Maddahi, Yzhar Charuzi, Lawrence O'Connor, William Shell, Michael C. Fishbein, Robert Kass, Alfonso Miyamoto, and H. J. C. Swan. "Intracoronary Thrombolysis in Evolving Myocardial Infarction." *American Heart Journal* 101 (Jan. 1981): 4–13.

Gardner, Timothy J., Peter J. Horneffer, Teri A. Manolio, Thomas A. Pearson, Vincent L. Gott, William A. Baumgartner, A. Michael Borkon, Levi Watkins, and Bruce A. Reitz. "Stroke Following Coronary Artery Bypass Grafting: A Ten-Year Study." *Annals of Thoracic Surgery* 40 (Dec. 1985): 574–81.

Gawande, Atul. *The Checklist Manifesto: How to Get Things Right.* New York: Metropolitan Books, 2010.

———. *Complications: A Surgeon's Notes on an Imperfect Science.* New York: Metropolitan Books, 2002.

———. "The Cost Conundrum: What a Texas Town Can Teach Us about Health Care." *New Yorker,* 1 June 2009, 36–44.

Gaziano, Thomas A. "Reducing the Growing Burden of Cardiovascular Disease in the Developing World." *Health Affairs* 26 (Jan.–Feb. 2007): 13–24.

"Gene Austin Undergoes THI's First Coronary Artery Bypass." *THI Today* (Summer 1987). THI Papers, box 3, folder "Texas Heart Institute: 25th Anniversary."

General Assembly of the United Nations. "Political Declaration of the High-Level Meeting of the General Assembly on the Prevention and Control of Non-communicable Diseases." Draft Resolution, 16 Sept. 2011. Accessed 30 Oct. 2011. www.un.org/en/ga/ncdmeeting2011.

Gilberstadt, Harold, and Yoshio Sako. "Intellectual and Personality Changes Following Open-Heart Surgery." *Archives of General Psychiatry* 16 (Feb. 1967): 210–14.

Gillum, Richard F. "Coronary Artery Bypass Surgery and Coronary Angiography in the United States, 1979–1983." *American Heart Journal* 113 (1987): 1255–60.

Gillum, Richard F. "Coronary Heart Disease in Black Populations. I. Mortality and Morbidity." *American Heart Journal* 104 (Oct. 1982): 839–51.

Gilman, Sid. "Cerebral Disorders after Open-Heart Operations." *New England Journal of Medicine* 272 (11 March 1965): 489–98.

Gilston, Alan. "Brain Damage after Cardiac Surgery." *Lancet* 327 (7 June 1986): 1323.

Glagov, Seymour. Preface to *Pathobiology of the Human Atherosclerotic Plaque*, ed. Seymour Glagov, William P. Newman, and Sheldon A. Schaffer, ix–xiv. New York: Springer-Verlag, 1990.

Glover, J. Alison. "The Incidence of Tonsillectomy in School Children." *Proceedings of the Royal Society of Medicine* 31 (1938): 1219–36.

Glover, Robert P., Thomas J. E. O'Neill, and Charles P. Bailey. "Commissurotomy for Mitral Stenosis." *Circulation* 1 (March 1950): 329–42.

Gofman, John W., Frank Lindgren, Harold Elliott, William Mantz, John Hewitt, Beverly Strisower, Virgil Herring, and Thomas P. Lyon. "The Role of Lipids and Lipoproteins in Atherosclerosis." *Science* 111 (17 Feb. 1950): 166–71, 186.

Goldberg, Kenneth C., Arthur J. Hartz, Steven J. Jacobsen, Henry Krakauer, and Alfred A. Rimm. "Racial and Community Factors Influencing Coronary Artery Bypass Graft Surgery Rates for All 1986 Medical Patients." *JAMA* 267 (18 March 1992): 1473–77.

Goldstein, James A., Demetris Demetriou, Cindy L. Grines, Mark Pica, Mazen Shoukfeh, and William W. O'Neill. "Multiple Complex Coronary Plaques in Patients with Acute Myocardial Infarction." *New England Journal of Medicine* 343 (28 Sept. 2000): 915–22.

Good, Bryon J. "How Medicine Constructs Its Objects." In *Medicine, Rationality, and Experience: An Anthropological Perspective*, 65–87. Cambridge: Cambridge University Press, 1994.

Gorlin, Richard, Valentin Fuster, and John A. Ambrose. "Anatomic-Physiologic Links between Acute Coronary Syndromes." *Circulation* 74 (July 1986): 6–9.

Gorman, Christine. "Hearts and Minds." *Time*, 19 Feb. 2001, 58–59.

Gornick, Marian. "Medical Patients: Regional Differences in Length of Hospital Stays, 1969–71." *Social Security Bulletin* 38 (July 1975): 16–33.

Gott, Vincent L. "Oral History" (26 Aug. 1997). In *Pioneers of Cardiac Surgery*, ed. William S. Stoney, 332–40. Nashville, TN: Vanderbilt University Press, 2008.

Gottesman, Rebecca F., Maura A. Grega, Maryanne M. Bailey, Luu D. Pham, Scott L. Zeger, William A. Baumgartner, Ola A. Selnes, and Guy M. McKhann. "Delirium after Coronary Artery Bypass Graft Surgery and Late Mortality." *Annals of Neurology* 67 (2010): 338–44.

Goy, Jean-Jacques, and Eric Eeckhout. "Intracoronary Stenting." *Lancet* 351 (27 June 1998): 1943–49.

Graves, E. J., and M. F. Owings. *1996 Summary: National Hospital Discharge Survey*. Advance Data from Vital and Health Statistics, no. 301. Hyattsville, MD: National Center for Health Statistics, 1998.

Green, George E. "Microvascular Technique in Coronary Artery Surgery." *American Heart Journal* 79 (Feb. 1970): 276–79.

Green, George E., Simon H. Stertzer, Richard B. Gordon, and David A. Tice. "Anastomosis of the Internal Mammary Artery to the Distal Left Anterior Descending Coronary Artery." *Circulation* 41 suppl. 2 (May 1970): 79–84.

Greene, Jeremy. "The Abnormal and the Pathological: Cholesterol, Statins, and the Threshold of Disease." In *Medicating Modern America: Pharmaceutical Drugs in History*, ed. Andrea Tone and Elizabeth Watkins, 183–228. New York: New York University Press, 2007.

———. *Prescribing by Numbers: Drugs and the Definition of Disease.* Baltimore, MD: Johns Hopkins University Press, 2007.

———. "Releasing the Flood Waters: Diuril and the Transformation of Hypertension." *Bulletin of the History of Medicine* 79 (Winter 2005): 749–94.

Greene, Jeremy A., and Scott H. Podolsky. "Keeping Modern in Medicine: Pharmaceutical Promotion and Physician Education in Postwar America." *Bulletin of the History of Medicine* 83 (2009): 331–77.

Gregory, Elisha A. "Cell Antagonism." *JAMA* 8 (11 June 1887): 645–49.

Groopman, Jerome. "Heart Surgery, Unplugged." *New Yorker,* 11 Jan. 1999, 43–51.

———. *How Doctors Think.* Boston: Houghton Mifflin, 2007.

Groopman, Jerome, and Pamela Hartzband. *Your Medical Mind: How to Decide What Is Right for You.* New York: Penguin Press, 2011.

Grüntzig, Andreas. "Transluminal Dilatation of Coronary-Artery Stenosis." *Lancet* 311 (4 Feb. 1978): 263.

Grüntzig, Andreas R., Åke Senning, and Walter E. Siegenthaler. "Nonoperative Dilatation of Coronary-Artery Stenosis." *New England Journal of Medicine* 301 (12 July 1979): 61–68.

Gruppo Italiano per lo Studio della Streptochinasi nell'Infarto Miocardio (GISSI). "Effectiveness of Intravenous Thrombolytic Treatment in Acute Myocardial Infarction." *Lancet* 327 (22 Feb. 1986): 397–402.

Hacking, Ian. *Mad Travelers: Reflections on the Reality of Transient Mental Illness.* Cambridge, MA: Harvard University Press, 1998.

Hacking, Ian. *The Social Construction of What?* Cambridge, MA: Harvard University Press, 2000.

Hadler, Nortin M. *Worried Sick: A Prescription for Health in an Overtreated America.* Durham: University of North Carolina Press, 2008.

Haggerty, Robert J. "Effectiveness of Medical Care." *New England Journal of Medicine* 289 (16 Aug. 1973): 372–73.

Hale, Nathan G. *The Rise and Crisis of Psychoanalysis in the United States: Freud and the Americans, 1917–1985.* New York: Oxford University Press, 1995.

Hall, M. J., C. J. DeFrances, S. N. Williams, A. Golosinskiy, and A. Schwarzmann. *National Hospital Discharge Survey: 2007 Summary.* National Health Statistics Reports No 29. Hyattsville, MD: National Center for Health Statistics, 2010.

Hallin, Roger W., U. Scott Page, John C. Bigelow, and William R. Sweetman. "Revascularization of the Heart: AortocoronaryBypass in Sixty-three Patients." *American Journal of Surgery* 122 (Aug. 1971): 164–68.

Hamm, Christian W., Jacobus Reimers, Thomas Ischinger, Hans-Jürgen Rupprecht, Jürgen Berger, and Walter Bleifeld, for the Germany Angioplasty Bypass Surgery Investigation. "A Randomized Study of Coronary Angioplasty Compared with Bypass Surgery in Patients with Symptomatic Multivessel Coronary Disease." *New England Journal of Medicine* 331 (20 Oct. 1994): 1037–43.

Hamman, Louis. "The Symptoms of Coronary Occlusion." *Bulletin of the Johns Hopkins Hospital* 38 (1926): 273–319.

Hanlon, C. Rollins. "Oral History" (26 Oct. 2000). In *Pioneers of Cardiac Surgery*, ed. William S. Stoney, 159–72. Nashville, TN: Vanderbilt University Press, 2008.

Hannan, Edward L., Harold Kilburn, Joseph F. O'Donnell, Gary Lukacik, and Eileen P. Shields. "Interracial Access to Selected Cardiac Procedures for Patients Hospitalized with Coronary Artery Disease in New York State." *Medical Care* 29 (May 1991): 430–41.

Hannan, Edward L., Michael J. Racz, Gary Walford, Robert H. Jones, Thomas J. Ryan, Edward Bennett, Alfred T. Culliford, O. Wayne Isom, Jeffrey P. Gold, and Eric A. Rose. "Long-Term Outcomes of Coronary-Artery Bypass Grafting versus Stent Implantation." *New England Journal of Medicine* 352 (26 May 2005): 2174–83.

Hannan, Edward L., Chuntao Wu, Gary Walford, Alfred T. Culliford, Jeffrey P. Gold, Craig R. Smith, Robert S. D. Higgins, Russell E. Carlson, and Robert H. Jones. "Drug-Eluting Stents vs. Coronary-Artery Bypass Grafting in Multivessel Disease." *New England Journal of Medicine* 358 (24 Jan. 2008): 331–41.

Harken, Alden H. "Oral History" (5 July 2001). In *Pioneers of Cardiac Surgery*, ed. William S. Stoney, 341–53. Nashville, TN: Vanderbilt University Press, 2008.

Harken, Dwight E., and Harrison Black. "Improved Valvuloplasty for Mitral Stenosis." *New England Journal of Medicine* 253 (20 Oct. 1955): 669–78.

Harken, Dwight E., and Lon E. Curtis. "Heart Surgery—Legend and a Long Look." *American Journal of Cardiology* 19 (March 1967): 393–400.

Harken, Dwight E., Lewis Dexter, Laurence B. Ellis, Robert E. Farrand, and James F. Dickson. "The Surgery of Mitral Stenosis. III. Finger-Fracture Valvuloplasty." *Annals of Surgery* 134 (Oct. 1951): 722–42.

Harken, Dwight E., Laurence B. Ellis, Paul F. Ware, and Leona R. Norman. "The Surgical Treatment of Mitral Stenosis. I. Valvuloplasty." *New England Journal of Medicine* 239 (25 Nov. 1948): 801–9.

Harken, Dwight E., and Paul M. Zoll. "Foreign Bodies in and in Relation to the Thoracic Blood Vessels and Heart." *American Heart Journal* 32 (July 1946): 1–16.

Harris, Gardiner. "Doctor Faces Suits Over Cardiac Stents." *New York Times*, 5 Dec. 2010.

Harrison, Donald C., and Norman Shumway. "Evaluation and Surgery for Impending Myocardial Infarction." *Hospital Practice* 9 (Dec. 1972): 49–58.

Harrison, T. R., and William H. Resnick. "Etiologic Aspects of Heart Disease (Including Treatment of the Different Etiologic Types)." In *Principles of Internal Medicine*, ed. T. R. Harrison, Paul B. Beeson, William H. Resnick, George W. Thorn, and M. M. Wintrobe, 1274–1311. Philadelphia, PA: Blakiston, 1950.

Havlik, Richard, and Manning Feinleib, eds. *Proceedings of the Conference on the Decline in Coronary Heart Disease Mortality*. National Heart, Lung, and Blood Institute, NIH Publication No. 79-1610. Bethesda: National Institutes of Health, 1979.

Healy, David. *The Antidepressant Era*. Cambridge, MA: Harvard University Press, 1997.

———. *The Creation of Psychopharmacology*. Cambridge, MA: Harvard, University Press, 2002.

"Heart Attack Deaths (2002)." *Worldmapper: The World As You've Never Seen It*. Map 454. Accessed 14 Feb. 2012. www.worldmapper.org/.

Heberden, William. "Some Account of a Disorder of the Breast." *Medical Transactions* 2 (1772): 59–67.

Hedgecoe, Adam M. "Terminology and the Construction of Scientific Disciplines: The Case of Pharmacogenomics." *Science, Technology and Human Values* 28 (2003): 513–37.

Heller, S., D. S. Kornfeld, K. A. Frank, and J. Barsa. "Delirium after Coronary Artery Bypass Surgery." In *Psychic and Neurological Dysfunctions after Open-Heart Surgery*, ed. Hubert Speidel and Georg Rodewald, 156–61. Stuttgart: Georg Thieme Verlag, 1980.

Hennekens, Charles H., Julie E. Buring, Peter Sandercock, Rory Collins, and Richard Peto. "Aspirin and Other Antiplatelet Agents in the Secondary and Primary Prevention of Cardiovascular Disease." *Circulation* 80 (1989): 749–56.

Herper, Matthew, and Robert Langreth. "Dangerous Devices." Forbes.com, 27 Nov. 2006. Accessed 28 Dec. 2011. www.forbes.com/forbes/2006/1127/094.html.

Herrick, James B. "Clinical Features of Sudden Obstruction of the Coronary Arteries." *JAMA* 49 (7 Dec. 1912): 2015–22.

———. "Thrombosis of the Coronary Arteries." *JAMA* 72 (8 Feb. 1919): 387–90.

Hiatt, Howard H. "Protecting the Medical Commons: Who Is Responsible?" *New England Journal of Medicine* 293 (31 July 1975): 235–41.

Hill, J. Donald, Mary Jane Aguilar, Arnold Baranco, Primal de Lanerolle, and Frank Gerbode. "Neuropathological Manifestations of Cardiac Surgery." *Annals of Thoracic Surgery* 7 (May 1969): 409–19.

Hiremath, J. S. "Future of Thrombolytic Therapy—An Indian Context." *Journal of the Association of Physicians of India* suppl. 59 (Dec. 2011): 49–50.

"His Joy Killed Him: Supposed Leper on Being Released from Infected Colony in San Juan Has Fatal Heart Attack." *New York Times*, 2 Sept. 1903, 1.

Hlatky, Mark A., Constance Bacon, Derek Boothroyd, Elizabeth Mahanna, J. G. Reves, Mark F. Newman, Iain Johnstone, Carla Winston, Maria Mori Brooks, Allan D. Roses, Daniel B. Mark, Bertram Pitt, William Rogers, Thomas Ryan, Robert Wiens, and James A. Blumenthal. "Cognitive Function 5 Years After Randomization to Coronary Angioplasty or Coronary Artery Bypass Graft Surgery." *Circulation* 96, suppl. 2 (4 Nov. 1997): 11–14.

Hoerr, S. O., to Members of the Division of Surgery, 9 July 1969. CCF Archives, 21-TA, "Surgical Committee Minutes, 1966–1969," folder 1/8-6/16/69 (#2303).

Holahan, John, Robert A. Berenson, Peter G. Kachavos. "Area Variations in Selected Medicare Procedures." *Health Affairs* 9 (Winter 1990): 166–75.

Howell, Joel. *Heart Attack: A Biography.* Baltimore, MD: Johns Hopkins University Press, forthcoming.

———. *Technology in the Hospital: Transforming Patient Care in the Early Twentieth Century.* Baltimore, MD: Johns Hopkins University Press, 1995.

Howso, Christopher P., K. Srinath Reddy, Thomas J. Ryan, and Judith R. Bale, eds., for the Institute of Medicine. *Control of Cardiovascular Diseases in Developing Countries: Research, Development, and Institutional Strengthening.* Washington, DC: National Academies Press, 1998.

Hultgren, Herbert N., Timothy Takaro, and Katherine Detre. "Medical versus Surgical

Treatment of Stable Angina Pectoris: Progress Report of a Large Scale Study." *Postgraduate Medical Journal* 52 (Dec. 1976): 757–64.

Iglehart, John K. "From the Editor." *Health Affairs* 3 (Summer 1984).

Ingelfinger, Julie R. "Bariatric Surgery in Adolescents." *New England Journal of Medicine* 365 (2011): 1365–67.

In re: Denton A. Cooley, d/b/a Cardiovascular Associations, d/b/a Southwest Apartment Homes, d/b/a/ Point of Southwest, d/b/a Southwestern Plaza, d/b/a Texas American Bank Bldg., and d/b/a Cool Acres Ranch, debtor. Case No. 88-00154-H5-11 (Chapter 11). United States Bankruptcy Court for the Southern District of Texas, Houston Division. "First Amended Disclosure Statement." 28 March 1988. DC Papers, box "Chapter 11 Court Records."

International Study of Comparative Health Effectiveness with Medical and Invasive Approaches [ISCHEMIA Trial]. Accessed 1 Feb. 2012. www.ischemiatrial.org/ischemia trial.html

Isaacs, David, and Dominic Fitzgerald. "Seven Alternatives to Evidence Based Medicine." *BMJ* 319 (1999): 618.

Ischinger, Thomas, Andreas R. Gruentzig, Jay Hollman, Spencer King III, John Douglas, Bernhard Meier, James Bradford, and Rose Tankersley. "Should Coronary Arteries with Less Than 60% Diameter Stenosis Be Treated by Angioplasty?" *Circulation* 68 (July 1983): 148–54.

Iskowitz, Marc. "Cardiovascular—Therapeutic Focus." *Medical Marketing and Media* 46 (Feb. 2011): 38–39.

Ivert, Torbjörn. "Coronary Bypass Surgery—a Five-Year Follow-Up." *Scandinavian Journal of Thoracic and Cardiovascular Surgery* suppl. 28 (1981): 5–24.

Jasanoff, Sheila. "The Songlines of Risk." *Environmental Values* 8 (1999): 135–52.

Jauhar, Sandeep. "Saving the Heart Can Sometimes Mean Losing the Memory." *New York Times,* 19 Sept. 2000, F1, F4.

Javid, Hushang, Henry M. Tufo, Hassan Najafi, William S. Dye, James A. Hunter, and Ormand C. Julian. "Neurological Abnormalities Following Open-Heart Surgery." *Journal of Thoracic and Cardiovascular Surgery* 58 (Oct. 1969): 502–9.

Jeffrey, Kirk. *Machines in Our Hearts: The Cardiac Pacemaker, the Implantable Defibrillator, and American Health Care.* Baltimore, MD: Johns Hopkins University Press, 2001.

Johnson, W. Dudley, Robert J. Flemma, and Derward Lepley. "Direct Coronary Surgery Utilizing Multiple-Vein Bypass Grafts." *Annals of Thoracic Surgery* 5 (May 1970): 436–44.

Johnson, W. Dudley, and Kenneth L. Kayser. "An Expanded Indication for Coronary Surgery." *Annals of Thoracic Surgery* 16 (July 1973): 1–6.

Jones, David S. "The Health Care Experiments at Many Farms: The Navajo, Tuberculosis, and the Limits of Modern Medicine, 1952–1962." *Bulletin of the History of Medicine* 76 (Winter 2002): 749–90.

———. *On the Origins of Therapies.* In preparation.

———. "The Persistence of American Indian Health Disparities." *American Journal of Public Health* 96 (Dec. 2006): 2122–34.

———. *Rationalizing Epidemics: Meanings and Uses of American Indian Mortality since 1600.* Cambridge, MA: Harvard University Press, 2004.

——. "Virgin Soils Revisited." *William and Mary Quarterly* 60 (Oct. 2003): 703–42.

——. "Visions of a Cure: Visualization, Clinical Trials, and Controversies in Cardiac Therapeutics, 1968–1998." *Isis* 91 (Sept. 2000): 504–41.

Jones, David S., and Jeremy A. Greene. "The Fall and Rise of Heart Disease: A Study in Public Health Millennialism." In preparation.

——. "The Contributions of Prevention and Treatment to the Decline in Cardiovascular Mortality: Lessons from a Forty-Year Debate." *Health Affairs* 31 (2012): forthcoming.

Jones, David S., Scott H. Podolsky, and Jeremy A. Greene. "The Burden of Disease and the Changing Task of Medicine." *New England Journal of Medicine* 366 (21 June 2012): 2333–38.

Jones, Robert H., Eric J. Velazquez, Robert E. Michler, George Sopko, Jae K. Oh, Christopher M. O'Connor, James A. Hill, Lorenzo Menicanti, Zygmunt Sadowski, Patrice Desvigne-Nickens, Jean-Lucien Rouleau, and Kerry L. Lee, Ph.D., for the STICH Hypothesis 2 Investigators. "Coronary Bypass Surgery with or without Surgical Ventricular Reconstruction." *New England Journal of Medicine* 360 (23 April 2009): 1705–17.

Julian, Desmond G. "Myocardial Infarction." In *Cecil: Textbook of Medicine*, 15th ed., ed. Paul B. Beeson, Walsh McDermott, and James B. Wyngaarden, 1229–37. Philadelphia, PA: W. B. Saunders, 1979.

Kaiser, David. "Stick-Figure Realism: Conventions, Reification, and the Persistence of Feynman Diagrams, 1948–1964." *Representations* 70 (2000): 49–86.

Kaplan, Norman M. "The Support of Continuing Medical Education by Pharmaceutical Companies." *New England Journal of Medicine* (25 Jan. 1979): 194–96.

Kaplan, Stanley M. "Psychological Aspects of Cardiac Disease." *Psychosomatic Medicine* 18 (1956): 221–33.

Karagoz, Haldun, Murat Kurtoglu, Beyhan Bakkaloglu, Beril Somnez, Taner Cetintas, and Kemal Bayazit. "Coronary Artery Bypass Grafting in the Awake Patient: Three Years' Experience in 137 Patients." *Journal of Thoracic and Cardiovascular Surgery* 125 (June 2003): 1401–4.

Karagoz, Haldun Y., Beril Sönmez, Beyhan Bakkaloglu, Murat Kortoglu, Melih Erdinc, Aylin Türkeli, and Kemal Bayazit. "Coronary Artery Bypass Grafting in the Conscious Patient without Endotracheal General Anesthesia." *Annals of Thoracic Surgery* 70 (July 2000): 91–96.

Kattus, Albert A. "Physiologic Management of Coronary Artery Disease." In *Proceedings of the 4th Annual Texas Heart Institute Symposium: Physiological Approach to Clinical Cardiology*, 31 May–2 June 1973, THI Papers, box 1.

Kawachi, Ichiro, and Peter Conrad. "Medicalization and the Pharmacological Treatment of Blood Pressure." In *Contested Ground: Public Purpose and Private Interest in the Regulation of Prescription Drugs*, ed. Peter Davis, 26–41. New York: Oxford University Press, 1996.

Kee, Frank, Penny McDonald, and Brian Gaffney. "Risks and Benefits of Coronary Angioplasty: The Patient's Perspective: A Preliminary Study." *Quality in Health Care* 6 (1997): 131–39.

Kennedy, Janet A., and Hyman Bakst. "The Influence of Emotions on the Outcome of

Cardiac Surgery: A Predictive Study." *Bulletin of the New York Academy of Medicine* 42 (Oct. 1966): 811–45.

Khan, A. Hafrez, and Julian Haywood. "Myocardial Infarction in Nine Patients with Radiologically Patent Coronary Arteries." *New England Journal of Medicine* 291 (29 Aug. 1974): 427–31.

Khan, Natasha E., Anthony De Souza, Rebecca Mister, Marcus Flather, Jonathan Clague, Simon Davies, Peter Collins, Duolao Wang, Ulrich Sigwart, and John Pepper. "A Randomized Comparison of Off-Pump and On-Pump Multivessel Coronary-Artery Bypass Surgery." *New England Journal of Medicine* 350 (1 Jan. 2004): 21–28.

Killip, Thomas. "Angina Pectoris." In *Cecil-Loeb Textbook of Medicine*, 14th ed., ed. Paul B. Beeson and Walsh McDermott, 994–1006. Philadelphia, PA: W. B. Saunders, 1975.

Kim, Jim Yong, and Paul Farmer. "AIDS in 2006—Moving Toward One World, One Hope?" *New England Journal of Medicine* 355 (17 Aug. 2006): 645–47.

King, Spencer B. "The Development of Interventional Cardiology." *Journal of the American College of Cardiology* 31 suppl. (15 March 1998): 64B–88B.

King, Spencer B., Nicholas J. Lembo, William S. Weintraub, Andrzej S. Kosinski, Huiman X. Barnhart, Michael H. Kutner, Naomi P. Alazraki, Robert A. Guyton, and Xue-Qiao Zhao, for the Emory Angioplasty versus Surgery Trial (EAST). "A Randomized Trial Comparing Coronary Angioplasty with Coronary Bypass Surgery." *New England Journal of Medicine* 331 (20 Oct. 1994): 1044–50.

Kirklin, John W., chair. "ACC/AHA Guidelines and Indications for Coronary Artery Bypass Graft Surgery. A Report of the American College of Cardiology / American Heart Association Task Force on Assessment of Diagnostic and Therapeutic Cardiovascular Procedures (Subcommittee on Coronary Artery Bypass Graft Surgery)." *Circulation* 83 (March 1991): 1125–73.

Klaidman, Stephen. *Coronary: A True Story of Medicine Gone Awry*. New York: Scribner, 2007.

Kleinman, Arthur. *What Really Matters: Living a Moral Life amidst Uncertainty and Danger*. New York: Oxford University Press, 2006.

———. *Writing at the Margin: Discourse Between Anthropology and Medicine*. Berkeley: University of California Press, 1995.

Knowles, John H. "The Responsibility of the Individual." *Daedalus* 106 (1977): 57–80.

Kolata, Gina. "Considering When It Might Be Best Not to Know about Cancer." *New York Times*, 29 Oct. 2011.

———. "How It Happens: It's Not a 'Plumbing Problem.'" *New York Times*, 8 April 2007.

———. "New Heart Studies Question the Value of Opening Arteries." *New York Times*, 21 March 2004, 1.

Konnagan, Robert D., Memo to Russell J. Blattner and William D. Seybold, "Patient Consent to Anesthesia and Consent to Surgery Forms," 6 March 1970. WS Papers, series V, box 3, folder 35, "Operating Room Committee, 1956–1966."

Konstantinov, Igor E., Nicolai Mejevoi, and Nikolai M. Anichkov. "Nikolai N. Anichkov and His Theory of Atherosclerosis." *Texas Heart Institute Journal* 33 (2006): 417–23.

Kornfeld, Donald S. "The Influence of Emotions on the Outcome of Cardiac Surgery." *Bulletin of the New York Academy of Medicine* 42 (Oct. 1966): 846–49.

Kornfeld, Donald S., Stanley S. Heller, Kenneth A. Frank, and Reed Moskowitz. "Personality and Psychological Factors in Postcardiotomy Delirium." *Archives of General Psychiatry* 31 (Aug. 1974): 249–53.

Kornfeld, Donald S., Sheldon Zimberg, and James R. Malm. "Psychiatric Complications of Open-Heart Surgery." *New England Journal of Medicine* 273 (5 Aug. 1965): 287–92.

Kramer, John R., Hidemasa Kitazume, William L. Proudfit, Yasuo Matsuda, Marlene Goormastic, George W. Williams, and F. Mason Sones. "Segmental Analysis of the Rate of Progression in Patients with Progressive Coronary Atherosclerosis." *American Heart Journal* 106 (Dec. 1983): 1427–31.

Kuhn, Evelyn M., Arthur J. Hartz, and Mario Baras. "Correlation of Rates of Coronary Artery Bypass Surgery, Angioplasty, and Cardiac Catheterization in 305 Large Communities for Persons Age 65 and Older." *Health Services Research* 30 (Aug. 1995): 425–36.

Kuhn, Thomas. *The Structure of Scientific Revolutions.* Chicago: University of Chicago Press, 1962.

Kulik, Alexander, William H. Shrank, Raisa Levin, and Niteesh K. Choudry. "Adherence to Statin Therapy in Elderly Patients after Hospitalization for Coronary Revascularization." *American Journal of Cardiology* 107 (2011): 1409–14.

Kuroda, Yasuhiro, Ryogo Uchimoto, Reiji Kaieda, Reiko Shinkura, Kouichi Shinohara, Shigeru Miyamoto, Shuzo Oshita, and Hiroshi Takeshita. "Central Nervous System Complications after Cardiac Surgery: A Comparison between Coronary Artery Bypass Grafting and Valve Surgery." *Anesthesia and Analgesia* 76 (1993): 222–27.

Kurusz, M., C. W. Shaffer, E. W. Christman, and G. F Tyers. "Runaway Pump Head: New Cause of Gas Embolism during Cardiopulmonary Bypass." *Journal of Thoracic and Cardiovascular Surgery* 77 (May 1979): 792–95.

Lange, Ramon L., Michael S. Reid, Donald D. Tresch, Michael H. Keelan, Victor M. Bernhard, and George Coolidge. "Nonatheromatous Ischemic Heart Disease Following Withdrawal from Chronic Industrial Nitroglycerin Exposure." *Circulation* 46 (Oct. 1972): 666–78.

Lasby, Clarence G. *Eisenhower's Heart Attack: How Ike Beat Heart Disease and Held on to the Presidency.* Lawrence: University of Kansas Press, 1997.

Latour, Bruno. "Drawing Things Together." In *Representation in Scientific Practice,* ed. Michael Lynch and Steve Woolgar, 19–68. Cambridge, MA: MIT Press, 1990.

———. *The Pasteurization of France.* Cambridge, MA: Harvard University Press, 1988.

———. *Reassembling the Social: An Introduction to Actor-Network-Theory.* Oxford: Oxford University Press, 2007.

———. *What Is the Style of Matters of Concern?: Two Lectures on Empirical Philosophy.* Amsterdam: Vangorcum, 2005.

———. "Why Has Critique Run Out of Steam? From Matters of Fact to Matters of Concern." *Critical Inquiry* 30 (Winter 2004): 225–48.

Latour, Bruno, and Steve Woolgar. *Laboratory Life: The Construction of Scientific Facts.* Beverly Hills, CA: Sage Publications, 1979; Princeton, NJ: Princeton University Press, 1986.

Lawrence, L., and M. J. Hall. *1997 Summary: National Hospital Discharge Survey.* Advance

Data from Vital and Health Statistics, no. 308. Hyattsville, MD: National Center for Health Statistics, 1999.

Layne, Ottis L., and Stuart Yudofsky. "Postoperative Psychosis in Cardiotomy Patients: The Role of Organic and Psychiatric Factors." *New England Journal of Medicine* 294 (11 March 1971): 518–20.

Lazarus, Herbert R., and Herome H. Hagens. "Prevention of Psychosis Following Open-Heart Surgery." *American Journal of Psychiatry* 124 (March 1968): 1190–95.

Leary, Timothy. "Cocoanut Grove Club," no date. TL Papers, box 4.

———. "Coronary Spasm as a Possible Factor in Producing Sudden Death." *American Heart Journal* 10 (Feb. 1935): 338–44.

———. "Disasters," no date. TL Papers, box 4.

———. "Experimental Atherosclerosis in the Rabbit Compared with Human (Coronary) Atherosclerosis." *Archives of Pathology* 17 (April 1934): 453–92.

———. "Mallory Institute Dedication," 13 Dec. 1933. TP Papers, box 4.

———. "Pathology of Coronary Sclerosis." *American Heart Journal* 10 (Feb. 1935): 328–37.

———. "Research in the Hospital," Feb. 1939. TL Papers, box 4.

———. "Vascularization of Atherosclerotic Lesions." *American Heart Journal* 16 (Nov. 1938): 549–54.

Lederer, Susan. *Subjected to Science: Human Experimentation in America before the Second World War.* Baltimore, MD: Johns Hopkins University Press, 1995.

Lee, William H., Mary P. Brady, Junius M. Rowe, and William C. Miller. "Effects of Extracorporeal Circulation upon Behavior, Personality, and Brain Function: Part II, Hemodynamic, Metabolic, and Psychometric Correlations." *Annals of Surgery* 173 (June 1971): 1013–23.

Lee, William H., W. Miller, J. Rowe, P. Hairston, and M. P. Brady. "Effects of Extracorporeal Circulation on Personality and Cerebration." *Annals of Thoracic Surgery* 7 (June 1969): 562–70.

Leeder, Stephen, Susan Raymond, and Henry Greenberg. *Race Against Time: The Challenge of Cardiovascular Disease in Developing Countries.* New York: Trustees of Columbia University, 2004.

Leinbach, Robert C., and Herman K. Gold. "Coronary Angiography during Acute Myocardial Infarction: A Search for Spasm." *American Heart Journal* 103 (April 1982): 769–72.

Lembcke, Paul A. "Measuring the Quality of Medical Care through Vital Statistics Based on Hospital Service Areas: 1. Comparative Study of Appendectomy Rates." *American Journal of Public Health* 42 (March 1952): 276–86.

Lemmon, William M., J. Stauffer Lehman, and Randal A. Boyer. "Suprasternal Transaortic Coronary Arteriography." *Circulation* 19 (Jan. 1959): 47–54.

Lendon, Corinne L., M. J. Davies, G. V. R. Born, and P. D. Richardson. "Atherosclerotic Plaque Caps Are Locally Weakened When Macrophages Density Is Increased." *Atherosclerosis* 87 (1991): 87–90.

Lenzer, Jeanne, and Shannon Brownlee. "Reckless Medicine." *Discover* (Nov. 2010): 65–70, 76.

Lerner, Barron H. *The Breast Cancer Wars: Hope, Fear, and the Pursuit of a Cure in Twentieth-Century America.* New York: Oxford University Press, 2001.

———. "The Perils of 'X-Ray Vision': How Radiographic Images Have Historically Influenced Perception." *Perspectives in Biology and Medicine* 35 (1992): 382–97.

"Letters to the Editors: Houston Doctors' Feud." *Life,* 1 May 1970.

Levine, Samuel A. *Clinical Heart Disease.* Philadelphia, PA: W. B. Saunders, 1937.

Lewis, Charles E. "Variations in the Incidence of Surgery." *New England Journal of Medicine* 281 (16 Oct. 1969): 880–84.

Lewis, H. Daniel, James W. Davis, Donald G. Archibald, William E. Steinke, Thomas C. Smitherman, James E. Doherty, Harold W. Schnaper, Martin M. LeWinter, Esteban Linares, J. Maurice Pouget, Subhash C. Sabharwal, Elliot Chesler, and Henry DeMots. "Protective Effects of Aspirin against Acute Myocardial Infarction and Death in Men with Unstable Angina: Results of a Veterans Administration Cooperative Study." *New England Journal of Medicine* 309 (18 Aug. 1983): 396–403.

Li, Jie Jack. *Triumph of the Heart: The Story of Statins.* New York: Oxford University Press, 2009.

Libby, Peter. "Atherosclerosis." In *Harrison's Principles of Internal Medicine,* 14th edition, ed. Anthony S. Fauci, Eugene Braunwald, Kurt J. Isselbacher, Jean D. Wilson, Joseph B. Martin, Dennis L. Kasper, Stephen L. Hauser, and Dan L. Longo, 1345–52. New York: McGraw-Hill, 1998.

———. "Current Concepts of the Pathogenesis of the Acute Coronary Syndromes." *Circulation* 104 (21 July 2001): 365–72.

———. "Inflammation in Atherosclerosis." *Nature* 420 (19–26 Dec. 2002): 868–74.

———. "Molecular Bases of the Acute Coronary Syndromes." *Circulation* 91 (1995): 2844–50.

Libby, Peter, Paul M. Ridker, and Attilio Maseri. "Inflammation and Atherosclerosis." *Circulation* (5 March 2002): 1135–43.

Libby, Peter, Uwe Schoenbeck, Francois Mach, Andrew P. Selwyn, and Peter Ganz. "Current Concepts in Cardiovascular Pathology: The Role of LDL Cholesterol in Plaque Rupture and Stabilization." *American Journal of Cardiology* 104 suppl. 2A (Feb. 1998): 14S–18S.

Libby, Peter, and Pierre Theroux. "Pathophysiology of Coronary Artery Disease." *Circulation* 111 (28 June 2005): 3481–88.

Liddicoat, John E., Szabolcs Bekassy, Arthur C. Beall, Donald H. Glaeser, and Michael E. DeBakey. "Membrane vs. Bubble Oxygenator: Clinical Comparison." *Annals of Surgery* 181 (May 1975): 747–52.

Liebman, Milton. "DTC's Role in the Statin Bonanza." *Medical Marketing and Media* 36 (Nov. 2001): 86–92.

Likoff, William, Bernard L. Segal, and Hratch Kasparian. "Paradox of Normal Selective Coronary Arteriograms in Patients Considered to Have Unmistakable Coronary Heart Disease." *New England Journal of Medicine* 276 (11 May 1967): 1063–66.

Likosky, Donald S., Charles A. S. Marrin, Louis R. Caplan, Yvon R. Baribeau, Jeremy R. Morton, Ronald M. Weintraub, Gregg S. Hartman, Felix Fernandez, Stephen P. Braff, David C. Charlesworth, David J. Malenka, Cathy S. Ross, and Gerald T. O'Connor, for the Northern New England Cardiovascular Disease Study Group. "Determination of Etiologic Mechanisms of Strokes Secondary to Coronary Artery Bypass Graft Surgery." *Stroke* 34 (Dec. 2003): 2830–34.

Lillehei, C. Walton. "Oral History" (4 May 1998). In *Pioneers of Cardiac Surgery*, ed. William S. Stoney, 83–99. Nashville, TN: Vanderbilt University Press, 2008.

Lillehei, C. Walton, Morley Cohen, Herbert E. Warden, Newell R. Ziegler, and Richard L. Varco. "The Results of Direct Vision Closure of Ventricular Septal Defects in Eight Patients by Means of Controlled Cross Circulation." *Surgery, Gynecology and Obstetrics* 101 (Oct. 1955): 446–66.

Lin, Grace A., R. Adams Dudley, and Rita F. Redberg. "Cardiologists' Use of Percutaneous Coronary Interventions for Stable Coronary Artery Disease." *Archives of Internal Medicine* 167 (Aug. 13–27 2007): 1604–9.

———. "Why Physicians Favor Use of Percutaneous Coronary Intervention to Medical Therapy: A Focus Group Study." *Journal of General Internal Medicine* 23 (2008): 1458–63.

Little, William C., Martin Constantinescu, Robert J. Applegate, Michael A. Kutcher, Mark T. Burrows, Frederic R. Kahl, and William P. Santamore. "Can Coronary Angiography Predict the Site of a Subsequent Myocardial Infarction in Patients with Mild-to-Moderate Coronary Artery Disease?" *Circulation* 78 (Nov. 1988): 1157–66.

Little, William C., Nelson S. Gwinn, Mark T. Burrows, Michael A. Kutchner, Frederic R. Kahl, and Robert J. Applegate. "Cause of Acute Myocardial Infarction Late After Successful Coronary Artery Bypass Grafting." *American Journal of Cardiology* 65 (15 March 1990): 808–10.

Littmann, David. "The Second Eugene C. Eppinger Lecture." 24 March 1972. *Fifty-Eighth Annual Report, 1970–1971*, Peter Bent Brigham Hospital (1971), 90–103, BWH Archives.

Lock, Margaret. *Twice Dead: Organ Transplants and the Reinvention of Death*. Berkeley: University of California Press, 2002.

Long, Franklin A., and Alexandra Oleson, eds. *Appropriate Technology and Social Values— A Critical Appraisal*. Cambridge, MA: Ballinger, 1980.

Loop, Floyd D., Delos M. Cosgrove, Bruce W. Lytle, Robert L. Thurer, Conrad Simpfendorfer, Paul C. Taylor, and William L. Proudfit. "An 11-Year Evolution of Coronary Arterial Surgery." *Annals of Surgery* 190 (Oct. 1979): 444–54.

Loop, Floyd D., James Szabo, Richard D. Rowlinson, and Kenneth Urbanek. "Events Related to Microembolism during Extra-corporeal Perfusion in Man: Effectiveness of In-line Filtration Recorded by Ultrasound." *Annals of Thoracic Surgery* 21 (May 1976): 412–20.

Loscalzo, Joseph. "Regression of Coronary Atherosclerosis." *New England Journal of Medicine* 323 (8 Nov. 1990): 1337–39.

"Louisiana Cardiologist Sentenced in Federal Court for Performing Unnecessary Medical Procedures." United States Attorney's Office, Western District of Louisiana, press release, 4 June 2009.

Lunbeck, Elizabeth. *The Psychiatric Persuasion: Knowledge, Gender, and Power in Modern America*. Princeton, NJ: Princeton University Press, 1994.

Lynch, Michael. "Discipline and the Material Form of Images: An Analysis of Scientific Visibility." *Social Studies of Science* 15 (1985): 37–66.

———. "The Externalized Retina: Selection and Mathematization in the Visual Documentation of Objects in the Life Sciences." *Human Studies* 11 (April–June 1988): 201–34.

———. "Science in the Age of Mechanical Reproduction: Moral and Epistemic Relations between Diagrams and Photographs." *Biology and Philosophy* 6 (April 1991): 205–26.

Lynn, Geoffrey M., Karen Stefanko, James F. Reed, William Gee, and Gary Nicholas. "Risk Factors for Stroke after Coronary Artery Bypass." *Journal of Thoracic and Cardiovascular Surgery* 104 (Dec. 1992): 1518–23.

Lytle, Bruce W. "Surgical Treatment of Coronary Artery Disease." In *Cecil Textbook of Medicine*, 22nd ed., ed. Lee Goldman and Dennis Ausiello, 428–31. Philadelphia, PA: W. B. Saunders, 2004.

MacKenzie, Rachel. "Risk." *New Yorker*, 21 Nov. 1970, 56–101.

———. *Risk*. New York: Viking Press, 1971.

Madden, Terri, Kathy Blankenhorn, and Nancy Duckwitz. "Business Watch 2000 in Review: Growth Slows, But Goes On." *Medical Marketing and Media* 36 (May 2001): 74–90.

Magovern, George J. "Oral History" (4 Sept. 1997). In *Pioneers of Cardiac Surgery*, ed. William S. Stoney, 297–307. Nashville, TN: Vanderbilt University Press, 2008.

Mangano, Dennis T. "Cardiovascular Morbidity and CABG Surgery—A Perspective." *Journal of Cardiac Surgery* 10, suppl. 4 (1995): 366–68.

Mark, Daniel B., C. David Naylor, Mark A. Hlatky, Robert M. Califf, Eric J. Topol, Christopher B. Granger, J. David Knight, Charlotte L. Nelson, Kerry L. Lee, Nancy E. Clapp-Channing, Wanda Sutherland, Louise Pilote, and Paul W. Armstrong. "Use of Medical Resources and Quality of Life after Acute Myocardial Infarction in Canada and the United States." *New England Journal of Medicine* 331 (27 Oct. 1994): 1130–35.

Marks, Harry. *The Progress of Experiment: Science and Therapeutic Reform in the United States, 1900–1990*. Cambridge: Cambridge University Press, 1997.

———. "What Does Evidence Do? Histories of Therapeutic Research." In *Harmonizing Drugs: Standards in 20th-Century Pharmaceutical History*, edited by Christian Bonah, Christophe Masutti, Anne Rasmussen, and Jonathan Simon, 81–100. Paris: Editions Glyphe, 2009.

Marseille, Elliot, Paul B. Hofmann, and James G. Kahn. "HIV Prevention before HAART in sub-Saharan Africa." *Lancet* 359 (25 May 2002): 1851–56.

Martin, David S. "From Omnivore to Vegan: The Dietary Education of Bill Clinton." CNN, 18 Aug. 2011. Accessed 30 Sept. 2011. www.cnn.com/2011/HEALTH/08/18/bill.clinton .diet.vegan/index.html.

Marx, Leo. "Technology: The Emergence of a Hazardous Concept." *Technology and Culture* 51 (July 2010): 561–77.

Maseri, Attilio, and Valentin Fuster. "Is There a Vulnerable Plaque?" *Circulation* 107 (29 April 2003): 2068–71.

Mason, George A. "Myocardial Ischemia and Its Surgical Relief." *Lancet* 257 (17 Feb. 1951): 359–67.

Master, Arthur M. "The Role of Effort and Occupation (Including Physicians) in Coronary Occlusion." *JAMA* 174 (22 Oct. 1960): 942–48.

Master, Arthur M., Simon Dack, and Harry L. Jaffe. "Activities Associated with the Onset of Acute Coronary Artery Occlusion." *American Heart Journal* 18 (1939): 434–43.

Master, Arthur M., Harry L. Jaffe, Simon Dack, and Arthur Grishman. "Coronary Occlusion, Coronary Insufficiency, and Angina Pectoris." *American Heart Journal* 27 (June 1944): 803–16.

Mateen, Farrah J., Rajanandini Muralidharan, Russell T. Shinohara, Joseph E. Parisi, Gregory J. Schears, Eelco F. M. Wijdicks. "Neurological Injury in Adults Treated with Extracorporeal Membrane Oxygenation." *Archives of Neurology* 68 (2011): 1543–49.

Mathers, Colin D., and Dejan Loncar. "Projections of Global Mortality and Burden of Disease from 2002 to 2030." *PLoS Medicine* 3 (Nov. 2006): 2011–30.

May, Angelo M., William E. Neville, and William Rumel, for the American College of Chest Physicians. "1971 Reflection of 1970 Statistics." *Chest* 6 (May 1972): 475–77.

Maynard, Charles, Lloyd D. Fisher, Eugene R. Passamani, and Thomas Pullum. "Blacks in the Coronary Artery Surgery Study (CASS): Race and Clinical Decision Making." *American Journal of Public Health* 76 (Dec. 1986): 1446–48.

McDermott, Walsh, Kurt W. Deuschle, and Clifford R. Barnett. "Health Care Experiment at Many Farms." *Science* 175 (7 Jan. 1972): 23–31.

McDougall, John. "Bill Clinton's Madness: A Consequence of Heart-Bypass Surgery Brain Damage." Accessed 12 June 2008. www.drmcdougall.com/misc/20080ther/080412 clinton.htm.

McDougall, John, to William Jefferson Clinton, ca. 6 Sept. 2004. Accessed 14 Feb. 2012. www.nealhendrickson.com/mcdougall/20041bn/040904pfnews.htm

McFadden, Robert D. "Clinton Suffers Pains in Chest; Bypass Surgery Is Scheduled." *New York Times,* 4 Sept. 2004.

McFadden, Robert D., and Lawrence K. Altman. "Clinton Is Given Bypass Surgery for 4 Arteries." *New York Times,* 7 Sept. 2004.

McGlynn, Elizabeth A., Steven M. Asch, John Adams, Joan Keesey, Jennifer Hicks, Alison DeCristofaro, and Eve A. Kerr. "The Quality of Health Care Delivered to Adults in the United States." *New England Journal of Medicine* 348 (26 June 2003): 2635–45.

McGlynn, Elizabeth A., C. David Naylor, Geoffrey M. Anderson, Lucian L. Leape, Rolla Edward Park, Lee H. Hilborne, Steven J. Bernstein, Bernard S. Goldman, Paul W. Armstrong, Joan W. Keesey; Laurie McDonald, S. Patricia Pinfold, Cheryl Damberg, Marjorie J. Sherwood, Robert H. Brook. "Comparison of Appropriateness of Coronary Angiography and Coronary Artery Bypass Graft Surgery between Canada and New York State." *JAMA* 272 (28 Sept. 1994): 934–40.

McGoon, Dwight C., Thomas H. Allen, and Emerson A. Moffitt. "Decreased Mortality Rate in Open-Heart Surgery." *Archives of Surgery* 88 (April 1964): 681–88.

McIntosh, Henry D., and Jorge A. Garcia. "The First Decade of Aortocoronary Bypass Grafting, 1967–1977: A Review." *Circulation* 57 (1978): 405–31.

McKhann, Guy M., Maura A. Goldsborough, Louis M. Borowicz, E. David Mellits, Ronald Brookmeyer, Shirley A. Quaskey, William A. Baumgartner, Duke E. Cameron, R. Scott Stuart, and Timothy J. Gardner. "Predictors of Stroke Risk in Coronary Artery Bypass Patients." *Annals of Thoracic Surgery* 63 (Feb. 1997): 516–21.

McKhann, Guy M., Maura A. Goldsborough, Louis M. Borowicz, Ola A. Selnes, E. David Mellits, Cheryl Enger, Shirley A. Quaskey, William A. Baumgartner, Duke E. Cameron, R. Scott Stuart, and Timothy J. Gardner. "Cognitive Outcome after Coronary Artery Bypass: A One-Year Prospective Study." *Annals of Thoracic Surgery* 63 (Feb. 1997): 510–15.

McKhann, Guy M., Maura A. Grega, Louis M. Borowicz, William A. Baumgartner, and Ola A. Selnes. "Stroke and Encephalopathy after Cardiac Surgery: An Update." *Stroke* 37 (Feb. 2006): 562–71.

McKinlay, John B. "From 'Promising Report' to 'Standard Procedure': Seven Stages in the Career of a Medical Innovation." *Milbank Quarterly* 59 (1981): 374–411.

McPherson, David D., Loren F. Hiratzka, Wade C. Lamberth, Berkeley Brandt, Michelle Hunt, Robert A. Kieso, Melvin L. Marcus, and Richard E. Kerber. "Delineation of the Extent of Coronary Atherosclerosis by High-Frequency Epicardial Echocardiography." *New England Journal of Medicine* 316 (5 Feb. 1987): 304–9.

McPherson, Klim, P. M. Strong, Arnold Epstein, and Lesley Jones. "Regional Variations in the Use of Common Surgical Procedures: Within and Between England and Wales, Canada, and the United States of America." *Social Science & Medicine* 15A (1981): 273–88.

Mearns, John G. "Effler and Sones and the Boys at Cleveland Clinic Have Taken the You-Bet-Your-Life Out of Heart Surgery." *Cleveland Magazine* (Nov. 1973): 2–10.

Merwin, Susan L., and Harry S. Abram. "Psychologic Responses to Coronary Artery Bypass." *Southern Medical Journal* 70 (Feb. 1977): 153–55.

Meyendorf, R. "The Causes of Pre-operative Psychopathology in Cardiac Surgery Patients." In *Cerebral Damage before and after Cardiac Surgery*, ed. Allen Willner, 3–14. Dordrecht: Kluwer Academic Publishers, 1993.

Michaels, Leon. *The Eighteenth-Century Origins of Angina Pectoris: Predisposing Causes, Recognition, and Aftermath.* London: Wellcome Trust, 2002.

Miller, Donald W., Eugene A. Hessel, Loren C. Winterscheid, K. Alvin Merendino, and David H. Dillard. "Current Practice of Coronary Artery Bypass Surgery: Results of a National Survey." *Journal of Thoracic and Cardiovascular Surgery* 73 (Jan. 1977): 75–83.

Miller, G. Wayne. *King of Hearts: The True Story of the Maverick Who Pioneered Open Heart Surgery.* New York: Random House, 2000.

Miller, Henry. "Some Neurological Complications of Surgical Treatment." *Proceedings of the Royal Society of Medicine* 57 (Feb. 1964): 143–46.

Miller, Judith. "Anti-Arms Groups Rebut Plans for MX." *New York Times,* 24 Nov. 1982.

Miller, R. Drew, Howard B. Burchell, and Jesse E. Edwards. "Myocardial Infarction with and without Coronary Occlusion." *Archives of Internal Medicine* 88 (1951): 597–604.

Millman, Joyce. "The Top 10 Reasons David Letterman's Heart Bypass Operation Was a Good Thing." Salon.com, 20 March 2000. Accessed 1 Feb. 2012. www.salon.com/entertainment/col/mill/2000/03/20/letterman

Millman, Marcia. *The Unkindest Cut: Life in the Backrooms of Medicine.* New York: Morrow, 1977.

Minard, Eliza. "Pathological Anteflexion of the Uterus." *JAMA* 16 (13 June 1891): 846–48.

Mitchell, J. R. A., and C. J. Schwartz. "The Relation between Myocardial Lesions and Coronary Artery Disease. II. A Selected Group of Patients with Massive Cardiac Necrosis or Scarring." *British Heart Journal* 25 (1963): 1–24.

Mol, Anne-Marie. *The Body Multiple: Ontology in Medical Practice.* Durham, NC: Duke University Press, 2002.

Monagan, David. "American Cardiac Surgeons, Yesterday's Idols, Watch Jobs Go Begging." theheart.org, 5 Dec. 2002. Accessed 3 Jan. 2011. www.theheart.org/article/131813.do.

Monagan, David, with David O. Williams. *Journey into the Heart: A Tale of Pioneering Doctors and Their Race to Transform Cardiovascular Medicine.* New York: Gotham Books, 2007.

Moody, D. M., M. A. Bell, V. R. Challa, W. E. Johnston, and D. S. Prough. "Brain Microemboli during Cardiac Surgery or Aortography." *Annals of Neurology* 28 (Oct. 1990): 477–86.

Moore, Francis D. "What Puts the Surge in Surgery?" *New England Journal of Medicine* 282 (15 Jan. 1970): 162–64.

Moore, Norman. "Rheumatic Fever and Valvular Disease." *Lancet* 173 (24 April 1909): 1159–64.

Mora Mangano, Christina T. "Risky Business." *Journal of Thoracic and Cardiovascular Surgery* 125 (June 2003): 1204–7.

Moreno, Pedro R., Erling Falk, Igor F. Palacios, John B. Newell, Valentin Fuster, and John T. Fallon. "Macrophage Infiltration in Acute Coronary Syndromes." *Circulation* 90 (Aug. 1994): 775–78.

Morgan, Matthew W., Raisa B. Deber, Hilary A. Llewellyn-Thomas, Peter Gladstone, R. J. Cusimano, Keith O'Rourke, George Tomlinson, and Allan S. Detsky. "Randomized, Controlled Trial of an Interactive Videodisc Decision Aid for Patients with Ischemic Heart Disease." *Journal of General Internal Medicine* 15 (Oct. 2000): 752–54.

Mount Sinai Hospital. "Ironic that a Plumber Came to Us to Help Him Remove a Clog." *New York Times Magazine,* 17 April 2011, back cover. [Advertisement]

Muanya, Chukwuma. "Nigerian Scores First in Cardiac Angioplasty, Stenting Surgery." *Online Nigeria,* 8 July 2009. http://news.onlinenigeria.com/templates/?a=5714&z=12.

Mueller, Richard L., Todd K. Rosengart, and O. Wayne Isom. "The History of Surgery for Ischemic Heart Disease." *Annals of Thoracic Surgery* 63 (1997): 869–78.

Muller, James E. "Diagnosis of Myocardial Infarction: Historical Notes from the Soviet Union and the United States." *American Journal of Cardiology* 40 (1977): 269–71.

Muller, James E., George S. Abela, Richard W. Nesto, and Geoffrey H. Tofler. "Triggers, Acute Risk Factors and Vulnerable Plaques: The Lexicon of a New Frontier." *Journal of the American College of Cardiology* 23 (1 March 1994): 809–13.

Muller, James E., Peter H. Stone, Zoltan G. Turi, John D. Rutherford, Charles A. Czeisler, Corette Parker, W. Kenneth Poole, Eugene Passamani, Robert Roberts, Thomas Robertson, Burton E. Sobel, James T. Willerson, Eugene Braunwald, and the MILIS Study Group. "Circadian Variations in the Frequency of Onset of Acute Myocardial Infarction." *New England Journal of Medicine* 313 (21 Nov. 1985): 1315–22.

Muller, James E., Ahmed Tawakol, Sekar Kathiresan, and Jagat Narula. "New Opportunities for Identification and Reduction of Coronary Risk: Treatment of Vulnerable Patients, Arteries, and Plaques." *Journal of the American College of Cardiology* 47 suppl. C (18 April 2006): C2–C6.

Muller, James E., and Geoffrey H. Tofler. "Triggering and Hourly Variation of Onset of Arterial Thrombosis." *Annals of Epidemiology* 2 (July 1992): 393–405.

Muller, James E., Geoffrey H. Tofler, and Peter H. Stone. "Circadian Variation and Triggers of Onset of Acute Cardiovascular Disease." *Circulation* 79 (1989): 733–43.

Mulley, Albert G. "The Need to Confront Variation in Practice." *BMJ* 339 (31 Oct. 2009): 1007–9.

Mulley, Albert G., and Kim A. Eagle. "What Is Inappropriate Care?" *JAMA* 260 (22–29 July 1988): 540–41.

Müllges, Wolfgang, Jörg Babin-Ebell, Wiklo Reents, and Klaus V. Tokya. "Cognitive Performance after Coronary Artery Bypass Grafting: A Follow-up Study." *Neurology* 59 (Sept. 2002): 741–43.

Murkin, John M. "Introduction: Protecting the Brain during Cardiac Surgery." *Journal of Cardiothoracic and Vascular Anesthesia* 10 (Jan. 1996): 2.

Murkin, John M., W. Douglas Boyd, Suganthan Ganapathy, Sandra J. Adams, and Rhonda C. Peterson. "Beating Heart Surgery: Why Expect Less Central Nervous System Morbidity?" *Annals of Thoracic Surgery* 68 (Oct. 1999): 1498–1501.

Murphy, Marvin L., Herbert N. Hultgren, Katherine Detre, James Thomsen, Timothy Takaro, and Participants of the Veterans Administration Cooperative Study. "Treatment of Chronic Stable Angina: A Preliminary Report of Survival Data of the Randomized Veterans Administration Cooperative Study." *New England Journal of Medicine* 297 (22 Sept. 1977): 621–27.

Mustard, J. Fraser. "Platelets and Thrombosis in Acute Myocardial Infarction." *Hospital Practice* 7 (Jan. 1972): 115–28.

Naghavi, Morteza, Peter Libby, Erling Falk, S. Ward Casscells, Silvio Litovsky, John Rumberger, Juan Jose Badimon, Christodoulos Stefanadis, Pedro Moreno, Gerard Pasterkamp, Zahi Fayad, Peter H. Stone, Sergio Waxman, Paolo Raggi, Mohammad Madjid, Alireza Zarrabi, Allen Burke, Chun Yuan, Peter J. Fitzgerald, David S. Siscovick, Chris L. de Korte, Masanori Aikawa, K. E. Juhani Airaksinen, Gerd Assmann, Christoph R. Becker, James H. Chesebro, Andrew Farb, Zorina S. Galis, Chris Jackson, Ik-Kyung Jang, Wolfgang Koenig, Robert A. Lodder, Keith March, Jasenka Demirovic, Mohamad Navab, Silvia G. Priori, Mark D. Rekhter, Raymond Bahr, Scott M. Grundy, Roxana Mehran, Antonio Colombo, Eric Boerwinkle, Christie Ballantyne, William Insull, Jr, Robert S. Schwartz, Robert Vogel, Patrick W. Serruys, Goran K. Hansson, David P. Faxon, Sanjay Kaul, Helmut Drexler, Philip Greenland, James E. Muller, Renu Virmani, Paul M Ridker, Douglas P. Zipes, Prediman K. Shah, and James T. Willerson. "From Vulnerable Plaque to Vulnerable Patient: A Call for New Definitions and Risk Assessment Strategies: Part 1." *Circulation* 108 (7 Oct. 2003): 1664–72.

Naghavi, Morteza, Peter Libby, Erling Falk, S. Ward Casscells, Silvio Litovsky, John Rumberger, Juan Jose Badimon, Christodoulos Stefanadis, Pedro Moreno, Gerard Pasterkamp, Zahi Fayad, Peter H. Stone, Sergio Waxman, Paolo Raggi, Mohammad Madjid, Alireza Zarrabi, Allen Burke, Chun Yuan, Peter J. Fitzgerald, David S. Siscovick, Chris L. de Korte, Masanori Aikawa, K. E. Juhani Airaksinen, Gerd Assmann, Christoph R. Becker, James H. Chesebro, Andrew Farb, Zorina S. Galis, Chris Jackson, Ik-Kyung Jang, Wolfgang Koenig, Robert A. Lodder, Keith March, Jasenka Demirovic, Mohamad Navab, Silvia G. Priori, Mark D. Rekhter, Raymond Bahr, Scott M. Grundy, Roxana Mehran, Antonio Colombo, Eric Boerwinkle, Christie Ballantyne, William Insull, Jr, Robert S. Schwartz, Robert Vogel, Patrick W. Serruys, Goran K. Hansson,

David P. Faxon, Sanjay Kaul, Helmut Drexler, Philip Greenland, James E. Muller, Renu Virmani, Paul M Ridker, Douglas P. Zipes, Prediman K. Shah, and James T. Willerson. "From Vulnerable Plaque to Vulnerable Patient: A Call for New Definitions and Risk Assessment Strategies: Part II." *Circulation* 108 (14 Oct. 2003): 1772–78.

Narayan, K. M. Venkat, Mohammed K. Ali, and Jeffrey P. Koplan. "Global Noncommunicable Diseases—Where Worlds Meet." *New England Journal of Medicine* 363 (23 Sept. 2010): 1196–98.

Narayan, K. M. Venkat, Mohammed K. Ali, Carlos del Rio, Jeffrey P. Koplan, and James Curran. "Global Noncommunicable Diseases—Lessons from the HIV-AIDS Experience." *New England Journal of Medicine* 365 (8 Sept. 2011): 876–78.

Nathoe, Hendrik M., Diederik van Dijk, Erik W. L. Jansen, Willem J. L. Suyker, Jan C. Diephuis, Wim-Jan van Boven, Aart Brutel de la Rivière, Cornelius Borst, Cor J. Kalkman, Diederick E. Grobbee, Erik Buskens, and Peter P. T. de Jaegere, for the Study Group. "A Comparison of On-Pump and Off-Pump Coronary Bypass Surgery in Low-Risk Patients." *New England Journal of Medicine* 348 (30 Jan. 2003): 394–402.

Neema, Praveen Kumar, Sameet Pathak, Praveen Kerala Verma, Sethuraman Manikandan, Ramesh Chandra Rathod, Deepak K. Tempe, and Avery Tung. "Systemic Air Embolization after Termination of Cardiopulmonary Bypass." *Journal of Thoracic and Cardiovascular Anesthesia* 21 (April 2007): 288–97.

Nelkin, Dorothy, and M. Susan Lindee. *The DNA Mystique: The Gene as a Cultural Icon.* 2nd ed. Ann Arbor: University of Michigan Press, 2004.

Newman, Mark F., Jerry L. Kirchner, Barbara Phillips-Bute, Vincent Gaver, Hilary Grocott, Robert H. Jones, Daniel B. Mark, Joseph G. Reves, James A. Blumenthal, for The Neurological Outcome Research Group and the Cardiothoracic Anesthesiology Research Endeavors Investigators. "Longitudinal Assessment of Neurocognitive Function after Coronary-Artery Bypass Grafting." *New England Journal of Medicine* 344 (8 Feb. 2001): 395–402.

Newman, Stanton P., and Michael J. G. Harrison, with David A. Stump, Peter Smith, and Key Taylor, eds. *The Brain and Cardiac Surgery: Causes of Neurological Complications and Their Prevention.* Amsterdam: Harwood Academic Publishers, 2000.

Newman, Stanton, Louise Klinger, Graham Venn, Peter Smith, Michael Harrison, and Tom Treasure. "Subjective Reports of Cognition in Relation to Assessed Cognitive Performance Following Coronary Artery Bypass Surgery." *Journal of Psychosomatic Research* 33 (1989): 227–33.

Nissen, Steven E., Cindy L. Grines, John C. Gurley, Kevin Sublett, David Haynie, Cheryl Diaz, David C. Booth, and Anthony N. DeMaria. "Application of a New Phased-Array Ultrasound Imaging Catheter in the Assessment of Vascular Dimensions: In Vivo Comparison to Cineangiography." *Circulation* 81 (Feb. 1990): 660–66.

[Note, no identifiers], 29 Dec. 1970. HB Papers, series V, box 6.

Nutton, Vivian. "From Medical Certainty to Medical Amulets: Three Aspects of Ancient Therapeutics." *Clio Medica* 22 (1991): 13–22.

"Obama's Health Cost Illusion: The President's Main Case for Reform Is Rooted in False Claims and Little Evidence." *Wall Street Journal*, 8 June 2009.

Oberman, Albert, and Gary Cutter. "Issues in the Natural History and Treatment of Coronary Heart Disease in Black Populations: Surgical Treatment." *American Heart Journal* 108 (Sept. 1984): 688–94.

Ochsner, John L. "Oral History" (1 May 2000). In *Pioneers of Cardiac Surgery*, ed. William S. Stoney, 391–404. Nashville, TN: Vanderbilt University Press, 2008.

Ochsner, John L., and Noel L. Mills. *Coronary Artery Surgery*. Philadelphia, PA: Lea and Febiger, 1978.

O'Connor, Christopher M., Michael W. Dunne, Marc A. Pfeffer, Joseph B. Muhlestein, Louis Yao, Sandeep Gupta, Rebecca J. Benner, Marian R. Fisher, Thomas D. Cook, for the Investigators in the WIZARD Study. "Azithromycin for the Secondary Prevention of Coronary Heart Disease Events. The WIZARD Study: A Randomized Controlled Trial." *JAMA* 290 (17 Sept. 2003): 1459–66.

Oliva, Philip B., and John C. Breckinridge. "Arteriographic Evidence of Coronary Arterial Spasm in Acute Myocardial Infarction." *Circulation* 56 (Sept. 1977): 366–74.

Oliva, Philip B., Daniel E. Potts, and Richard G. Pluss. "Coronary Arterial Spasm in Prinzmetal Angina: Documentation by Coronary Arteriography." *New England Journal of Medicine* 288 (12 April 1973): 745–51.

Olsen, Axel K., Charles P. Bailey, and Kenneth Keown. "Cerebral Embolization in Intra-Cardiac Surgery Prevention and Treatment." *Transactions of the American Neurological Association* 77 (1952): 171–74.

O'Neill, William, Gerald C. Timmis, Patrick D. Bourdillon, Peter Lai, V. Ganghadarhan, Joseph Walton, Renato Ramos, Nathan Laufer, Seymor Gordon, Anthony Schork, and Bertram Pitt. "A Prospective Randomized Clinical Trial of Intracoronary Streptokinase versus Coronary Angioplasty for Acute Myocardial Infarction." *New England Journal of Medicine* 314 (27 March 1986): 812–18.

Oppenheimer, Gerald M. "Becoming the Framingham Study, 1947–1950." *American Journal of Public Health* 95 (April 2005): 602–10.

Orenstein, Jan Marc, Noriko Sato, Benjamin Aaron, Bryan Buckholz, and Sherman Bloom. "Microemboli Observed in Deaths Following Cardiopulmonary Bypass Surgery: Silicone Antifoam Agents and Polyvinyl Chloride Tubing as Sources of Emboli." *Human Pathology* 13 (Dec. 1982): 1082–90.

Organization for Economic Co-operation and Development. *Health at a Glance 2011: OECD Indicators*. OECD Publishing, 2011. www.oecd.org/dataoecd/6/28/49105858.pdf.

O'Shaughnessy, Laurence. "Future of Cardiac Surgery." *Lancet* 234 (4 Nov. 1939): 969–71.

Osler, William. "Angina Pectoris." *Lancet* 175 (12 March 1910): 697–702.

———. *Principles and Practice of Medicine*. New York: D. Appleton and Company, 1892.

Ozner, Michael. *The Great American Heart Hoax: Lifesaving Advice Your Doctor Should Tell You About Heart Disease Prevention (But Probably Never Will)*. Dallas, TX: BenBella Books, 2008.

Padmavati, S. "Interventional Cardiology: An Indian Perspective." *Journal of Interventional Cardiology* 8 (1995): 39–42.

Parisi, Alfred F., Peter Peduzzi, Katherine Detre, Gerald Shugoll, Herbert N. Hultgren, and Timothy Takaro. "Characteristics and Outcomes of Medical Nonadherers in the

Veterans Administration Cooperative Study of Coronary Artery Surgery." *American Journal of Cardiology* 53 (Jan. 1984): 23–28.

Park, Rolla Edward, Arlene Fink, Robert H. Brook, Mark R. Chassin, Katherine L. Kahn, Nancy J. Merrick, Jacqueline Kosecoff, and David H. Solomon. "Physician Ratings of Appropriate Indications for Six Medical and Surgical Procedures." *American Journal of Public Health* 76 (July 1986): 766–72.

Park, Seung-Jung, Young-Hak Kim, Duk-Woo Park, Sung-Cheol Yun, Jung-Min Ahn, Hae Geun Song, Jong-Young Lee, Won-Jang Kim, Soo-Jin Kang, Seung-Whan Lee, Cheol Whan Lee, Seong-Wook Park, Cheol-Hyun Chung, Jae-Won Lee, Do-Sun Lim, Seung-Woon Rha, Sang-Gon Lee, Hyeon-Cheol Gwon, Hyo-Soo Kim, In-Ho Chae, Yangsoo Jang, Myung-Ho Jeong, Seung-Jea Tahk, and Ki Bae Seung. "Randomized Trial of Stents versus Bypass Surgery for Left Main Coronary Artery Disease." *New England Journal of Medicine* 364 (2011): 1718–27.

Paterson, J. C. "Capillary Rupture with Intimal Haemorrhage as a Causative Factor in Coronary Thrombosis." *Archives of Pathology* 25 (1938): 474–87.

———. "Capillary Rupture with Intimal Hemorrhage as a Cause of Pulmonary Thrombosis." *American Heart Journal* 18 (1939): 451–57.

———. "Vascularization and Hemorrhage of the Intima of Arteriosclerotic Coronary Arteries." *Archives of Pathology* 22 (Sept. 1936): 313–24.

Patrick, Robert T., John W. Kirklin, and Richard A. Theye. "The Effects of Extracorporeal Circulation on the Brain." In *Extracorporeal Circulation,* ed. J. Garrott Allen, Francis D. Moore, Andrew G. Morrow, and Henry Swan, 272–78. Springfield, IL: Charles C Thomas, 1958.

Patterson, Russel H., Jeffry S. Wasser, and Robert S. Porro. "The Effect of Various Filters on Microembolic Cerebrovascular Blockage Following Cardiopulmonary Bypass." *Annals of Thoracic Surgery* 17 (May 1974): 464–73.

Paul, Charlotte. "The Relentless Therapeutic Imperative." *BMJ* 329 (18–25 Dec. 2004): 1457–59.

Paul-Shaheen, Pamela, Jane Deane Clark, and Daniel Williams. "Small Area Analysis: A Review and Analysis of the North American Literature." *Journal of Health Politics, Policy, and Law* 12 (Winter 1987): 741–809.

Pear, Robert. "Health Care Spending Disparities Stir a Fight." *New York Times,* 9 June 2009.

Pearson, R. John C., Björn Smedby, Ragnar Berfenstam, Robert F. L. Logan, Alex Burgess, and Osler L. Peterson. "Hospital Caseloads in Liverpool, New England and Uppsala. An International Comparison." *Lancet* 292 (7 Sept. 1968): 559–66.

Pellegrino, Edmund D. "The Sociocultural Impact of Twentieth-Century Therapeutics." In *The Therapeutic Revolution,* ed. Morris Vogel and Charles E. Rosenberg, 245–66. Philadelphia: University of Pennsylvania Press, 1979.

Penry, J. Kiffin, A. Robert Cordell, Frank R. Johnston, and Martin G. Netsky. "Cerebral Embolism by Antifoam in a Bubble Oxygenator System: An Experimental and Clinical Study." *Surgery* 47 (May 1960): 784–94.

Pernick, Martin S. *A Calculus of Suffering: Pain, Professionalism, and Anesthesia in Nineteenth-Century America.* New York: Columbia University Press, 1985.

Petryna, Adriana. *Life Exposed: Biological Citizens after Chernobyl.* Princeton, NJ: Princeton University Press, 2002.

Phibbs, Brendan. "The Abuse of Coronary Arteriography." *New England Journal of Medicine* 301 (20 Dec. 1979): 1394–96.

Phillips, Margaret A., Richard G. A. Feachem, and Jeffrey P. Koplan. "The Emerging Agenda for Adult Health." In *The Health of Adults in the Developing World,* ed. Richard G. A. Feachem, Tord Kjellstrom, Christopher J. L. Murray, Mead Over, and Margaret A. Phillips, 261–94. Washington, DC: World Bank, 1992.

Pillsbury, R. Cree, Richard R. Lower, Edward J. Hurle, Eugene Dong, and Norman E. Shumway. "Four Hundred and Fifty Consecutive Open Heart Operations." *California Medicine* 102 (March 1965): 181–84.

Pilote, Louise, Robert M. Califf, Shelly Sapp, Dave P. Miller, Daniel B. Mark, W. Douglas Weaver, Joel M. Gore, Paul W. Armstrong, E. Magnus Ohman, and Eric J. Topol, for the GUSTO-1 Investigators. "Regional Variation Across the United States in the Management of Acute Myocardial Infarction." *New England Journal of Medicine* 333 (31 Aug. 1995): 565–72.

Pokar, H. "A Study of Equipment for and Performance of Extracorporeal Circulation." In *Impact of Cardiac Surgery on the Quality of Life,* ed. Allen E. Willner and Georg Rodewald, 247–54. New York: Plenun Press, 1990.

Pollock, Anne. "The Internal Cardiac Defibrillator." In *The Inner History of Devices,* ed. Sherry Turkle, 98–111. Cambridge, MA: MIT Press, 2008.

Porter, Theodore M. "Life Insurance, Medical Testing, and the Management of Mortality." In *Biographies of Scientific Objects,* ed. Lorraine Daston, 226–46. Chicago: University of Chicago Press, 2000.

Potter, G. G., B. L. Plassman, M. J. Helms, D. C. Steffens, and K. A. Welsh-Bohmer. "Age Effects of Coronary Artery Bypass Graft on Cognitive Status Change among Elderly Male Twins." *Neurology* 63 (Dec. 2004): 2245–49.

Pressman, Jack. *Last Resort: Psychosurgery and the Limits of Medicine.* Cambridge: Cambridge University Press, 1998.

Preston, Thomas A. "The Hazards of Poorly Controlled Studies in the Evaluation of Coronary Artery Surgery." *Chest* 73 (April 1978): 441–42.

———. "Marketing an Operation." *Atlantic Monthly* 254 (Dec. 1984): 32–40.

Priest, Walter S., Misha S. Zaks, George K. Yacorzynski, and Benjamin Boshes. "The Neurologic, Psychiatric, and Psychologic Aspects of Cardiac Surgery." *Medical Clinics of North America* 41 (1957): 155–69.

Prinzmetal, Myron, Rexford Kennamer, Rueben Merliss, Takashi Wada, and Naci Bor. "Angina Pectoris: I. A Variant Form of Angina Pectoris, Preliminary Report." *American Journal of Medicine* 27 (Sept. 1959): 375–88.

Prioreschi, Plinio. "Myocardial Infarction, Experimental Cardiac Necroses, and Potassium." *Perspectives in Biology and Medicine* 9 (Spring 1966): 369–75.

Proctor, Robert. "Agnotology: A Missing Term to Describe the Cultural Production of Ignorance (and Its Study)." In *Agnotology: The Making and Unmaking of Ignorance,* ed. Robert Proctor and Londa Schiebinger, 1–33. Stanford, CA: Stanford University Press, 2008.

Proctor, Robert, and Londa Schiebinger, eds. *Agnotology: The Making and Unmaking of Ignorance.* Stanford, CA: Stanford University Press, 2008.

Proudfit, William L. "John Hunter: On Heart Disease." *British Heart Journal* 56 (1986): 109–14.

———. "Origin of Concept of Ischaemic Heart Disease." *British Heart Journal* 50 (1983): 209–12.

Proudfit, William L., to the Board of Governors, 6 Jan. 1971. CCF Archives, A08-27/1, "Cardiology Annual Reports, 1975–1985" (#1554).

Purdum, Todd S. "Bubba Trouble: The Comeback Id." *Vanity Fair,* July 2008.

Rabbani, Ramin, and Eric J. Topol. "Strategies to Achieve Coronary Arterial Plaque Stabilization." *Cardiovascular Research* 41 (1 Feb. 1999): 402–17.

Rabiner, Charles J., Allen E. Willner, and Jirina Fishman. "Psychiatric Complications Following Coronary Bypass Surgery." *Journal of Nervous and Mental Disease* 160 (1975): 342–48.

Radner, Stig. "An Attempt at the Roentgenologic Visualization of Coronary Blood Vessels in Man." *Acta Radiologica* 26 (1945): 497–502.

Rafflenbeul, Wolf, Lloyd R. Smith, William J. Rogers, John A. Mantle, Charles E. Rackley, and Richard O. Russell. "Quantitative Coronary Arteriography: Coronary Anatomy of Patients with Unstable Angina Pectoris Reexamined 1 Year After Optimal Medical Therapy." *American Journal of Cardiology* 43 (April 1979): 699–707.

Ramlawi, Basel, Hasan Out, James L. Rudolph, Shigetoshi Mieno, Isaac S. Kohane, Handan Can, Towia A. Libermann, Edward R. Marcantonio, Cesario Bianchi, and Frank W. Sellke. "Genomic Expression Pathways Associated with Brain Injury after Cardiopulmonary Bypass." *Journal of Thoracic and Cardiovascular Surgery* 134 (Oct. 2007): 996–1005.

Ramlawi, Basel, James L. Rudolph, Shigetoshi Mieno, Jun Feng, Munir Boodhwani, Kamal Khabbaz, Sue E. Levkoff, Edward R. Marcantonio, Cesario Bianchi, and Frank W. Sellke. "C-Reactive Protein and Inflammatory Response Associated to Neurocogntive Decline following Cardiac Surgery." *Surgery* 221 (Aug. 2006): 221–26.

"Randomised Trial of Intravenous Streptokinase, Oral Aspirin, Both, or Neither among 17,187 Cases of Suspected Acute Myocardial Infarction: ISIS-2." *Lancet* 332 (1988): 349–60.

Rapp, Rayna. *Testing Women, Testing the Fetus: The Social Impact of Amniocentesis in America.* New York: Routledge, 1999.

Rasmussen, Nicolas. *Picture Control: The Electron Microscope and the Transformation of Biology in America, 1940–1960.* Stanford, CA: Stanford University Press, 1997.

Ratcliff, J. D. "The 'Stopped Heart' Operation: New Era in Surgery?" *Reader's Digest* (Aug. 1956): 29–33.

Reddy, K. Srinath. Foreword to *Race Against Time: The Challenge of Cardiovascular Disease in Developing Countries,* by Stephen Leeder, Susan Raymond, and Henry Greenberg, vi–vii. New York: Trustees of Columbia University, 2004.

Reiser, Stanley Joel. "The Intensive Care Unit: The Unfolding Ambiguities of Survival Therapy." *International Journal of Technology Assessment in Health Care* 8 (1992): 382–94.

Rentrop, K. P., H. Blanke, K. R. Karsch, H. Kaiser, H. Köstering, and K. Leitz. "Selective Intracoronary Thrombolysis in Acute Myocardial Infarction and Unstable Angina Pectoris." *Circulation* 63 (Feb. 1981): 307–17.

Rentrop, K. P., H. Blanke, K. R. Karsch, and H. Kreuzer. "Initial Experience with Transluminal Recanalization of the Recently Occluded Infarct-Related Coronary Artery in Acute Myocardial Infarction—Comparison with Conventionally Treated Patients." *Clinical Cardiology* 2 (April 1979): 92–105.

Restrepo, Lucas, Robert J. Wityk, Maura A. Grega, Lou Borowicz, Peter B. Barker, Michael A. Jacobs, Norman J. Beauchamp, Argye E. Hillis, and Guy M. McKhann. "Diffusion- and Perfusion-Weighted Magnetic Resonance Imaging of the Brain Before and After Coronary Artery Bypass Grafting Surgery." *Stroke* 33 (Dec. 2002): 2909–15.

Reul, George J., Denton A. Cooley, Don C. Wukasch, E. Ross Kyger, Frank M. Sandiford, Grady L. Hallman, and John C. Norman. "Long-Term Survival Following Coronary Artery Bypass: Analysis of 4,522 Consecutive Patients." *Archives of Surgery* 110 (1975): 1419–24.

Richards, Evelleen. *Vitamin C and Cancer: Medicine or Politics?* London: Macmillan, 1991.

Richards, Norman V. *Heart to Heart: A Cleveland Clinic Guide to Understanding Heart Disease and Open Heart Surgery.* New York: Atheneum, 1987.

Ridker, Paul M. "A Tale of Three Labels: Translating the JUPITER Trial Data into Regulatory Claims." *Clinical Trials* 8 (2011): 417–22.

Ridker, P. M., J. E. Manson, J. M. Gaziano, J. E. Buring, and C. H. Hennekens. "Low-Dose Aspirin Therapy for Chronic Stable Angina: A Randomized, Placebo-Controlled Clinical Trial." *Annals of Internal Medicine* 114 (1991): 835–39.

Rief, Winfried, Jerry Avorn, and Arthur J. Barsky. "Medication-Attributed Adverse Effects in Placebo Groups." *Archives of Internal Medicine* 166 (23 Jan. 2006): 155–60.

Rief, Winfried, Yvonne Nestorius, Anna von Lilienfeld-Toal, Imis Dogan, Franziska Schreiber, Stefan G. Hofmann, Arthur J. Barsky, and Jerry Avorn. "Differences in Adverse Effect Reporting in Placebo Groups in SSRI and Tricyclic Antidepressant Trials: A Systematic Review and Meta-Analysis." *Drug Safety* 32 (Nov. 2009): 1041–56.

Roach, Gary W., Marc Kanchuger, Christina Mora Mangano, Mark Newman, Nancy Nussmeier, Richard Wolman, Anil Aggarwal, Katherine Marschall, Steven H. Graham, Catherine Ley, Gerard Ozanne, Dennis T. Mangano, Ahvie Herskowitz, Vera Katseva, Rita Sears, for The Multicenter Study of Perioperative Ischemia Research Group and the Ischemia Research and Education Foundation Investigators. "Adverse Cerebral Outcomes after Coronary Bypass Surgery." *New England Journal of Medicine* 335 (19 Dec. 1996): 1857–63.

Robbins, Stanley L. "Cardiac Pathology—A Look at the Last Five Years." *Human Pathology* 5 (Jan. 1974): 9–24.

———. "Heart." In *Pathology*, 3rd ed., 511–66. Philadelphia, PA: W. B. Saunders, 1967.

Robbins, Stanley L., and Ramzi S. Cotran. "The Heart." In *Pathologic Basis of Disease*, 2nd ed., 643–711. Philadelphia, PA: W. B. Saunders, 1979.

Roberts, James C., Robert C. Boice, Robert L. Brownell, and David H. Brown. "Spontaneous Atherosclerosis in Pacific Toothed and Baleen Whales." In *Comparative Atherosclerosis: The Morphology of Spontaneous and Induced Atherosclerotic Lesions in Animals and*

Humans in Relation to Disease, ed. James C. Roberts, Reuben Straus, and Miriam S. Cooper, 151–55. New York: Harper and Row, 1965.

Roberts, James C., Reuben Strauss, and Miriam S. Cooper, eds. *Comparative Atherosclerosis: The Morphology of Spontaneous and Induced Atherosclerotic Lesions in Animals and Humans in Relation to Disease.* New York: Harper and Row, 1965.

Roberts, William C. "Coronary Arteries in Fatal Acute Myocardial Infarction." *Circulation* 45 (Jan. 1972): 215–30.

Roberts, William C., and L. Maximilian Buja. "The Frequency and Significance of Coronary Arterial Thrombi and Other Observations in Fatal Acute Myocardial Infarction: A Study of 107 Necropsy Patients." *American Journal of Medicine* 52 (April 1972): 425–43.

Robinson, George, and Avraham Merav. "Informed Consent: Recall by Patients Tested Postoperatively." *Annals of Thoracic Surgery* 22 (Sept. 1976): 209–12.

Romaine-Davis, Ada. *John Gibbon and His Heart-Lung Machine.* Philadelphia: University of Pennsylvania Press, 1991.

Roos, Noralou P. "Hysterectomy: Variations in Rates Across Small Areas and Across Physicians' Practices." *American Journal of Public Health* 74 (April 1984): 327–35.

Rosenberg, Charles E. "Disease and Social Order in America: Perceptions and Expectations." *Milbank Quarterly* 64, suppl. 1 (1986): 34–55.

———. *Our Present Complaint: American Medicine, Then and Now.* Baltimore, MD: Johns Hopkins University Press, 2007.

———. "Pathologies of Progress: The Idea of Civilization as Risk." *Bulletin of the History of Medicine* 72 (1998): 714–30.

———. "The Therapeutic Revolution: Medicine, Meaning, and Social Change in Nineteenth-Century America." *Perspectives in Biology and Medicine* 20 (1977): 485–506.

———. "The Tyranny of Diagnosis: Specific Entities and Individual Experience." *Milbank Quarterly* 80 (2002): 237–60.

Ross, Richard S. "Ischemic Heart Disease." In *Harrison's Principle's of Internal Medicine,* 7th ed., ed. Maxwell M. Wintrobe, George W. Thorn, Raymond D. Adams, Eugene Braunwald, Kurt J. Isselbacher, and Robert G. Petersdorf, 1194–1210. New York: McGraw-Hill, 1974.

Rothberg, Michael B., Senthil K. Sivalingam, Javed Ashraf, Paul Visintainer, John Joelson, Reva Kleppel, Neelima Vallurupalli, and Marc J. Schweiger. "Patients' and Cardiologists' Perceptions of the Benefits of Percutaneous Coronary Intervention for Stable Coronary Disease." *Annals of Internal Medicine* 153 (2010): 307–13.

Rothstein, William J. *Public Health and the Risk Factor: A History of an Uneven Medical Revolution.* Rochester, NY: University of Rochester Press, 2003.

Rouleau, Jean L., Lemuel A. Moye, Marc A. Pfeffer, J. Malcolm O. Arnold, Victoria Bernstein, Thomas E. Cuddy, Gilles R. Dagenais, Edward M. Geltman, Steven Goldman, David Gordon, Peggy Hamm, Marc Klein, Gervasio A. Lamas, John McCans, Patricia McEwan, Francis J. Menapace, John O. Parker, Francois Sestier, Bruce Sussex, and Eugene Braunwald, for the SAVE Investigators. "A Comparison of Management Patterns after Acute Myocardial Infarction in Canada and the United States." *New England Journal of Medicine* 328 (18 March 1993): 779–84.

"Roundtable Discussion." *American Journal of Cardiology* 75 suppl. 1 (23 Feb. 1995): 93B–97B.

Rozhon, Tracie. "Pfizer to Buy Maker of Promising Cholesterol Drug." *New York Times*, 22 Dec. 2003, C2.

Rubens, Fraser D., and Howard Nathan. "Lessons Learned on the Path to a Healthier Brain: Dispelling the Myths and Challenging the Hypotheses." *Perfusion* 22 (May 2007): 153–60.

Rutkow, Ira M., Alan M. Gittelsohn, and George D. Zuidema. "Surgical Decision Making: The Reliability of Clinical Judgment." *Annals of Surgery* 190 (Sept. 1970): 409–19.

Sabiston, David C. "The Coronary Circulation." In *Davis-Christopher Textbook of Surgery: The Biological Basis of Modern Surgical Practice*, 10th ed., ed. David C. Sabiston, 2025–41. Philadelphia, PA: W. B. Saunders, 1972.

———. "The Coronary Circulation." *Johns Hopkins Medical Journal* 134 (June 1974): 314–29.

Sanders, Kenneth. "The Inner World of Some Patients with Coronary Thrombosis." *Proceedings of the Royal Society of Medicine* 55 (1962): 691–93.

Sanghavi, Darshak. "Plumber's Butt? The Right and Wrong Way to Think about Heart Attacks." *Slate*, 8 May 2007.

Saunders, Barry F. *CT Suite: The Work of Diagnosis in the Age of Noninvasive Cutting*. Durham, NC: Duke University Press, 2008.

Schaar, Johannes A., James E Muller, Erling Falk, Renu Virmani, Valentin Fuster, Patrick W Serruys, Antonio Colombo, Christodoulos Stefanadis, S. Ward Casscells, Pedro R Moreno, Attilio Maseri and Anton F. W. van der Steen. "Terminology for High-Risk and Vulnerable Coronary Plaques." *European Heart Journal* 25 (June 2004): 1077–82.

Schlich, Thomas. "The Emergence of Modern Surgery." In *Medicine Transformed: Health, Disease and Society in Europe, 1800–1930*, ed. Deborah Brunton, 61–91. Manchester: Manchester University Press, 2004.

Schoen, Frederick J. "The Heart." In *Robbins Pathologic Basis of Disease*, 6th ed., ed. Ramzi S. Cotran, Vinay Kumar, and Tucker Collins, 543–600. Philadelphia, PA: W. B. Saunders, 1999.

———. "The Heart." In *Robbins and Cotran Pathologic Basis of Disease*, 7th ed., ed. Vinay Kumar, Abul K. Abbas, and Nelson Fausto, 555–618. Philadelphia, PA: Elsevier Saunders, 2005.

Schoen, Frederick J., and Ramzi S. Contran. "Blood Vessels." In *Robbins Pathologic Basis of Disease*, 6th ed., ed. Ramzi S. Cotran, Vinay Kumar, and Tucker Collins, 493–541. Philadelphia, PA: W. B. Saunders, 1999.

Schön, Donald A. *The Reflective Practitioner: How Professionals Think in Action*. New York: Basic Books, 1983.

Schulman, Kevin A., Jesse A. Berlin, William Harless, Jon F. Kerner, Shyrl Sistrunk, Bernard J. Gersh, Ross Dubé, Christopher K. Taleghani, Jennifer E. Burke, Sankey Williams, John M. Eisenberg, and José J. Escarce. "The Effect of Race and Sex on Physicians' Recommendations for Cardiac Catheterization." *New England Journal of Medicine* 340 (25 Feb. 1999): 618–26.

Schwartz, Lisa M., Steven Woloshin, and H. Gilbert Welch. "Misunderstandings about the Effects of Race and Sex on Physicians' Referrals for Cardiac Catheterization." *New England Journal of Medicine* 341 (22 July 1999): 279–83.

Scull, Andrew T. *Madhouse: A Tragic Tale of Megalomania and Modern Medicine.* New Haven, CT: Yale University Press, 2005.

Secord, James. *Victorian Sensation: The Extraordinary Publication, Reception, and Secret Authorship of Vestiges of the Natural History of Creation.* Chicago: University of Chicago Press, 2000.

Selby, Joe V., Bruce H. Fireman, Robert J. Lundstrom, Bix E. Swain, Alison F. Truman, Candice C. Wong, Erika S. Froelicher, Hal V. Barron, and Mark A. Hlatky. "Variation among Hospitals in Coronary-Angiography Practices and Outcomes after Myocardial Infarction in a Large Health Maintenance Organization." *New England Journal of Medicine* 335 (19 Dec. 1996): 1888–96.

Selnes, Ola A., Maura A. Goldsborough, Louis M. Borowicz, Cheryl Enger, Shirley A. Quaskey, and Guy M. McKhann. "Determinants of Cognitive Change after Coronary Artery Bypass Surgery: A Multifactorial Problem." *Annals of Thoracic Surgery* 67 (June 1999): 1669–76.

Selnes, Ola A., Rebecca F. Gottesman, Maura A. Grega, William A. Baumgartner, Scott L. Zeger, and Guy M. McKhann. "Cognitive and Neurologic Outcomes after Coronary-Artery Bypass Surgery." *New England Journal of Medicine* 366 (2012): 250–57.

Selnes, Ola A., Maura A. Grega, Maryanne M. Bailey, Luu D. Pham, Scott L. Zeger, William A. Baumgartner, and Guy M. McKhann. "Cognition 6 Years after Surgical or Medical Therapy for Coronary Artery Disease." *Annals of Neurology* 63 (May 2008): 581–90.

———. "Do Management Strategies for Coronary Artery Disease Influence 6-Year Cognitive Outcomes?" *Annals of Thoracic Surgery* 88 (2009): 445–54.

Selnes, Ola A., and Guy M. McKhann. "Neurocognitive Complications after Coronary Artery Bypass Surgery." *Annals of Neurology* 57 (May 2005): 615–21.

Selwyn, Andrew P., and Eugene Braunwald. "Ischemic Heart Disease." In *Harrison's Principles of Internal Medicine,* 14th ed., ed. Anthony S. Fauci, Eugene Braunwald, Kurt J. Isselbacher, Jean D. Wilson, Joseph B. Martin, Dennis L. Kasper, Stephen L. Hauser, and Dan L. Longo, 1365–75. New York: McGraw-Hill, 1998.

Sepucha, Karen, and Albert G. Mulley. "A Perspective on the Patient's Role in Treatment Decisions." *Medical Care Research and Review* 66 suppl. (Feb. 2009): 53S–74S.

Serruys, Patrick W., Hector M. Garcia-Garcia, and Evelyn Regar. "From Postmortem Characterization to the In Vivo Detection of Thin-Capped Fibroatheromas: The Missing Link Toward Percutaneous Treatment." *Journal of the American College of Cardiology* 50 (2007): 950–52.

Serruys, Patrick W., Marie-Claude Morice, A. Pieter Kappetein, Antonio Colombo, David R. Holmes, Michael J. Mack, Elisabeth Ståhle, Ted E. Feldman, Marcel van den Brand, Eric J. Bass, Nic Van Dyck, Katrin Leadley, Keith D. Dawkins, and Friedrich W. Mohr, for the SYNTAX Investigators. "Percutaneous Coronary Intervention versus Coronary-Artery Bypass Grafting for Severe Coronary Artery Disease." *New England of Medicine* 360 (5 March 2009): 961–72.

Serruys, Patrick W., Felix Unger, J. Eduardo Sousa, Adib Jatene, Hans J. R. M. Bonnier,

Jacques P. A. M. Schönberger, Nigel Buller, Robert Bonser, Marcel J. B. van sen Brand, Lex A. Van Herwerden, Marie-Angèle M. Morel, and Ben A. van Hout, for the Arterial Revascularization Therapies Study Group. "Comparison of Coronary-Artery Bypass Surgery and Stenting for the Treatment of Multivessel Disease." *New England Journal of Medicine* 344 (12 April 2001): 1117–24.

Seung, Ki Bae, Duk-Woo Park, Young-Hak Kim, Seung-Whan Lee, Cheol Whan Lee, Myeong-Ki Hong, Seong-Wook Park, Sung-Cheol Yun, Hyeon-Cheol Gwon, Myung-Ho Jeong, Yangsoo Jang, Hyo-Soo Kim, Pum Joon Kim, In-Whan Seong, Hun Sik Park, Taehoon Ahn, In-Ho Chae, Seung-Jea Tahk, Wook-Sung Chung, and Seung-Jung Park. "Stents versus Coronary-Artery Bypass Grafting for Left Main Coronary Artery Disease." *New England Journal of Medicine* 358 (24 April 2008): 1781–92.

Sewell, William H. "Coronary Spasm as a Primary Cause of Myocardial Infarction: A Preliminary Report." *Angiology* 17 (Jan. 1966): 1–8.

Shahian, David M., Fred H. Edwards, Victor A. Ferraris, Constance K. Haan, Jeffrey B. Rich, Sharon-Lise T. Normand, Elizabeth R. DeLong, Sean M. O'Brien, Cynthia M. Shewan, Rachel S. Dokholyan, and Eric D. Peterson. "Quality Measurement in Adult Cardiac Surgery: Part 1—Conceptual Framework and Measure Selection." *Annals of Thoracic Surgery* 83 (2007): S3–S12.

Shapin, Steven. *A Social History of Truth: Civility and Science in Seventeenth-Century England.* Chicago: University of Chicago Press, 1995.

Shapin, Steven, and Simon Schaeffer. *Leviathan and the Air-Pump: Hobbes, Boyle, and the Experimental Life.* Princeton, NJ: Princeton University Press, 1985.

Shapiro, Barbara. *A Culture of Fact: England, 1550–1720.* Cornell, NY: Cornell University Press, 2000.

Shaw, Pamela J., David Bates, Niall E. F. Cartlidge, Joyce M. French, David Heaviside, Desmond G. Julian, and David A. Shaw. "Early Intellectual Dysfunction Following Coronary Bypass Surgery." *Quarterly Journal of Medicine* 58 (Jan. 1986): 59–68.

———. "Neurologic and Neuropsychological Morbidity Following Major Surgery: Comparison of Coronary Artery Bypass and Peripheral Vascular Surgery." *Stroke* 18 (July–Aug. 1987): 700–707.

Shaw, Pamela J., David Bates, Niall E. F. Cartlidge, David Heaviside, Joyce M. French, Desmond G. Julian, and David A. Shaw. "Neurological Complications of Coronary Artery Bypass Graft Surgery: Six-Month Follow-up Study." *BMJ* 293 (19 July 1986): 165–67.

Shaw, Pamela J., David Bates, Niall E. F. Cartlidge, David Heaviside, Desmond G. Julian, and David A. Shaw. "Early Neurological Complications of Coronary Artery Bypass Surgery." *BMJ* 291 (16 Nov. 1985): 1384–87.

Sheldon, William. "1977 Annual Report, Department of Cardiology," 1 April 1978. CCF Archives, A08-27/1, "Cardiology Annual Reports, 1975–1985" (#1554).

———. "1979 Annual Report, Department of Cardiology," 17 April 1980. CCF Archives, A08-27/1, "Cardiology Annual Reports, 1975–1985" (#1554).

———. *Pathfinders of the Heart: The History of Cardiology at the Cleveland Clinic.* Bloomington, IN: Xlibris: 2008.

Sheldon, William, to Richard G. Farmer, 19 Nov. 1979. CCF Archives, 3-PR10, "Grüntzig, Andreas R."

Sheps, Mindel C., and Alvin P. Shapiro. "The Physician's Responsibility in the Age of Therapeutic Plenty." *Circulation* 25 (Feb. 1962): 399–407.

Shirey, Earl K. "1978 Annual Report, Department of Cardiology," 26 Feb. 1979. CCF Archives, A08-27/1, "Cardiology Annual Reports, 1975–1985" (#1554).

Shroyer, A. Laurie, Frederick L. Grover, Brack Hattler, Joseph F. Collins, Gerald O. McDonald, Elizabeth Kozora, John C. Lucke, Janet H. Baltz, and Dimitri Novitzky, for the Veterans Affairs Randomized On/Off Bypass (ROOBY) Study Group. "On-Pump versus Off-Pump Coronary-Artery Bypass Surgery." *New England Journal of Medicine* 361 (5 Nov. 2009): 1827–37.

Shub, Clarence, Ronald E. Vlietstra, Hugh C. Smith, Richard E. Fulton, and Lila R. Elveback. "The Unpredictable Progression of Symptomatic Coronary Artery Disease." *Mayo Clinic Proceedings* 56 (March 1981): 155–60.

Shumacker, Harris B. *A Dream of the Heart: The Life of John H. Gibbon, Jr., Father of the Heart-Lung Machine.* Santa Barbara: Fithian Press, 1999.

———. *The Evolution of Cardiac Surgery.* Bloomington: Indiana University Press, 1992.

Sicherman, Barbara. "The Uses of a Diagnosis: Doctors, Patients, and Neurasthenia." *Journal of the History of Medicine and Allied Sciences* 32 (1977): 33–54.

Sikri, Nikhil, and Amit Bardia. "A History of Streptokinase Use in Acute Myocardial Infarction." *Texas Heart Institute Journal* 34 (2007): 318–27.

Silverstein, Allen, and Howard P. Krieger. "Neurologic Complications of Cardiac Surgery." *Transactions of the American Neurological Association* 85 (1960): 151–54.

Skinner, Jonathan, and Elliott Fisher. "Regional Disparities in Medicare Expenditures: An Opportunity for Reform." *National Tax Journal* 1 (1997): 413–25.

Sloan, Herbert, Joe D. Morris, James Mackenzie, and Aaron Stern. "Open Heart Surgery: Results in 600 Cases." *Thorax* 17 (1962): 128–38.

Smedley, Brian D., Adrienne Y. Stith, and Alan R. Nelson, ed., for the Institute of Medicine. *Unequal Treatment: Confronting Racial and Ethnic Disparities in Health Care.* Washington, DC: National Academies Press, 2003.

Smith, Ben. "Clinton Attacks Vanity Fair." Politico.com, 1 June 2008. Accessed 12 June 2008. www.politico.com/blogs/bensmith/0608/Clinton_attacks_Vanity_Fair.html.

Smith, Merritt Roe. "Technological Determinism in American Culture." In *Does Technology Drive History? The Dilemma of Technological Determinism,* ed. Merritt Roe Smith and Leo Marx, 1–35. Cambridge, MA: MIT Press, 1994.

Smith, Peter L. C. "Cerebral Dysfunction after Cardiac Surgery: Closing Address." *Annals of Thoracic Surgery* 59 (May 1995): 1359–62.

Smith, Sidney C. "Risk-Reduction Therapy: The Challenge to Change." *Circulation* 93 (15 June 1996): 2205–11.

Sobel, Burton E. "Acute Myocardial Infarction." In *Cecil: Textbook of Medicine,* 19th ed., ed. James B. Wyngaarden, Lloyd H. Smith, and J. Claude Bennett, 304–18, Philadelphia, PA: W. B. Saunders, 1992.

Solomon, Daniel H., Sebastian Schneeweiss, Robert J. Glynn, Yuka Kiyota, Raisa Levin, Helen Mogun, and Jerry Avorn. "Relationship Between Selective Cyclooxygenase-2 Inhibitors and Acute Myocardial Infarction in Older Adults." *Circulation* 109 (2004): 2068–73.

Sones, F. Mason. "1963 Annual Report, Department of Pediatric Cardiology and the Cardiac Laboratory" (no date). CCF Archives, A08-27/1, "Cardiology Annual Reports, 1975–1985" (#1554).

Sones, F. Mason, to E. P. Roy, 12 Dec. 1963. CCF Archives, A08-27/1, "Cardiology Annual Reports, 1975–1985" (#1554).

Sones, F. Mason, and Earl K. Shirey. "Cine Coronary Angiography." *Modern Concepts of Cardiovascular Disease* 31 (July 1962): 735–38.

Sotaniemi, Kyösti. "Cerebral Outcome after Extracorporeal Circulation: Comparison Between Prospective and Retrospective Evaluations." *Archives of Neurology* 40 (Feb. 1983): 75–77.

Souttar, Henry, to Dwight E. Harken, 22 Sept. 1961. DH Papers, box 63, folder 6.

Souttar, H. S. "The Surgical Treatment of Mitral Stenosis." *BMJ* 3 (3 Oct. 1925): 603–6.

Spain, David M., and Victoria A. Bradess. "The Relationship of Coronary Thrombosis to Coronary Atherosclerosis and Ischemic Heart Disease: A Necropsy Study Covering a Period of 25 Years." *Amercan Journal of Medical Science* 240 (Dec. 1960): 701–10.

———. "Sudden Death from Coronary Heart Disease: Survival Time, Frequency of Thrombi, and Cigarette Smoking." *Chest* 58 (Aug. 1970): 107–10.

Spencer, Merrill P., G. Hugh Lawrence, George I. Thomas, and Lester R. Sauvage. "The Use of Ultrasonics in the Determination of Arterial Aeroembolism During Open-Heart Surgery." *Annals of Thoracic Surgery* 8 (Dec. 1969): 489–97.

Spodick, David H. "Percutaneous Transluminal Coronary Angioplasty: Opportunity Fleeting." *JAMA* 242 (12 Oct. 1979): 1658–59.

———. "Revascularization of the Heart—Numerators in Search of Denominators." *American Heart Journal* 81 (Feb. 1971): 149–57.

Sprague, Howard B. " 'Heart Attacks.' " *Boston Medical and Surgical Journal* 1967 (24 March 1927): 472–76.

Stallones, Reuel A. "The Rise and Fall of Ischemic Heart Disease." *Scientific American* 243 (Nov. 1980): 53–59.

Stammers, Alfred H. "Historical Aspects of Cardiopulmonary Bypass: From Antiquity to Acceptance." *Journal of Cardiothoracic and Vascular Anesthesia* 11 (May 1997): 266–74.

———. "Trends in Extracorporeal Circulation for the 1990s: Renewed Interest and Advancing Technologies." *Journal of Thoracic and Cardiovascular Anesthesia* 6 (April 1992): 226–37.

Starr, Paul. *The Social Transformation of American Medicine: The Rise of a Sovereign Profession and the Making of a Vast Industry.* New York: Basic Books, 1982.

Steinberg, Daniel. "An Interpretive History of the Cholesterol Controversy, Part I." *Journal of Lipid Research* 45 (2004): 1583–93.

———. "An Interpretive History of the Cholesterol Controversy, Part V: The Discovery of the Statins and the End of the Controversy." *Journal of Lipid Research* 47 (2006): 1339–51.

Stone, Gregg W., Akiko Maehara, Alexandra J. Lansky, Bernard de Bruyne, Ecaterina Cristea, Gary S. Mintz, Roxana Mehran, John McPherson, Naim Farhat, Steven P. Marso, Helen Parise, Barry Templin, Roseann White, Zhen Zhang, and Patrick W. Serruys, for the PROSPECT Investigators. "A Prospective Natural-History Study of Coronary Atherosclerosis." *New England Journal of Medicine* 364 (20 Jan. 2011): 226–35.

Stoney, William S., ed. *Pioneers of Cardiac Surgery.* Nashville, TN: Vanderbilt University Press, 2008.

Strauss, Reuben, Harry Sobel, S. K. Abul-Haj, and Robert J. Kositchek. "Spontaneous Myocardial Infarcts in Treated and Nontreated Animals." In *Comparative Atherosclerosis: The Morphology of Spontaneous and Induced Atherosclerotic Lesions in Animals and Humans in Relation to Disease,* ed. James C. Roberts, Reuben Strauss, and Miriam S. Cooper, 186–95. New York: Harper and Row, 1965.

Stump, David A. "Selection and Clinical Significance of Neuropsychologic Tests." *Annals of Thoracic Surgery* 59 (May 1995): 1340–44.

Stutz, Bruce. "Pump Head." *Scientific American* 289 (July 2003): 76–81.

Stygall, Jan, Stanton P. Newman, Geraldine Fitzgerald, Liz Steed, Kathleen Mulligan, Joseph E. Arrowsmith, Wilfred Pugsley, Steve Humphries, and Michael J. Harrison. "Cognitive Change 5 Years after Coronary Artery Bypass Surgery." *Health Psychology* 22 (Nov. 2003): 579–86.

Sullivan, Robert B. "Sanguine Practices: A Historical and Historiographic Reconsideration of Heroic Therapy in the Age of Rush." *Bulletin of the History of Medicine* 68 (Summer 1994): 211–34.

"Surgery in the Heart." *Time,* 30 April 1956.

"Surgery: The Texas Tornado." *Time,* 28 May 1965.

Surgical Committee, "Minutes," 16 June 1969. CCF Archives, 21-TA, "Surgical Committee Minutes, 1966–1969," folder 1/8-6/16/69 (#2303).

Swank, Roy L., and Raymond F. Hain. "The Effect of Different Sized Emboli on the Vascular System and Parenchyma of the Brain." *Journal of Neuropathology and Experimental Neurology* 11 (1952): 280–99.

Sweeney, Patrick J., and Frederick Lautzenheiser. *Neuroscience at the Cleveland Clinic: The Early Years, a Short History.* Cleveland, OH: Cleveland Clinic Foundation Press, 2006.

Szilagyi, D., Emerick, Richard T. McDonald, and Lloyd C. France. "The Applicability of Angioplastic Procedures in Coronary Atherosclerosis: An Estimate through Postmortem Injection Studies." *Annals of Surgery* 148 (Sept. 1958): 447–59.

Taggart, David P., Stuart M. Browne, Peter W. Halligan, and Derick T. Wade. "Is Cardiopulmonary Bypass Still the Cause of Cognitive Dysfunction after Cardiac Operations?" *Journal of Thoracic and Cardiovascular Surgery* 118 (Sept. 1999): 414–21.

Taggart, David P., and Stephen Westaby. "Neurological and Cognitive Disorders after Coronary Artery Bypass Grafting." *Current Opinion in Cardiology* 16 (Sept. 2001): 271–76.

Takaro, Timothy, Herbert N. Hultgren, Martin J. Lipton, and Katherine M. Detre. "The VA Cooperative Randomized Study of Surgery for Coronary Arterial Occlusive Disease. II. Subgroup with Significant Left Main Lesions." *Circulation* 54 suppl. 3 (Dec. 1976): 107–17.

Takaro, Timothy, Herbert N. Hultgren, David Littmann, and Elizabeth C. Wright. "An Analysis of Deaths Occurring in Association with Coronary Arteriography." *American Heart Journal* 86 (Nov. 1973): 587–97.

Tardiff, Barbara E., Mark F. Newman, Ann M. Saunders, Warren J. Strittmatter, James A.

Blumenthal, William D. White, Narda D. Croughwell, R. Duane Davis, Allen D. Roses, Joseph G. Reves, and the Neurologic Outcome Research Group of the Duke Heart Center. "Preliminary Report of a Genetic Basis for Cognitive Decline after Cardiac Operations." *Annals of Thoracic Surgery* 64 (Sept. 1997): 715–20.

Taubes, Gary. *Good Calories, Bad Calories: Challenging the Conventional Wisdom on Diet, Weight Control, and Disease.* New York: Anchor Books, 2007.

Taussig, Michael T. "Reification and the Consciousness of the Patient." *Social Science and Medicine* 148 (1980): 3–13.

Taylor, C. Bruce. "Experimentally Induced Arteriosclerosis in Nonhuman Primates." In *Comparative Atherosclerosis: The Morphology of Spontaneous and Induced Atherosclerotic Lesions in Animals and Humans in Relation to Disease,* ed. James C. Roberts, Reuben Straus, and Miriam S. Cooper, 215–43. New York: Harper and Row, 1965.

Taylor, C. Bruce, George E. Cox, Marjorie Counts, and Nelson Yogi. "Fatal Myocardial Infarction in the Rhesus Monkey with Drug-Induced Hypercholesterolemia." *American Journal of Pathology* 35 (May–June 1959): 674.

Taylor, C. Bruce, George E. Cox, Pacita Manalo-Estrella, and Jane Southworth. "Atherosclerosis in Rhesus Monkeys. II. Arterial Lesions Associated with Hypercholesterolemia Induced by Dietary Fat and Cholesterol." *Archives of Pathology* 64 (July 1962): 16–34.

Taylor, K. M. "Cardiac Surgery and the Brain: An Introduction." In *Cardiac Surgery and the Brain,* ed. Peter L. Smith and K. M. Taylor, 1–14. London: Edward Arnold, 1993.

"Tenet Physicians Settle Case Over Unnecessary Heart Procedures at Redding Medical Center." *Medical News Today,* 17 Nov. 2005. www.medicalnewstoday.com/articles/33643.php.

"Therapeutics: Disturbances of the Heart." *JAMA* (14 Dec. 1912): 2151–53.

"Therapeutics: Disturbances of the Heart." *JAMA* (21 Dec. 1912): 2255–57.

Theye, Richard A., Robert T. Patrick, and John W. Kirklin. "The Electro-encephalogram in Patients Undergoing Open Intracardiac Operations with the Aid of Extracorporeal Circulation." *Journal of Thoracic Surgery* 34 (Dcember 1957): 709–17.

Thomas, Lewis. *The Youngest Science: Notes of a Medicine-Watcher.* New York: Viking Press, 1983.

Thompson, Charis. *Making Parents: The Ontological Choreography of Reproductive Technologies.* Cambridge, MA: MIT Press, 2005.

Thompson, Samuel A., and Aaron Plachta. "Experiences with Cardiopericardiopexy in the Treatment of Coronary Disease." *JAMA* (20 June 1953): 678–81.

Thompson, Thomas. "The Texas Tornado vs. Doctor Wonderful." *Life,* 10 April 1970, 62B–74.

Thorn, George A. "Report of the Physician in Chief." *Fifty-Fourth Annual Report, 1966–1967,* Peter Bent Brigham Hospital (1967), BWH Archives.

———"Report of the Physician in Chief." *Fifty-Sixth Annual Report, 1968–1969,* Peter Bent Brigham Hospital (1969), BWH Archives.

——— "Report of the Physician in Chief." *Fifty-Seventh Annual Report, 1969–1970,* Peter Bent Brigham Hospital (1970), BWH Archives.

Timmermans, Stefan, and Marc Berg. *The Gold Standard: The Challenge of Evidence-Based Medicine.* Philadelphia, PA: Temple University Press, 2003.

Timmins, Edward Francis, to Timothy Leary, 26 July 1934. TL Papers, box 2.

Tomes, Nancy. *The Gospel of Germs: Men, Women, and the Microbe in American Life.* Cambridge, MA: Harvard University Press, 1998.

Topol, Eric J. "Coronary Angioplasty for Acute Myocardial Infarction." *Annals of Internal Medicine* 109 (1988): 970–80.

Topol, Eric J., and Steven E. Nissen. "Our Preoccupation with Coronary Luminology: The Dissociation between Clinical and Angiographic Findings in Ischemic Heart Disease." *Circulation* 92 (Oct. 1995): 2333–42.

Torch, Suzanne M., Jay S. Greenspan, Michael S. Kornhauser, John P. O'Connor, David B. Nash, and Alan R. Spitzer. "The Timing of Neonatal Discharge: An Example of Unwarranted Variation?" *Pediatrics* 107 (Jan. 2001): 73–77.

Tu, Jack V., Chris L. Pashos, C. David Naylor, Erluo Chen, Sharon-Lise Normand, Joseph P. Newhouse, and Barbara J. McNeil. "Use of Cardiac Procedures and Outcomes in Elderly Patients with Myocardial Infarction in the United States and Canada." *New England Journal of Medicine* 336 (22 May 1997): 1500–1505.

Urschel, Harold C., Maruf A. Razzuk, Martin J. Nathan, Ervin R. Miller, Donald M. Nicholson, and Donald L. Paulson. "Combined Gas (CO_2) Endarterectomy and Vein Bypass Graft for Patients with Coronary Artery Disease." *Annals of Thoracic Surgery* 10 (Aug. 1970): 119–31.

van den Brand, M., and the European Angioplasty Survey Group. "Utilization of Coronary Angioplasty and Cost of Angioplasty Disposables in 14 Western European Countries." *European Heart Journal* 14 (1993): 391–97.

van der Mast, Rose C., and Frits H. J. Roest. "Delirium after Cardiac Surgery: A Critical Review." *Journal of Psychosomatic Research* 41 (July 1996): 13–30.

van Schaik, Katherine D. " 'Taking' a History." *JAMA* 304 (15 Sept. 2010): 1159–60.

Vedantam, Shankar. "Clinton's Heart Bypass Surgery Called a Success." *Washington Post,* 7 Sept. 2004, A1.

Velazquez, Eric J., Kerry L. Lee, Marek A. Deja, Anil Jain, George Sopko, Andrey Marchenko, Imtiaz S. Ali, Gerald Pohost, Sinisa Gradinac, William T. Abraham, Michael Yii, Dorairaj Prabhakaran, Hanna Szwed, Paolo Ferrazzi, Mark C. Petrie, Christopher M. O'Connor, Pradit Panchavinnin, Lilin She, Robert O. Bonow, Gena Roush Rankin, Robert H. Jones, and Jean-Lucien Rouleau, for the STICH Investigators. "Coronary-Artery Bypass Surgery in Patients with Left Ventricular Dysfunction." *New England Journal of Medicine* 364 (2011): 1607–16.

Vineberg, Arthur, to Donald Effler, 21 Dec. 1960. AV Fonds, P126/C8, folder 78, "Effler."

Vlodaver, Zeev, Robert Frech, Robert A. Van Tassel, and Jesse E. Edwards. "Correlation of the Antemortem Coronary Arteriogram and the Postmortem Specimen." *Circulation* 47 (Jan. 1973): 162–69.

Vontz, Frederick K., Sam E. Myrick, John R. Ibach, and Larry H. Birch. "One Hundred Consecutive Cases of Aorto-coronary Bypass Surgery." *Journal of the Florida Medical Association* 5 (May 1972): 29–35.

Wailoo, Keith. *Dying in the City of Blues: Sickle Cell Anemia and the Politics of Race and Health.* Chapel Hill: University of North Carolina Press, 2000.

"Warfare, Medicine Converge in Naming Heart Condition." *Kerrville Daily Times* (Texas), 11 Jan. 1999, 6A.

Warner, John Harley. *The Therapeutic Perspective: Medical Practice, Knowledge, and Identity in America, 1820–1885.* Cambridge, MA: Harvard University Press, 1986.

Watkins, Elizabeth S. *The Estrogen Elixir: A History of Hormone Replacement Therapy in America.* Baltimore, MD: Johns Hopkins University Press, 2007.

Waxman, Sergio, Fumiyuki Ishibashi, and James E. Muller. "Detection and Treatment of Vulnerable Plaques and Vulnerable Patients: Novel Approaches to Prevention of Coronary Events." *Circulation* 114 (2006): 2390–2411.

Weindling, Paul. "Origins of Informed Consent: The International Scientific Commission on Medical War Crimes and the Nuremberg Code." *Bulletin of the History of Medicine* 75 (Spring 2001): 37–71.

Weisse, Allen B., ed. *Heart to Heart: The Twentieth Century Battle Against Cardiac Disease.* New Brunswick: Rutgers University Press, 2002.

Weisz, George. "The Origins of Medical Ethics in France: The International Congress of Morale Médicale of 1955." In *Social Science Perspectives on Medical Ethics,* ed. George Weisz, 145–61. Dordrecht: Kluwer Academic Publishers, 1990.

Weisz, George, Alberto Cambrosio, Peter Keating, Loes Knaapen, Thomas Schlich, and Virginie J. Tournay. "The Emergence of Clinical Practice Guidelines." *Milbank Quarterly* 85 (Dec. 2007): 691–727.

Wennberg, David, John Dickens, David Soule, Mirle Kellett, David Malenka, John Robb, Thomas Ryan, William Bradley, Paul Vaitkus, Michael Hearne, Gerald O'Connor, and Robert Hillman. "The Relationship between the Supply of Cardiac Catheterization Laboratories, Cardiologists and the Use of Invasive Cardiac Procedures in Northern New England." *Journal of Health Services Research and Policy* 2 (April 1997): 75–80.

Wennberg, David E., Merle A. Kellett, John D. Dickens, David J. Malenka, Leonard M. Keilson, and Robert B. Keller. "The Association between Local Diagnostic Testing Intensity and Invasive Cardiac Procedures." *JAMA* 275 (17 April 1996): 1161–64.

Wennberg, John E. *Tracking Medicine: A Researcher's Quest to Understand Health Care.* New York: Oxford University Press, 2010.

———. "Unwarranted Variations in Healthcare Delivery: Implications for Academic Medical Centres." *BMJ* 325 (26 Oct. 2002): 961–64.

———. "Which Rate Is Right?" *New England Journal of Medicine* 314 (30 Jan. 1986): 310–11.

Wennberg, John E., Benjamin A. Barnes, and Michael Zubkoff. "Professional Uncertainty and the Problem of Supplier-Induced Demand." *Social Science & Medicine* 16 (1982): 811–24.

Wennberg, John, and Alan Gittelsohn. "Health Care Delivery in Maine. I. Patterns of Use of Common Surgical Procedures." *Journal of the Maine Medical Association* 66 (May 1975): 123–30, 149.

———. "Small Area Variations in Health Care Delivery." *Science* 182 (14 Dec. 1973): 1102–8.

Wenneker, Mark B., and Arnold M. Epstein. "Racial Inequalities in the Use of Procedures for Patients with Ischemic Heart Disease in Massachusetts." *JAMA* 261 (13 Jan. 1989): 253–57.

Wharton, Thomas P., Victor A. Umans, and Hans O. Peels. "PCI for Stable Coronary Disease." *New England Journal of Medicine* 357 (26 July 2007): 415.

White, Carl W., Creighton B. Wright, Donald B. Doty, Loren F. Hiratza, Charles L. Eastham, David G. Harrison, and Melvin L. Marcus. "Does Visual Interpretation of the Coronary Arteriogram Predict the Physiologic Importance of a Coronary Stenosis?" *New England Journal of Medicine* 310 (29 March 1984): 819–24.

Whittle, Jeff, Joseph Conigliaro, C. B. Good, and Monica Joswiak. "Do Patient Preferences Contribute to Racial Differences in Cardiovascular Procedure Use?" *Journal of General Internal Medicine* 12 (May 1997): 267–73.

Whittle, Jeff, Joseph Conigliaro, Chester B. Good, Mary E. Kelley, and Melissa Skanderson. "Understanding of the Benefits of Coronary Revascularization Procedures among Patients who Are Offered Such Procedures." *American Heart Journal* 154 (Oct. 2007): 662–68.

Wi, Stephenson L. "History of Cardiac Surgery." In *Cardiac Surgery in the Adult*, ed. L. H. Cohn and L. H. Edmunds, 3–29. New York: McGraw-Hill, 2003.

Widimsky, Petr, William Wijns, Jean Fajadet, Mark de Belder, Jiri Knot, Lars Aaberge, George Andrikopoulos, Jose Antonio Baz, Amadeo Betriu, Marc Claeys, Nicholas Danchin, Slaveyko Djambazov, Paul Erne, Juha Hartikainen, Kurt Huber, Petr Kala, Milka Klinčeva, Steen Dalby Kristensen, Peter Ludman, Josephina Mauri Ferre, Bela Merkely, Davor Miličić, Joao Morais, Marko Noć, Grzegorz Opolski, Miodrag Ostojić, Dragana Radovanović, Stefano De Servi, Ulf Stenestrand, Martin Studenčan, Marco Tubaro, Zorana Vasiljević, Franz Weidinger, Adam Witkowski, and Uwe Zeymer on behalf of the European Association for Percutaneous Cardiovascular Interventions. "Reperfusion Therapy for ST Elevation Acute Myocardial Infarction in Europe: Description of the Current Situation in 30 Countries." *European Heart Journal* 31 (2010): 943–57.

Wilde, Sally. "Truth, Trust, and Confidence in Surgery, 1890–1910: Patient Autonomy, Communication, and Consent." *Bulletin of the History of Medicine* 83 (Summer 2009): 302–30.

Williams, Guy H., to the Board of Governors, 5 Feb. 1971. CCF Archives, A93-11/55, folder "1970, Annual Reports—Acknowledged, Div. of Med." (#2301).

Williams, Isla M. "Intravascular Changes in the Retina during Open-Heart Surgery." *Lancet* 298 (25 Sept. 1971): 688–91.

Willner, Allen E. "The Use of Neuropsychological Tests as Criteria of Brain Dysfunction in Cardiac Surgery Research." In *Cerebral Damage before and after Cardiac Surgery*, ed. Allen E. Willner, 195–202. Dordrecht: Kluwer Academic Publishers, 1993.

Wilson, Duff. "Side Effects May Include Lawsuits." *New York Times,* 2 Oct. 2010.

Wilson, Hugh E., Martin L. Dalton, Ridlon J. Kiphart, and Walter M. Allison. "Increased Safety of Aorto-coronary Artery Bypass Surgery with Induced Ventricular Fibrillation to Avoid Anoxia." *Journal of Thorac and Cardiovascular Surgery* 64 (Aug. 1972): 193–202.

Winslow, C.-E. A. "An Outbreak of Tonsillitis or Septic Sore Throat in Eastern Massachusetts and Its Relation to an Infected Milk Supply." *Boston Medical and Surgical Journal* 165 (14 Dec. 1911): 899–904.

Winslow, Constance Monroe, Jacqueline B. Kosecoff, Mark Chassin, David E. Kanouse, and Robert H. Brook. "The Appropriateness of Performing Coronary Artery Bypass Surgery." *JAMA* 260 (22–29 July 1988): 505–9.

Winter, Alison. *Mesmerized: Powers of Mind in Victorian Britain.* Chicago: University of Chicago Press, 2000.

Woods, Hiram. "Use of Diaphoresis and Diaphoretic Agents in Ophthalmic Therapeutics." *JAMA* 43 (24 Dec. 1904): 1957–60.

Wooler, Geoffrey. "Oral History" (10 June 1997). In *Pioneers of Cardiac Surgery,* ed. William S. Stoney, 138–47. Nashville, TN: Vanderbilt University Press, 2008.

World Bank. *Investing in Health: World Development Report 1993.* New York: Oxford University Press, 1993.

Yach, Derek. Foreword to *Race Against Time: The Challenge of Cardiovascular Disease in Developing Countries,* by Stephen Leeder, Susan Raymond, and Henry Greenberg, iii–v. New York: Trustees of Columbia University, 2004.

Yudofsky, Stuart. "Hilde, the Teacher." *JAMA* 257 (20 Feb. 1987): 975.

Zaks, Misha S., Jordan Lachman, George K. Yacorzynski, and Benjamin Boshes. "The Neuropsychiatric and Psychologic Significance of Cerebrovascular Damage (Strokes) Following Rheumatic Heart Surgery." *American Journal of Cardiology* 5 (June 1960): 768–76.

Zhang, Yan, and Yong Huo. "Early Reperfusion Strategy for Acute Myocardial Infarction: A Need for Clinical Implementation." *Journal of Zhejiang University—Science B (Biomedicine and Biotechnology)* 12 (2011): 629–32.

Zimmerman, Henry A., and Martial A. Demany. "Coronary Artery Visualization and Coronary Surgery—A Word of Caution." *American Heart Journal* 71 (June 1966): 725–26.

Zuckerman, Stephen, Timothy Waidmann, Robert Berenson, and Jack Hadley. "Clarifying Sources of Geographic Differences in Medicare Spending." *New England Journal of Medicine* 363 (2010): 54–62.

Index